Androids in the Enlightenment

Androids in the Enlightenment

Mechanics, Artisans, and Cultures of the Self

ADELHEID VOSKUHL

The University of Chicago Press   Chicago and London

The University of Chicago Press, Chicago 60637
The University of Chicago Press, Ltd., London
© 2013 by The University of Chicago
All rights reserved. Published 2013.
Paperback edition 2015
Printed in the United States of America

24 23 22 21 20 19 18 17 16 15   2 3 4 5 6

ISBN-13: 978-0-226-03402-7 (cloth)
ISBN-13: 978-0-226-03416-4 (paper)
ISBN-13: 978-0-226-03433-1 (e-book)
10.7208/chicago/9780226034331.001.0001

Library of Congress Cataloging-in-Publication Data

Voskuhl, Adelheid.
   Androids in the Enlightenment : mechanics, artisans, and cultures
of the self / Adelheid Voskuhl
      pages. cm.
   Includes bibliographical references and index.
   ISBN 978-0-226-03402-7 (cloth : alk. paper)—ISBN 978-0-226-03416-4
(pbk. : alk. paper)—ISBN 978-0-226-03433-1 (e-book) 1. Robots—History.
2. Mechanical engineering—History. 3. Artisans—Europe—History.
4. Enlightenment—Europe. I. Title.
   TJ211.V675 2013
      629.8—dc23   2013005611

♾ This paper meets the requirements of ANSI/NISO Z39.48-1992
(Permanence of Paper).

To the memory of my mother, again.

Contents

Illustrations

Acknowledgments

This book would not have been possible without the help of numerous people and institutions. Peter Dear, at Cornell's Department of Science and Technology Studies, guided it with extraordinary patience and inspiration from its earliest ideas and research, as did Ron Kline and Itsie Hull. The Cornell campus as a whole provided an exceptionally welcoming and stimulating environment. I was privileged to be part of it and am grateful to Steve Hilgartner, Mike Lynch, Trevor Pinch, Michael Dennis, Peter Hohendahl, Michael Steinberg, Dominick LaCapra, Peter Holquist, and Anette Schwartz. My friends in Ithaca were and have continued to be the source of unending joy and support. I thank in particular Emily Dolan, Josh Greenberg, Francesca Wong, Gary Tsifrin, Camille Robcis, Chris Bilodeaux, Tracie Matysik, Hilary Emmett, Kevin Lamb, Ute Tellmann, Ryan Plumley, Marianne Tettlebaum, Dorian Stuber, Suman Seth, Mindy Smith, John Kim, Lara Kelingos, Shobita Parthasarathy, Cyrus Mody, and Javier Lezaun.

Over the years, I received support from Cornell University, the National Science Foundation, the Thyssen Foundation, the Deutsches Museum in Munich, the Musée international d'horlogerie in La Chaux-de-Fond, and the Department of History and Philosophy of Science at Cambridge University. At these places, I received gracious hospitality from Bettine Menke, Hans Medick, Julika Funk, Dietmar Schmidt, Helmuth Trischler, Hartmut Petzold, Silke Berdux, Ludwig Oechslin, Nick Jardine, Simon Schaffer, and Adam Mosley. Olaf Breidbach hosted me for an entire year at the Friedrich-Schiller-Universität at Jena, and he and Thomas Bach, Gerhard Wiesenfeldt, Jan Frercks, and Heiko Weber made my time there both pleasant and productive. Other museums and archives I visited include the Technisches Museum in Vienna, the Conservatoire national des arts et métiers in Paris, the Mathematisch-Physikalischer Salon in Dresden, the Musée d'art et d'histoire at Neuchâtel, the Staatsarchiv in Basel, the Bibliothèque de Genève and the Archives d'Etat in Geneva, the Metropolitan Museum of Art in New York City, the Archiv der Brüder-Unität in Saxony, the Fürstlich-Wiedisches Archiv in Neuwied, the Bibliothèque de la Ville La Chaux-de-Fonds, the Musée internationale d'art et d'histoire in Neuchâtel, and the Forschungsbibliothek Gotha in Schloss Friedenstein. At all these places, I received warm welcome and crucial help in exploring artifact and manuscript collections, notably from Otmar Moritsch, Mirko Herzog, Ghislaine Aung Ko, Rüdiger Kröger, Peter Plaßmeyer, and Annette Gerlach. I am particularly grateful to the staff of the Conservatoire national des arts et métiers in Paris and the Musée d'art et d'histoire at Neuchâtel for help with the images and permissions to reproduce them in this book, and to Daniela Helbig for her help with the logistics. I also thank Elsevier for permission to use material that appeared in a previous article of mine in *Studies in History and Philosophy of Science*.

I have presented the material in this book at annual meetings of the Society for the History of Technology, the History of Science Society, the Society for Social Studies of Science, and the Society for Philosophy and Technology, and at colloquia and talks at Bates College, the University of Pennsylvania, MIT, Virginia Tech, Georgetown University, Cornell University, the University of Minnesota, and the UCLA Center for Seventeenth- and Eighteenth-Century Studies. There and on other occasions, I received very helpful criticism from Jessica Riskin, Matt Jones, John Tresch, Matt Wisnioski, Roz Williams, Myles Jackson, Ann Johnson, Roe Smith, Rebecca Herzig, Lynda Coon, Lorraine Daston, Debbie Coen, Jan Golinski, Mary Terrall, Pamela Smith, Leonard Rosenband, David Mindell, and Deborah Harkness.

In recent years, the Department of the History of Science at Harvard University has been my intellectual and personal home. I am very grateful to its students, faculty, and affiliates for ideas, criticism, and companionship, in particular Katy Park, Steven Shapin, Peter Galison, Allan Brandt, Mario Biagioli, Anne Harrington, Charles Rosenberg, Janet Brown, Sheila Jasanoff, Jimena Canales, Jeremy Greene, Sophia Roosth, Rebecca Lemov, Elly Truitt, Nasser Zakariya, and Nico Wey-Gomez. Most of this book was written while I spent a very happy year on leave at the Newhouse Center for the Humanities at Wellesley College. Its director Tim Peltason and fellow fellows, in particular Daniel Ussishkin and John Carson, made it a wonderful time. My friends in the Boston area, in Germany, and in other places have helped me through many stages of life and work over the last years, and I owe special thanks to Shu-Yi Oei, Andy Jewett, Sindhu Revuluri, Robin Bernstein, Maya Jasanoff, Ian Miller, Judith Surkis, Shauna Helton, Natalie Wagner, Mary Ann Cicala, Rebecca Dornin, Susanne Jacob, Frauke Sterwerf, Silke Baehr, Falk Müller, and Uljana Feest.

Karen Merikangas Darling and Amy Krynak provided tireless help in putting this book into production. The Press's two anonymous referees made me rethink the manuscript as a whole and provoked me into sharpening its crucial and most difficult points. Sonam Singh gets a special call-out for diligent and brilliant editorial help. All remaining errors are, of course, my own.

ONE

Introduction: Androids, Enlightenment, and the Human-Machine Boundary

Android automata, robots, and mechanical humans have been central to our understanding of the relationship between humans and machines. They are designed to look and move as human beings do and perform motions and techniques such as walking, writing, or music-making. For a spectator, it is often difficult to determine whether an android object is human or machine. Because androids so effectively destabilize our sense of the boundary between humans and machines and, by extension, our sense of our own constitution, they and their histories evoke a broad range of concerns, most significantly, perhaps, those related to the promises and perils of the modern industrial age.

Specifically, those android automata that were made during the Enlightenment have played an influential role in our understanding of modern industrial society and the human-machine boundary in it, either directly or through texts. Enlightenment automata were spectacular and innovative self-moving objects and, in regard to mechanical complexity, the most sophisticated of their kind compared to not only earlier but also later periods. They have attracted people's attention continuously from the time they were made to the present and have served throughout all stages of industrialization during the nineteenth, twentieth, and twenty-first centuries as eloquent metaphors for the social and economic changes, and novel human-machine encounters, brought about by the steam engine, the factory, and the electronic computer. The Enlightenment period as such is also widely recognized as a milestone in the making and theorizing of modern society. For this reason as well, Enlightenment automata are often taken to be forerunners and figureheads of the modern, industrial machine age, an age in which the economic, social, cultural, and aesthetic constitution of humans changed fundamentally and supposedly became "mechanized."[1]

In this book I discuss two android automata that were made in the Enlightenment. They both represent women playing a keyboard-type instrument (see figs. 1 and 2).[2] One is a harpsichord player that was built by Pierre and Henri-Louis Jaquet-Droz, father and son, who were clock-makers from the small town La Chaux-de-Fonds in the Neuchâtel mountains in the western part of Switzerland. They presented this remarkable, life-size automaton to the public in 1774. The seated figure is about four feet high and shows a young woman of about age fifteen in front of a small organ. She plays her instrument like a human by pushing keys on the keyboard with her fingers. The other automaton is a dulcimer player that was made by the cabinetmaker David Roentgen in collaboration with the clock-maker Peter Kinzing. David Roentgen, together with his father, ran a sizable furniture manufacture (a preindustrial type of factory) in the small principality of Neuwied in the Holy Roman Empire, about seventy miles south of Cologne on the Rhine. He offered this elegant, delicate automaton as a gift to Marie-Antoinette in 1785. The dulcimer player is smaller than the Jaquet-Droz family's harpsichord player: she is one foot and nine inches tall, is arranged on top of a table, and strikes the strings of her dulcimer with two hammers. She purportedly represents Marie-Antoinette herself, who was thirty years old when the automaton was made. The musical instrument is large in relation to the automaton's body: it is about two feet long by one foot wide, and the arrangement as a whole—consisting of

1. Cultural critics of the Frankfurt School, historians of the Industrial Revolution, historians of science, and intellectual historians are among those who take the Enlightenment to be a milestone in the emergence of modern industrial society. Outram, "Enlightenment Our Contemporary"; Daston, "Afterword"; Schaffer, "Enlightened Automata"; Alder, *Engineering the Revolution*, xi; and Allen, *British Industrial Revolution*. Minsoo Kang singles out the eighteenth century in the general history of automaton-making as a period of an "automaton craze." Kang, *Sublime Dreams*, 7.

2. I use the phrase "keyboard-type" as a heuristic, since the dulcimer is not, strictly speaking, a keyboard instrument. The crucial aspect for my purposes is the two instruments' shared geometry: it determines the construction of the clockworks inside of the automata, as I analyze in chapter 4.

FIGURE 1 The harpsichord player (called La musicienne). Made by Pierre and Henri-Louis Jaquet-Droz, La Chaux-de-Fonds, 1772–1774. Musée d'art et d'histoire, Neuchâtel (Switzerland); photograph S. Iori.

FIGURE 2 The dulcimer player (called La joueuse de tympanon). Made by David Roentgen and Peter Kinzing, Neuwied, 1785. © Musée des arts et métiers-Cnam, Paris/photo P. Faligot: Inv. 7501.

musician, dulcimer, and table—is about four feet wide, two feet deep, and four feet high. Both women automata, despite differences in design, therefore play music at the same height: the eye level of a seated spectator. They produce similarly captivating scenarios of music-making with different aesthetic means. The harpsichord player shows a musical performance in real-life dimensions with realistic motions of arms, hands, and fingers, while the dulcimer player's appeal resides in her miniature effect and the contrast in size between the instrument and the music-playing figure.[3]

The two mechanical androids are a subset of a group of about ten

3. Throughout the book I use the pronouns "she" and "her" to refer to the two women automata, for better readability. Detailed descriptions of the two automata are found in chapter 4.

that were made between 1730 and 1810, and they are exemplary of their genre: they realistically carry out sophisticated motions, they are appealing and charming in their performance, and their clockwork mechanisms are hidden entirely in their bodies.[4] The two also distinguish themselves from the others in important ways: they are the only women among the ten, the only musicians to play keyboard-type instruments, and the two (among pairs made by different artisans) with the greatest resemblance to each other. Being such an exemplary and distinguished pair within the larger set of Enlightenment automata, they allow us to draw several conclusions about mechanical arts, industrialism, and the human-machine boundary in the period on the cusp of the Industrial Revolution.

The other eight android automata that we know of from the eighteenth century display a variety of figures and activities. They were made by clock-makers, cabinetmakers, court mechanics, and other artisans who had the means, skills, and incentives to do so.[5] The French mechanic Jacques de Vaucanson presented three automata—a flute player, a galoubet player, and a duck—to the Académie des Sciences in Paris in 1738. He set the standard for mechanical androids for the following decades and continues to this day to be the best known and most-often-cited eighteenth-century automaton-maker. In 1760 Friedrich von Knaus, a mechanic employed at the imperial court in Vienna, built a writing machine for the mechanical cabinet of the Holy Roman Emperor Franz Stephan. It was the first mechanical device to implement the process of writing. Father and son Jaquet-Droz introduced in 1774 three android automata often considered to be the most spectacular and appealing of the entire period: a writer, a draftsman, and the harpsichord player. In 1785 Roentgen and Kinzing presented the dulcimer player to Marie-Antoinette as a gift. Finally, in 1810, the instrument-maker family Johann Gottfried Kaufmann from Dresden

---

4. Both automata are extant, functioning, and on regular display in their respective museum homes: the harpsichord player in the Musée d'art et d'histoire in Neuchâtel in Switzerland and the dulcimer player in the Conservatoire national des arts et métiers in Paris. There are several videos of both automata on the Internet, findable on YouTube by searching "La joueuse de tympanon" and "Jaquet-Droz androids."

5. Discussion continues about how many more automata were made during the eighteenth century in the prolific artisan communities in southern Germany, Switzerland, Hungary, and the areas around Paris, London, Vienna, and Berlin. There is no doubt that there was a great deal of skilled mechanical craftsmanship in European cities, courts, and the countryside, and many artisans tried to push the boundaries of their trade by building automata or related artifacts. Much artisan work was also being done in musical instrument–making, clock-making, and related trades. Textual evidence about automaton-making is often missing or, if it exists, is often unreliable. Beyer, *Faszinierende Welt der Automaten*.

built a trumpeter. It was reported to play tunes and fanfares in two voices and is today famed for inspiring E. T. A. Hoffmann's famous tale "The Sandman." I provide details about each of these automata on pages 29–36.[6]

I focus on the dulcimer player and the harpsichord player because they replicate in a more comprehensive and far-reaching way than the others, through their distinctive mechanical design, an entire setting of eighteenth-century cultural and political activity. They make manifest Enlightenment ideas about music-making that were at the core of efforts to constitute new types of selfhood in a new type of society. The automata teach us about both technical and cultural conditions that were central to the beginnings of modern society in the eighteenth century. In a novel way, they make processes of creating cultural and technical modernities visible, by cutting across material, intellectual, and political cultures that are often discussed separately.

The German and Swiss artisans built the two women automata so that they move not only their arms, hands, and fingers to play music; they also move their heads, eyes, and torsos in rhythm with the music they play: two-voice dance pieces that were standard at contemporary courts. Separate mechanisms independently drive the automata's motions of their heads, eyes, and, in the case of the harpsichord player, her torso to breathe and make a gracious bow at the end of each performance. These body motions were "extra" features, as it were, in addition to the music-making. The harpsichord player's mechanism for moving her head and eyes and for breathing, for example, can run for one and a half hours. The musician thus begins these movements long before she starts playing her instrument. Miniaturizing this type of mechanism to fit into the automaton was one of the Jaquet-Droz fam-

---

6. On Vaucanson, see Swoboda, *Der künstliche Mensch*, 91–96; and Heckmann, *Die andere Schöpfung*, 219–32. For critical work on Vaucanson and his importance in the development of research on artificial intelligence in the twentieth century, see Riskin, "Defecating Duck" and "Eighteenth-Century Wetware." Friedrich von Knaus called his machine an Alleschreibende Wundermaschine (Miracle writing machine). Folder "Friedrich von Knaus, Alleschreibende Wundermaschine" in the archives of the Technisches Museum in Vienna. The status of the three Jaquet-Droz automata as epitomes of automaton-making is demonstrated in the fact that the only illustration in the article "automaton" in the online *Encyclopædia Britannica* is of a Jaquet-Droz automaton. On E. T. A. Hoffmann's specific relationship to the Kaufmanns' trumpeter, see Krätz, "'Der makabre Android.'" Even though it is not a "real" android automaton, it is important to mention the chess-playing automaton by Wolfgang von Kempelen. It became famous even beyond the borders of Europe and functions to this day, despite its hoax character, as a signature automaton of the eighteenth century. See Schaffer, "Enlightened Automata"; and Standage, *The Turk*. Since it was a deliberate hoax and not an attempt to replicate human action by mechanical means, it is not part of my considerations here.

ily's most innovative feats, and it indicates their strong intent to make the automaton perform subtle and graceful body motions in addition to playing music.

The automata's bodily motions correspond to eighteenth-century performance techniques, as musicians at the time were expected to move their bodies while playing music to communicate affects and sentiments to the audience—affects and sentiments that they were meant to generate in themselves first, according to the program set forth by the musical piece. Cultivating sentiments was not only a feature of making and listening to music at the time; it was also part of a larger social movement in the European Enlightenment and was practiced in the reading and writing of literature, in the sciences and natural philosophy, in letter-writing and friendship, and in travel culture. Sentiments were taken as means to form new types of social interaction and used as the basis for a new, more equal and just social order: civil society. Civil society, made up of rational, sensible, and equal citizens, was meant to replace the traditional estate and court societies and be held together and sustained by, among other things, cultivated and shared sentiments.[7]

The artisan automaton makers thus manifested mechanically in their androids body practices that cultivated sociability, sentiments, and practices of subject- and society-formation.[8] The culture of sentimentality was not only a social and political affair, prescribing participation in other people's feelings and generating emancipated forms of social conduct such as in an audience or a reading club. It was also a practice of *self*-reflexivity in which feeling subjects were supposed

7. Accounts of these practices, and the experience of deliberate participation and society-building attendant upon them, are found in Dülmen, *Die Gesellschaft der Aufklärer*; and Hull, *Sexuality, State, and Civil Society*. On music-making and listening specifically, see Riley, *Musical Listening*; and Freia Hoffmann, *Instrument und Körper*. Jürgen Habermas, a major theorist of this eighteenth-century phenomenon, draws attention to the public sphere that emerged "between" state authority and citizens and was characterized by a strong sense of self-consciousness and deliberation on the part of its practitioners. Habermas, *Strukturwandel der Öffentlichkeit*; and Calhoun, *Habermas and the Public Sphere*, 4–14. In regard to understandings of feelings and bodily motions in earlier periods, see Dear, "A Mechanical Microcosm." Minsoo Kang is also interested in the sentimental aspect of eighteenth-century culture and its relevance for automata, focusing on medical and philosophical literature as well as fiction to emphasize the crises that strict mechanistic worldviews underwent in post-Cartesian debates. Complementing his work, I discuss, in particular in chapter 4, political aspects of sentimentality and their explicit manifestation in body techniques, both in human musicians and in the design of music-playing automata.

8. The two women automata that I discuss are not "technologies of the self" in the strict Foucauldian sense. Foucauldian technologies of the self comprise practices, methods, forms of knowledge, and a "way of being," as he says. I look at a mechanical replication of Foucault's technologies of the self. Foucault, "Technologies of the Self," 18. See also, in this regard, Schaffer, "Enlightened Automata."

to find and experience themselves, in a manner discussed in the era's cultural, political, and pedagogical literature.[9] The specific musical literature of the time made the same point about the relationship between musician and listeners in a musical performance. Music-playing women automata that communicate sentiments thus raise issues that touch simultaneously on mechanics, artisans, subject-formation, and social order.

The two automata replicate mechanically a comprehensive scenario of cultural and political activity of their time. The image of piano-playing women was a widespread stereotype: it embodied contested cultural and political gender issues and was the subject of pointed caricatures.[10] In my study, accordingly, I investigate the automata not only as symbols of industrial modernity, as they have been used widely, but also, and primarily, as products of their own period, the early, pre-industrial phases of modern society. Doing so, I bring the eighteenth century into a sustained dialogue with industrial modernity. Although the eighteenth century is rightfully considered to be a milestone in the making of the modern age, it still retained distinctly "early" modern features that set it apart from the industrial age. On the European continent, in particular, the eighteenth century was a period of largely pre- and protoindustrial production and was characterized by a social order that was culturally, politically, and legally organized as estates in a court society. Furthermore, the bourgeois classes were still in a very early phase of consolidating and emancipating, efforts that they based on sentimental practices among other things. Despite my two women automata's later association with industrial modernity, they and the textual commentary they occasioned are products of these emerging "early" modern practices: protoindustrial mechanical arts, novel techniques of bourgeois sociability in a courtly estate society, sentimental music-making, and novel types of literary production.

However, androids from the eighteenth century have sparked interest continuously since their creation, especially because the two hundred or so years between their making and the present span the entire history of industrialization in Europe and North America. Their continuous use as illustrations for the varied experiences in an industrializing world has kept the automaton motif alive to this day and has

9. Peter Hohendahl, *Der europäische Roman*, 2, calls the sentimental disposition a "dialectic" relationship of the subject with itself. I discuss this cultural, political, and pedagogical literature in chapter 4.

10. See caricatures, for example, in Rieger, *Frau, Musik und Männerherrschaft*, 37, 50. I analyze this image in depth in chapter 4.

maintained and perpetuated its credibility as an allegory for relationships between humans and machines. The automata that were made in the eighteenth century underwent a long and complex journey, literally and figuratively, between the Enlightenment and the end of the Cold War. The very idea of the "Enlightenment automaton" has been used in a wide range of ways and is not an unambiguous historical or conceptual category. Our current understanding of the Enlightenment automaton is preceded by a long and contested history of ideas about how android automata do, or do not, epitomize humans in the industrial age.[11]

My study raises the questions of how eighteenth-century automata became such successful symbols of a later and in many ways very different type of modernity and how their roots in their own time help us understand better this later industrial modernity for which we so readily make them a symbol. I uncover how protoindustrial conditions were already forming in the eighteenth century, as others have before me, but my case also prompts me to ask how objects rooted in a preindustrial court society became so successfully associated with industrial periods. We use eighteenth-century androids as emblems of human-machine problems that are situated firmly in the industrial revolutions of the nineteenth century and the major wars (including the Cold War) of the twentieth. I ask what this fact teaches us about our understanding of, on one hand, the relationship between humans and machines in the industrial age and, on the other, the objects that we choose to represent this relationship and make it comprehensible to ourselves.

The Book's Structure

I treat the range of themes that converge in the construction and interpretation of my two women automata by integrating the social and economic histories of artisan industries in western Switzerland and the Rhineland with cultural histories of sentimental body techniques and literary production on automata from 1740 to 1815. This effort occupies chapters 2 through 5. In chapter 6 I discuss these findings

11. Arguments similar to mine, developed around different technological artifacts, are found in Gerhard Dohrn-van Rossum's study about the clock and its relevance to our understanding of the industrial age and in David Mindell's study of the USS *Monitor*, in which he explains how its battle with the CSS *Virginia* "raised profound questions about the industrial world." Dohrn-van Rossum, *History of the Hour*, 14; Mindell, *War, Technology, and Experience*, 2.

in the context of the two-hundred-year history following my automata's construction. During this period Enlightenment automata were engaged, cited, and rendered symbols successively of each of the later stages of industrialization. The two women automata thus introduce us, on the one hand, to a multifaceted world in the eighteenth century that featured the design and construction of mechanical androids, music-making, nascent bourgeois self-awareness, and literary exploration of the human-machine boundary. On the other, they help us understand how objects that seem to successfully symbolize relationships between humans and machines travel equally successfully through later periods and get used in those later contexts as metaphors of later and later industrial modernities, often with an assumption that their symbolic connotations remain unchanged through different periods of human-machine relations.

My two women automata were made in artisan environments in the 1770s and 1780s that combined in unique and productive ways traditional and modern elements of craft and trade that were flourishing in protoindustrial early modern Europe. In regard to the history of economics and of technology, this was an exceptional period, on the cusp of the industrial revolution, that saw the last blossoming of the artisan and court worlds on the European continent. The body techniques that the automata display—mechanics built by artisans to perform cultures of the self —were a key practice in eighteenth-century struggles for a new type of society. The replication of these techniques in automata raised questions for contemporary critics of sentimental culture about the reliability and desirability of the idea of grounding a new social order on sentiments. Such critics exploited piano-playing women automata as a literary motif to express their skepticism about the feasibility of sentimental civil society, and I examine their texts to trace these critical strategies. I find in the convergence of eighteenth-century industry, musical culture, and literary commentary answers to the question of how Enlightenment automata help us understand human-machine boundaries in both preindustrial and industrial ages. The very self that was being challenged as having become "a machine" in the industrial age was being forged in the first place in eighteenth-century sentimental practices. The literary motif of the piano-playing woman automaton did not yet serve in the eighteenth century as an oracle for mechanization of bourgeois selfhood, as it did later in the nineteenth; instead, it served as an illustration of the initial, often apprehensively watched, constitution of bourgeois selfhood.

I begin my study by exploring in the next two chapters the artisan worlds in which the automata were made and the newspapers and letters in which contemporaries commented on them. Chapter 2 takes us to La Chaux-de-Fonds, a rural valley in western Switzerland that was densely populated with clock-makers, local dignitaries, natural philosophers, and aristocratic patrons. The village's social worlds provided unique opportunities for Pierre Jaquet-Droz and his son Henri-Louis to accumulate clock-making skills, education, capital, and patronage. I integrate Pierre Jaquet-Droz's training and early career as a clock-maker with the complex history of Europe's beginning industrialization and trace the channels through which he was introduced to the European court society as a young man. I investigate how he and his son designed and built their three android automata and showed them in a few exhibitions, and I look at how the Jaquet-Droz family's clock-making business evolved after they finished their automaton work, to uncover their self-understandings and ambitions as artisans and merchants. In the second half of chapter 2, I look at brief mentions of their automata that appeared between the 1770s and 1790s in letters and diaries of local dignitaries of La Chaux-de-Fonds and at printed texts in mass media such as newspapers, travel reports, and pocket calendars. The printed texts varied in length, served different functions, and were part of different contexts. Some were one-sentence mentions in travel reports, and others were short accounts of mechanical curiosities in newspapers or brief histories of Switzerland in pocket calendars. All descriptions of the two women automata were copied from one another, however, and did not engage the automata in a deliberate or detailed manner. I demonstrate how it was common and desirable in the media industry of the time to copy and paste texts across genres and how, consequently, these texts do not provide us insight into people's reactions to the Jaquet-Droz automata or information about how widely known they were as spectacles. I use these texts, instead, as a window into the economic and cultural conditions of print mass media at the time and the role of automaton spectacles in them.

Chapter 3 takes the journey down the Rhine to the furniture manufacture of Abraham and David Roentgen in Neuwied, about seventy miles south of Cologne. David Roentgen became the most distinguished and sought-after furniture-maker of his time, and his manufacture was the biggest and most productive of its kind in the entire Holy Roman Empire. He sold to the major European political and clerical courts, including Catherine the Great's, Louis XVI's, and Frederick the Great's.

Roentgen's success, like that of the Jaquet-Droz clock-making workshop in Switzerland, had its roots in a society in which early industrialization and the court culture's continuing need for luxury commodities coincided. David Roentgen was, furthermore, a member of a Pietist sect, which gave him access to mobile capital, skilled labor, and distinguished (that is, aristocratic) trade and business contacts, as well as exemption from guild regulation. The case of Roentgen's dulcimer player exemplifies the multifaceted social, spiritual, and political circumstances in the second half of the eighteenth century that facilitated the production of an android automaton. There was contemporary letter and periodical commentary on the dulcimer player, as on the harpsichord player. Since the dulcimer player was never put on public display, however, it is mentioned in only one letter and one newspaper article. I discuss the letter (by Marie Antoinette's physician, who was a witness of the first presentation of the dulcimer player at Louis XVI's court) and the newspaper article (in a German weekly magazine, about two months after the presentation in Paris) and find on the basis of those two and the texts on the Jaquet-Droz automata that contemporary descriptions of both automata emphasized clearly and consistently the additional motions of the automata's heads, eyes, and torsos. The artisans' strategy of furnishing their automata with such visible motions that corresponded to contemporary sentimental body techniques thus paid off. It was through such texts in contemporary media that this feature was transmitted to numerous audiences, including literary writers, in the following decades.

Chapter 4 analyzes in detail the clockwork mechanisms of the two automata and explains how they enable the women figurines to perform their minute sentimental-musical programs. I investigate the mechanisms in two ways. First I analyze the clockworks' individual parts to explain how mechanical modules correspond to motions of the figures, focusing specifically on the effects that these have on a spectator. This analysis illuminates other aspects of the automata's history as well, such as the exceptional skill that went into their design and construction, similarities and differences between the two mechanisms, choices that the artisans made about where to devote their technical skill and effort, conflicts between artistic and mechanical principles that the artisans faced, and the social and economic functions that the automata served in the respective artisans' lives. Second, I explore the women automata's mechanical-musical performance as a cultural scenario in its own right, with its own history and valences in the eighteenth century. Music-making was an important and visible activity at the time, with key social and economic functions, for women as well as for men. Through performances in salons and homes, the writing and printing of musical works and textbooks, and continuous dissemination of sentimental practices, the process of subject-formation was taking place in sustained and multiple ways. Music-playing women who moved their bodies to communicate sentiments were a common topic in musical theory, sentimental theory, social theory, and political philosophy. The chapter explores in detail the techniques of representation, expression, and subject-formation that were used in the music-making world, discussing how these were replicated in the two women automata.

In chapter 5 I focus on literary representations of the two women automata. I look at texts that use piano-playing women automata as motifs to develop broader intellectual and poetic agendas about the boundaries between humans and machines. In the years between 1750 and 1820, numerous writers concerned themselves with android automata and other artificial humans, and two of them used piano-playing women automata specifically: Johann Paul Friedrich Richter (who later in his life went by Jean Paul) in the 1780s and E. T. A. Hoffmann in the second decade of the next century. I read and discuss texts by Richter and Hoffmann that feature piano-playing women automata in relation to other texts about automata from the time. The two writers use the motif to elaborate on problems of the (sentimental) constitution of selfhood and society, playing at the same time with the human-machine boundary and making it disturbingly unstable for a spectator or reader. I argue based on my readings that piano-playing women automata raised questions not only about whether humans had become "like" machines in the modern age, but also about whether the initial formation of the modern subject—the one preceding its mechanization—was a real and reliable process. It is well known that critical stances toward ideas and practices of subject-formation emerged long before the industrial age. But this also means that accusing the bourgeois subject of "having become a machine" does not necessarily imply references to *industrial* machinery. The complaint that humans have become machines has its own history, precedes the industrial age, and is expressed in terms of a range of technologies, preindustrial and industrial.

The commentary produced in the eighteenth and early nineteenth centuries about the two women automata also brings to the fore issues of spectatorship in the history of the two automata. The commentary adds a spectator's gaze to the automata's mechanical performance, and it is this gaze that creates the epistemic uncertainty over the human-machine boundary in android automata's performances: it is a real or imagined spectator, after all, for whom android automata stabilize or destabilize the human-machine boundary.[12] Texts about the two women automata open up to us the historical space in which authors, readers, and spectators experienced the automata and the human-machine boundary in the eighteenth century.

Chapter 6 surveys how the two women automata and the larger set of Enlightenment automata and ideas about them "traveled" from their eighteenth-century origins through the various phases of industrialization in the nineteenth and twentieth centuries and were increasingly used over this period as symbols of industrial modernity overall. During the Industrial Revolution in the early and middle nineteenth century, interest in mechanical humans emerged repeatedly among both critics and advocates of industrialization. The literary writers of the romantic age, also in the first half of the nineteenth century, were fond of the motif and developed it in various directions, as did Victorian and modernist writers and the earliest authors of science fiction in the 1890s. The critical cultural theorists of the twentieth century, especially after each of the two world wars, also relied on ideas of mechanized humans and android automata. Their arguments were often driven by a desire to explain the catastrophes of twentieth-century industrial warfare and state action, including those catastrophes' relation to other groundbreaking changes in the interactions between humans, machines, and labor. Interest in automata flared up again during the Cold War with the coming of cybernetics, artificial intelligence, the heyday of science fiction, and the personal computer. My tour in this chapter through the industrial age demonstrates how widely eighteenth-century automata have been used and reveals the conduits through which they have remained visible and credible motifs throughout this long time window, to this day. I use representative texts from key episodes to provide an overview, treating some in greater detail to explain the origins of explicit and implicit assumptions about Enlightenment automata and their role in our imagination of the machine age.

12. Questions of gaze and spectatorship, as well as their fundamental relationship to epistemic uncertainties, have been discussed in theories of film and art, inspired, for example, by psychoanalysis and literary and feminist theory. See, e.g., Lacan, *Seminar of Jacques Lacan*, books 1 and 11; Rose, *Visual Methodologies*; and Mulvey, *Visual and Other Pleasures*.

Sentimental Androids in Larger Narratives of Modernity

My study is in dialogue with detailed and wide-reaching analyses of the emergence of modern society and its cultural, industrial, and philosophical components. The cultural practice (and political ambition) of sentimental music-making speaks to recent discussions on modern subject-formation and its gendered dimensions. Much of this literature builds on the work of twentieth-century social theorists who have posed questions about the relationship of cultural, social, and legal regulation of men's and women's bodies and behaviors (such as bodily comportment, sexual behavior, table manners, violence, rational conduct, and public comportment) to larger social and political transformations (such as the transition from the early modern court society to the modern nation-state). Andre Wakefield has aptly termed this literature the "disciplinary revolution."[13] Out of these concerns, many studies of the intertwined formation of the modern state and the modern subject have emerged.[14] Related questions about the origins of capitalism, scientific and technological practices, and modern rational types of behaviors are on the agendas of sociologists such as Max Weber and Robert Merton, who suggest that Protestant ascetic doctrines were the breeding grounds that enabled and justified such practices and behaviors.[15]

Studies such as Weber's and Merton's, and those of their successors, are also involved in the discussion about the emergence of that "other" part of modernity that I study in this book, the industrial machine age. Questions about the origins of industrial capitalism have generated answers that rely not only on shifts of individuals' behaviors or attitudes, but also of technological and economic conditions. In my study, I integrate artisans' life-worlds, industrialization, and widely shared cultural efforts at forging bourgeois subjects, and thus I rely on theories of "the modern" that speak to industrial aspects as well as to those of subject-formation. Among the crucial ingredients of accounts of the modern world is also the so-called mechanistic worldview of the early modern

13. Wakefield, *Disordered Police State*, 141.

14. Among the most important examples are the works by Michel Foucault and Norbert Elias and, subsequently, feminist theories of modernity and the embodied subject. Grosz, *Volatile Bodies*, vii; Williams, *Dream Worlds*, 197–214.

15. Merton, *Science, Technology, and Society*; Weber, *Protestant Ethic*; Cohen, *Puritanism*; Shapin, "Understanding the Merton Thesis."

period, nurtured and consolidated by philosophers such as Descartes and Newton and describing the world in terms of forces and matter. These theories are often thought of as having been in cross-pollination with the contemporaneous material and artisan cultures of clock- and instrument-making, not least because key mechanical artifacts (such as the clock and the orrery) embody mechanistic principles and, in turn, promote this worldview as well. Android automata from this period have been thought of as products of a marriage of philosophy and craftsmanship, too, since they mimic mechanically a key part of our cosmos, the human body. The intersection of philosophy and clock- and instrument-making—and this is where the above-outlined accounts of the emergence of modern society come full circle—is often considered to have been one of the hotbeds of the Industrial Revolution.[16]

The claims and findings of such large-scale narratives are at times in tension with more detailed small-scale historical studies. Microhistories of early modern Europe have in the past few decades complicated large-scale master narratives about the roots of modernity by carefully studying often marginal figures, attempting to capture their life-worlds and thus complementing broad and overarching assumptions with specific detail, differentiation, and previously unheard voices.[17] Before I describe my study's contributions to various accounts and theories of modernity, I want to discuss three assumptions that are often made about Enlightenment automata and that are intertwined with questions about the origins of modern society. First, Enlightenment automata, as I spelled out above, have been conceived as symbols or oracles of industrial modernity, despite their preindustrial origins, and their makers and audiences have been thought of as anticipating via them, in a more or less uninterrupted process, the perils of the industrial age. A related second assumption maintains that these automata were shown to large audiences in the eighteenth century and that a broad public knew them and shared the automaton makers' sensibilities, namely that automata are reminders of the general threat of social mechanization.[18] A third view states that automata were built

---

16. Bedini, "Role of Automata"; Solla Price, "Automata"; Schmidt-Biggemann, *Maschine und Teufel*; Mayr, *Authority*. On the relationships between artisanship and material and intellectual cultures, see also Smith, *Body of the Artisan*; Bennett, "Mechanics' Philosophy"; Jackson, *Harmonious Triads*.

17. Among the most influential works are Ginzburg, *Cheese and the Worms*; Davis, *Return of Martin Guerre*; and Darnton, *Great Cat Massacre*. See also Ginzburg, "Microhistory."

18. Ideas that large eighteenth-century audiences were captivated by automata's threatening mechanics originated in the nineteenth century. David Brewster, for example, wrote in 1832 that Vaucanson's celebrated duck "excited much interest throughout Europe" and that von Kem-

by artisans or philosophers as replicas of the human body in order to better understand its anatomy and physiology. Under this reasoning, automata served primarily epistemic functions to help us understand better the constitution of the human body and its status vis-à-vis the mind, thus yielding knowledge about the human or animal organic forms.[19]

My study's findings challenge these assumptions. The manufacture and display of androids was a strenuous mechanical, economic, and cultural affair in the eighteenth century. The automata had to be produced with considerable investment of time, money, and skill, often for capricious and obscure markets. This was possible only in circumstances in which unusual skill, labor, and a specific market coincided with each other and helped bring about a "business model" for proto-industrial artisanship.[20] The delicate automaton mechanisms, furthermore, had to be kept maintained, especially when transported in adverse conditions. One would have to dedicate one's entire business and life if one wanted to display them to broad audiences. Indeed, showing automata was only one of many options available to artisans, entrepreneurs, and impresarios, and not necessarily the most advantageous in any given set of cultural expectations and exigencies. In the cases of the two automata that I look at, both sets of makers had other business interests and ended up not pursuing automaton-making and -showing; they built no additional automata. Instead, they expanded their original businesses, clock-making and furniture-making. Their choices were

---

pelen's chess automaton was exhibited after its completion "to thousands." Brewster, *Letters*, 272. Hermann von Helmholtz claimed in 1865 that artisans in the eighteenth century built automata to replace human labor at the workplace. Helmholtz, *Wechselwirkung*, 6. A recent example is Gaby Wood's 2002 account of Enlightenment automata, in which she assumes that audiences reacted to them in coherent and universal ways, namely, with a fear that crossed time and space. Wood, *Living Dolls*, xiv. Frank Wittig's work on machine-men relies on the same combination of claims that large numbers of people watched automata with fear or apprehension. Wittig, *Maschinenmenschen*, 25. I explore this legend-building vis-à-vis "publics" in chapter 6.

19. Jessica Riskin, in "Defecating Duck," 610, calls eighteenth-century automata "mechanical experiments," naming Jacques de Vaucanson the founder of this approach. She also explicitly mentions the "epistemological utility" of Vaucanson's flute player (616). Martin Lister shares the view that automata simulated mechanistic physiology. Lister et al., *New Media*, 325. David Brewster and Hermann von Helmholtz emphasized in the mid-nineteenth century the same epistemic purpose. Brewster, *Letters*, 269; Helmholtz, *Wechselwirkung*, 6. This epistemic quality is rightfully ascribed to Jacques de Vaucanson and his automata. It is, however, often extended to all other eighteenth-century automata, and there is only weak evidence to justify this. An example for this faulty extension is found in Poritzky, *Julien Offray de Lamettrie*, 174–75.

20. I owe the label "business model" for my artisan workshops to Mario Biagioli, who used the term to refer to not only the plan that the artisans had in their minds to operate their firms, but also the larger set of conditions (clients and demand for and supply of commodities) that made possible this kind of business.

entirely unremarkable against the background of contemporary artisan entrepreneurship. This suggests that, for both artisan families, the epistemic relevance of automaton-building was at most a secondary motivating factor.

I also address the assumption that the automata of the eighteenth century were shown to a large number of people, that a large number of people knew about them, and that there was a substantial public whose imagination was captured by them in a sustained way. While there was certainly a pan-European spectacle culture in the late eighteenth century, of which android automata were a part, we cannot assume, based on that, that they were necessarily central to it.[21] I analyze the available evidence about what kinds of publics my two women automata were shown to and what kinds of people commented on them. Although some people's imaginations were definitely captured by them, so that they used piano-playing women automata as a motif for philosophical and poetic explorations, the generalization to a significant impact on a larger public is difficult to support.[22] Whatever exhibitions these automata were part of, they left behind only a small record on which we can base our understanding of the relationship between eighteenth-century automata and their contemporaneous "publics." In chapters 2 and 3 I analyze documentation about the automata and present crucial details about the automata's staging and performance, but there is no evidence for a large, coherent public that witnessed, through them, the mechanization of the human body and soul.

As I indicated, later renderings of eighteenth-century automata, presented in nineteenth- and twentieth-century texts, tended to portray the automata as objects of wide interest during exhibitions and as objects that captivated a large proportion of the general population. And this claim, that eighteenth-century automata provoked "much interest" wherever they went, has its own interesting history. Authors of these reports often do not cite sources to support claims for the automata's popularity, and I suspect that the authors projected their own contemporaneous experiences of large publics back into the eighteenth

21. There was a general emerging public sphere in the eighteenth century in which spectacles coincided with the discourse of natural philosophy. See Hochadel, *Öffentliche Wissenschaft*; Hankins and Silverman, *Instruments and the Imagination*, 46–81; and Evans and Marr, *Curiosity and Wonder*.

22. It is certainly true that some eighteenth-century automata attracted a great deal of attention. The best examples are Vaucanson's flute player, galoubet player, and duck, as well as von Kempelen's chess player. These automata were deliberately produced for this function, and their makers promoted public exhibitions of them.

century, a period during which this kind of public did not exist. This is one indication of the nineteenth- and twentieth-century legend-building around Enlightenment automata, which ties in with a great many misconceptions about the history of the emergence of publics themselves.[23]

It is the idea of a continuity connecting androids from the eighteenth century (or even from the ancient period) to the robots, science fiction figures, and artificial life of the present time that has inspired the related idea that there has *always* been a large and interested audience for robots and that this audience's concerns and preoccupations about machine-men coincide with our own current concerns over the status of humans in late-industrial societies.[24] The Enlightenment is an important period in the emergence of mass culture (and all other facets of modernity), but its "public sphere" is not comparable to what we see in the nineteenth century, and neither is its machine culture. Notwithstanding this, scholars have used automata to conceptualize problems of mass culture, industrial culture, and the perils of modernity.

Such ideas render Enlightenment automata more modern than they are. They turn them into symbols for concerns about industrial modernity and neglect to take sufficiently into account their eighteenth-century roots. The assumptions that I listed accompany the automaton all the way to the end of the twentieth century, and they are crucial to our understanding of androids and our ways of using them to understand the human condition in the industrial world. In contrast to such generalizing historical interpretations of eighteenth-century androids, I demonstrate for my two women automata how they embodied not a universal philosophical system but a cultural practice and political ambition specific to their own time. I also demonstrate, on the basis of other historical work, how the "roots" of industrialization are found not only in mechanistic philosophy but also in emerging novel business models, economic and labor practices, and domestic and international markets.[25]

My book makes a few interventions in this complex of themes and

23. See note 18.

24. This idea is, again, present in Gaby Wood's work. She draws a close connection between a staging of the Jaquet-Droz harpsichord player (described in note 18 above) and the founding of the first AI lab at MIT in the 1950s, arguing that "although what goes on at MIT is undeniably sophisticated and cutting-edge, the most extraordinary thing about it is how consistent it is with concerns that have been around for centuries." Wood, *Living Dolls*, xxv.

25. Hafter, *Women at Work*; Kriedte, Medick, and Schlumbohm, *Industrialization*; Mendels, "Proto-Industrialization"; Safley and Rosenband, *Workplace*; Berg, Hudson, and Sonenscher, *Manufacture in Town and Country*.

histories. I suggest, on the one hand, the need for more circumspection in the assumptions we make about automaton-makers, the people who saw the automata, and those who commented on them in writing. On the other hand, I offer more specific and detailed, and ultimately more far-reaching, claims about the lessons that these automata teach us, regarding two matters: first, how we, in an industrial and postindustrial age, try to make sense of our relationship to machines and, second, the artifacts and metaphors that we use to characterize this relationship and make it tangible for ourselves.

In this endeavor, I benefit from the guidance of historians of technology of the past fifty or so years who have navigated the assumptions, scattered records, and legends that present themselves to historians of the automaton. In 1964 Derek de Solla Price and Silvio Bedini contributed essays to an early issue of *Technology and Culture*, and they went on to be cited in every subsequent automaton study. They integrate automata more rigorously into intellectual and social contexts than did survey works from earlier in the century, but both essays also have distinctly universalist and teleological elements, which greatly influence their overall arguments. Their analyses provide little space to differentiate between automata's purposes and functions in different contexts and periods, or even within one and the same period.[26]

Jessica Riskin, Simon Schaffer, and Minsoo Kang have recently offered studies that avoid precisely this kind of teleology. Riskin and Schaffer study in historically specific ways connections between eighteenth-century automata and their original philosophical and political contexts. Riskin argues that automata took on new epistemic functions in the eighteenth century: as mechanical replicas of humans (and of animals), according to her, they provided knowledge about humans' and animals' anatomy and physiology. She contrasts Vaucanson's automata with automata outside of the eighteenth century, in particular the automata of the seventeenth century, whose builders merely aimed to provide mechanical representation or imitations that were "purely external." By the eighteenth century, she argues, Vaucanson made the transition from representation to simulation.[27] Riskin generalizes her finding and assumes that all eighteenth-century automata were of epistemic relevance to their makers and the public. This is where my account differs from hers and complements it by draw-

---

26. Bedini, "Role of Automata"; and Solla Price, "Automata."
27. Riskin, "Defecating Duck," 601–6 (quote on 601).

ing further distinctions between the automaton-makers of the eighteenth century: I understand the women automata by the Jaquet-Droz and Roentgen families to be not epistemically relevant simulations of live bodies, but rather mechanical replications of cultural and political body practices and ambitions.[28]

Relatedly, Simon Schaffer explores how leading figures of the European Enlightenment used automata and automaton metaphors to articulate key issues of the period, focusing in particular on ideas about social hierarchy and economics. He shows how pervasive "automaton-talk" was and how widely the motif was employed, using this proliferation to understand better the Enlightenment's affinity to metaphors of the machine.[29] Schaffer's studies focus less on the relation of specific, individual automata to the discourses generated around them, and more on the relation of eighteenth-century automata as a group to contemporaneous enlightened philosophies, mechanical principles of both nature and society, and materialism and engineering. Schaffer also addresses automata's status as commercial commodities.[30] I complement his work by putting a lens on two specific automata, their reception at the time of their creation, and the ways in which they were subsequently lumped together in the nineteenth and twentieth centuries with their fellow eighteenth-century automata as universal mechanical metaphors. Minsoo Kang has revived the much-needed genre of survey histories of automata, machine-men, and their conceptualization. He traces the automaton and the robot from the ancient period to the middle of the twentieth century, focusing in particular on the shifting meanings of these objects in intellectual and literary texts.[31] While my work, like his, relies on political philosophy and literary production, I select a particular period of automaton-making, protoindustrial and early modern in its character, and investigate through selected automata the relationship of this period to the modern machine age.

28. Riskin, "Eighteenth-Century Wetware," 101; and Riskin, "Defecating Duck," 599. Counterexamples to her claim are indeed the Jaquet-Droz automata, whom she takes to be "also anatomical and physiological simulations." "Eighteenth-Century Wetware," 102. My findings regarding the world of the Jaquet-Droz family do not support Riskin's suggestion. Father and son Jaquet-Droz were first and foremost clock-makers, and their automata are not designed to simulate internal parts of the human body. Instead, they are painstakingly designed to perform a specific cultural body technique.

29. Schaffer, "Enlightened Automata"; Schaffer, "Deus et machina."

30. The automaton that Schaffer deals with in the most depth, the famous chess automaton by Wolfgang von Kempelen, is an exception to this. Schaffer, "Enlightened Automata," 154–64.

31. Kang, *Sublime Dreams*, 11.

The Android Automaton in Early Modern Philosophy

I want to provide background for two more aspects of eighteenth-century automaton culture before I present my study of the two music-playing women automata. First, I describe a few key strands of eighteenth-century philosophy with which the automata are frequently associated. This association is not supported by my study, however. What I find is that the two women automata are closely related to early industrial cultures and early bourgeois emancipatory politics but that their connection to contemporary philosophical discussion is less close. It is only in later periods, once again, that they get lumped together into one category with other eighteenth-century automata and rendered embodiments of early modern mechanistic philosophy. Second, at the end of the chapter, I outline details about the construction of the other eighteenth-century automata, as well as about previous automaton-making, to stake out the cultural and economic horizon in which my two automata came into being and the purposes for which they were made. There are remarkable similarities and differences between automaton-makers, especially in the eighteenth century, and a comparative perspective allows for further insights into the variety of relationships that existed between automata, automaton-makers, and the original and subsequent audiences of automata.

One reason Enlightenment automata were so popular in their time, and continue to be so in our own, is that they speak to key philosophical discussions of the early modern and modern periods. Automata appear in discussions about the Cartesian mind-body dualism, the Newtonian "mechanistic" worldview, the question of human freedom in ethics, and the question of God and the clockwork structure of the universe that he created. Metaphysical as well as epistemic and political concerns are closely linked to human-machine questions and lend themselves to the automaton motif. In the Enlightenment, furthermore, there were stronger challenges to clerical, political, and epistemic authorities than in previous periods, and the themes inherited from the seventeenth century fermented intensely in the era and converged with one another.

Among the best-known philosophical works of the time relating to automata was *Man a Machine* (1748) by the philosopher and physician Julien Offray de La Mettrie. The evocative title might suggest that this work was related to mechanical arts, automaton-making, or other ways of thinking or building machine-men. However, the book was concerned primarily with Cartesian and post-Cartesian materialism: with the nature of the human mind and soul, their relationship to the body, and the relationship between materialism and atheism. La Mettrie did rely on Vaucanson's automata, which were by the 1750s well-known entities.[32] The relationships between motion, matter, and forces remained unresolved in the eighteenth century and afterward and continued to generate philosophical, theological, and political concerns. The Cartesian mechanistic ideal provided inadequate explanation for a wide range of phenomena in the biological and physical worlds, and antimaterialisms in various forms continued to attract advocates. The most pressing question was whether matter in itself could engender motion, either in biology or in physics.[33] The nineteenth and twentieth centuries inherited these debates, and there was a continuing stream of editions and translations of La Mettrie's *L'homme machine* into the 1920s. These books and the debates of the nineteenth and twentieth centuries touched on matters in physiology (also picking up on new findings in evolutionary biology), theology, and epistemology that continued to rely on the mechanical androids from the eighteenth century and thus kept them alive as a motif.

The android automata of the eighteenth century were not mere illustrations or metaphors in philosophical discussions. They were used as rhetorical instruments in social and political debates. In Newtonian thinking, for example, the inadequacies of the Cartesian system were interpreted in terms of theology and government and often expressed with automaton metaphors. The so-called Boyle lectures were a series of sermons endowed through the will of Robert Boyle (1627–1691) and were part of the larger project in the eighteenth century of popularizing Newtonian thinking. Automata played a key role in the lectures' attempt to connect the Francophone debate around materialism and atheism with the Anglophone world of Newtonianism.[34] One of the lectures, a confutation of atheism and challenge of materialism, was delivered by the Anglican clergyman and theologian Richard Bentley. In it, he reiterates materialist reasoning by explaining that material-

---

32. La Mettrie mentions Vaucanson's duck and the flutist, calling Vaucanson a "new Prometheus." La Mettrie, *L'homme machine* (1865), 140. Vaucanson's fame was primarily due to his 1738 book *Le mécanisme du fluteur automate*, which he promoted actively, and the article "L'automat" in Diderot's and d'Alembert's *Encyclopédie*.

33. Vartanian, "Biographical Sketch of La Mettrie," 18.

34. Another example of a philosopher who used automata to promote Newtonianism was John Theophilus Desaguliers. A distinguished natural philosopher, a skilled experimentalist, a popular lecturer, and a committed Newtonian, he translated Vaucanson's book on his flute-playing automaton into English in 1742.

ists consider mankind to be (merely) an engine of a "finer make and contexture" than animals; materialists point out, says Bentley, how closely animals apparently approach human reasoning merely through mechanisms in their bodies. He finishes by saying, "I must confess that the Cartesians and some others, men that have given no occasion to be suspected of irreligion, have asserted that brutes are mere *machines* and *automata*."[35]

The materialism debate also spawned prominent thought experiments about the mind-body problem, involving automata and automaton-related objects. Etienne Bonnot de Condillac, one of the leading figures of the French Enlightenment, published in 1754 a fundamental treatise on sensations, with a thought experiment about a statue at its center. The statue has no senses at first and then acquires senses one by one. With its help, Condillac illustrates the radical empiricist position that no cognitive abilities, such as will, desire, and judgment, were innate; instead, all derived exclusively from sensation.[36] He joined discussions on empiricism and human-animal distinctions in the 1740s and 1750s and frequented the Parisian salon society, and like many others at the time, he integrated John Locke's empiricism with the Newtonian search for fundamental principles in post-Cartesian French philosophy.

Animal, machine, God, and spirit—these entities and their philosophical analyses served as foils to discuss what it meant to be human, and it is not least for this reason that automata proved to be such persistent figures in these debates. I am dealing with objects and ideas around them that have broad meanings and are widely known. Few objects are so replete with meaning as automata are and have absorbed so many connotations over the past two centuries.

Automaton-Making in the Eighteenth Century

The eighteenth century was an unusually prolific period in the history of automaton-making, and the automata made during this period

35. Ibid., 28–29 (Bentley's emphasis).
36. Condillac, *Traité des sensations*. See also Condillac, *L'origine des connoissances* (1746); and Condillac, *Traité des animaux* (1755). The relationship between man and animal, and the metaphysical spin-offs of this question, had been at the center of French philosophy ever since Descartes had given the problem a foundation in the early seventeenth century. Coski, "Condillac," 57. Mike Lynch and Harry Collins pick up this theme in a special issue of *Science, Technology, and Human Values* on humans, animals, and machines in 1998 and emphasize likewise its Cartesian origins. Lynch and Collins, "Humans, Animals, and Machines," 371–72.

reflect the particular combination of skills, technologies, and market conditions that were present at the time. The engine that moved them was mechanical clockwork, hidden in their bodies, and they were individually produced as unique pieces. From a technical viewpoint, it is particularly difficult to make an automaton perform activities in the same way that a human would and at the same time make the mechanism small enough so that it can be hidden entirely in the automaton's body. The automata's minute and measured motions, combined with the constraint of limited space, made constructing such artifacts much more time-, skill-, and labor-intensive than building nonandroid mechanical devices that represent human activity.[37] The only people able to afford to construct such artifacts were well-to-do artisans, artisans in employment or under the patronage of a court, or artisans who could expect members of the nobility to buy their products.

The eighteenth century provided cultural and economic conditions under which such enormous efforts could go into producing mechanical replicas of human bodies and motions. Two factors in particular contributed to the extraordinary automaton-making: the final blossoming of traditional artisanship before the full onset of industrialization, and an intensification on the European continent of the ostentatious court culture instigated in the seventeenth century by Louis XIV.

Michael Stürmer has demonstrated how the final decades of Europe's ancien régime brought about conditions under which the traditional arts and crafts flourished for a final time. Less than thirty years later, Europe was in many places undergoing rapid industrialization.[38] In noting a bifurcation of preindustrial artisanship and industrial factory production in eighteenth-century Europe, Stürmer's interpretation not only suggests that these two modes of production overlapped with each other much more than the term *Industrial Revolution* would suggest; it also suggests that no obvious "decline" of traditional artisanship was occurring along with the "rise" of industrialism during the century. Specifically, clock-makers and furniture-makers in the eighteenth

37. To fulfill the standard criteria, an android automaton has to be a mechanical device that looks like a human, moves as a human does, and carries the driving engine hidden entirely within its body. Diderot's *Encyclopédie* of 1754 defines "automaton" as a "device that moves by itself, or a machine that carries in itself the principle of its motion." It defines "android" on this basis as an "automaton with a human figure, equipped through the means of a certain spring drive to act like and conduct functions which apparently resemble those of humans" [1:448]. The entry's author bases his text on contemporary android automata and talks at some length about Vaucanson's automata.

38. Stürmer, "Die Roentgen-Manufaktur," 26; Stürmer, *Handwerk*, 212–75; Stürmer, *Luxus*, 15; Stürmer, *Scherben*.

century were better trained and had more monetary resources and trade contacts than in earlier periods. There had been a great deal of change in mechanical artisanship since the prosperous clock-making eras in Nuremberg and Augsburg in the sixteenth and seventeenth centuries, when clock-making and the clock-making trade were more elitist. By the eighteenth century, mechanical trades were more widespread, involved more agents and more labor, and were geared toward foreign markets. In the period after 1810, to draw a contrast with a later era, the production of dolls, puppets, and other humanlike artifacts was industrialized, and their design, use, and consumption changed rapidly.[39]

There is no monolithic identity that we can assign to individual artisans or entrepreneurs in the eighteenth century. The specific relationship between the maker and the receiver of an automaton could take on a variety of forms. In some cases, automata were presented by their makers as gifts to princely rulers with the hope of receiving privileges or patronage in return. On other occasions, princely rulers themselves commissioned mechanics or clock-makers employed at their courts to produce automata either as gifts for other sovereigns or as part of the court's cabinet of curiosities.[40]

Throughout the early modern period, the court society and its spectacle culture were an influential and stimulating backdrop for the making and trading of automata. European courts had developed, from the fourteenth century, into a stage for high and low politics, a medium for the manifestation and legitimation of political power, and a community serving as a model for the society outside of the court. This variety of functions demanded rigid rules of etiquette and carefully negotiated hierarchies, and also ostentatious displays of power, spectacular events, ritual, theater, luxury commodities, and an audience.[41] Diplomacy took

---

39. The most authoritative and comprehensive survey works in automaton-making are by Alfred Chapuis and his collaborators: Chapuis and Droz, *Automata*; Chapuis and Droz, *Les Automates des Jaquet-Droz*; Chapuis and Gelis, *Le monde des automates*. See also Cipolla, *Clocks and Culture*, 61–62; Richter, *Wunderbares Menschenwerk*, 14–15; Sauer, *Marionetten*, 19; Swoboda, *Der künstliche Mensch*, 12. On the nineteenth century, see Bailly, *Automata*; Beyer, *Faszinierende Welt der Automaten*. On "courtly" consumption, see Williams, *Dream Worlds*; and on the recent scholarship on "things," see Daston, *Things that Talk*; and the special issue of *ISIS* from March 2007 (in particular Ken Alder's "Introduction").

40. Privileges could include exemption from guild regulation, financial support such as pensions, and permanent positions at courts. Swoboda, *Der künstliche Mensch*, 71. Both automata considered in this book are examples of this practice of gift-giving.

41. The absolutist court of Louis XIV was the model for other courts in Europe in the seventeenth and eighteenth centuries. Its organization was central to Louis's strategy of creating social order and cohesion in the rest of his kingdom. Between four thousand and five thousand

on a key role in the forging of hierarchies and pecking orders between courts. Envoys reported in detail to their home residences about the size and habits of the courts to which they were accredited, and these communications generated something like a pan-European "public sphere" whose purpose was to increase princes' prestige and reputation among the big courts and dynasties in Europe.[42] In this public and international court culture, automata and related artifacts were suitable items of wonder and spectacle; they provided incentive for mechanical innovation, and they were also suitable items for gift exchange among parties of political power. Technical ornamentation of palace parks and gardens was part of the same culture that inspired automaton-making.[43] It is no coincidence that the primary commodities that the Jaquet-Droz and Roentgen families produced in their respective workshops—clocks and furniture—were also primary objects for furnishing these court stages. The two women automata were part of a continuous spectrum of luxury commodities.

Automaton-Making before the Eighteenth Century

The production of automata goes back to antiquity, as seen in the examples of technical-mechanical displays of human action described in the first-century CE works of Hero of Alexandria.[44] Hero's works include

nobles attended the court at any given time. Sturdy, *Fractured Europe*, 292–95. On early modern court culture and its manifold interrelations with political power, the arts, and artisanship, see also Holme, *Princely Feasts and Festivals*; Apostolidès, *Le roi-machine*; Blanning, *Culture of Power*, 29–52.

42. Kunisch emphasizes the role of diplomacy in this intercourt competition, and he uses the term *Öffentlichkeit* to describe it. Kunisch, *Absolutismus*, 63–72. During the eighteenth century, the small courts of the minor German-speaking states rose to meet Louis XIV's standards, engaging in a strenuous competition to pursue their political agendas. As a result of this competition, costs of court households quintupled in the fifty years after 1672. Blanning, *Eighteenth Century*, 18–22, 140–44.

43. A specific example here is Vaucanson's flute player, which was supposedly modeled after a statue by Antoine Coysevox in the Tuileries. Coysevox was a sculptor at Louis XIV's courts. Prominent mechanics and scholars such as Caspar Schott or Salomon de Caus were in charge of musical and mechanical spectacles in royal gardens in the sixteenth and seventeenth centuries and thus influenced automaton-making. See Kircher, *Musurgia universalis* (1650); Schott, *Mechanica Hydraulico-pneumatica* (1657); Caus, *Les raisons des forces mouvantes* (1615). Historical analyses include Ord-Hume, *Clockwork Music*, 63; Protz, *Mechanische Musikinstrumente*, 22; and Chapuis, *Histoire de la boîte à musique*. See also *Wunder und Wissenschaft*.

44. My survey here is rather clichéd and opportunistic: I show in chapter 6 (notes 2, 7, 8, 15, and 17) how the same type of survey histories of automaton-making, covering the ancient period to the Enlightenment, came into being in the nineteenth century, was popularized in encyclopedias, dictionaries, and numerous other texts, and then became part of histories of industrialization or materialist philosophies. My survey is based on the same nineteenth-century

two treatises, *Pneumatica* and *Automata*, both of which report about automata, including water clocks, hydraulic devices depicting religious worship, and two automaton theaters. The *Pneumatica* and *Automata* influenced automaton-making for centuries to come. Hero's studies were first translated into Arabic, and it was in the Arabic-speaking world that the first automata were built. In Europe, automaton-building did not start until the thirteenth century, after Hero's work was translated into Latin in Sicily. Most of the automata built in the Renaissance and in the baroque period relied at least in part on Hero's instructions.[45]

Around the late Middle Ages and the early modern period, a number of legends appeared pertaining to the history of automata. Several prominent men, among them Gerbert d'Aurillac (ca. 950–1003), who later became Pope Sylvester II; Albertus Magnus (ca. 1193–1285); Roger Bacon (ca. 1214–1294); and Regiomontanus (1436–1476), are alleged to have built or owned automaton-like devices. Magnus's "Iron Man" worked as a guard at his door, asking visitors about their requests and then deciding, upon their answer, whether or not to grant them entrance. Although descriptions of this automaton are scarce and vary greatly, there is some consistency in the legend according to which Thomas Aquinas, Magnus's favorite pupil, demolished the automaton. Some sources say it was because the automaton's constant babbling disrupted Thomas's work; others report that Thomas recognized Satan himself in the automaton. D'Aurillac supposedly spent ten years building a talking head and consulted it for advice on difficult questions: the head was capable of answering questions clearly with "yes" or "no." Bacon is thought to have built a talking head made of bronze that advised him, upon inquiry, to build a wall of bronze around England to protect the country from attacks from the sea. Regiomontanus is reported to have constructed a mechanical fly as well as a mechanical flying eagle. On the occasion of Emperor Maximilian's arrival in Nürnberg in 1470, the eagle supposedly flew toward the emperor to greet him, circled around him once, and flew back to the castle wall.[46]

By the beginning of the fifteenth century, rapid developments in clock-making and the mechanical arts as well as increasing emphasis on ostentatious representation of political power at princely courts entailed that automata were no longer construed as magical and technical

---

encyclopedias and their twentieth-century renderings found in Heckmann, *Die andere Schöpfung;* Sauer, *Marionetten;* Beyer, *Faszinierende Welt der Automaten;* Swoboda, *Der künstliche Mensch.*

45. Heckmann, *Die andere Schöpfung,* 26–53.

46. Swoboda, *Der künstliche Mensch,* 46–66; Sauer, *Marionetten,* 18; Heckmann, *Die andere Schöpfung,* 87. On medieval technology, see also White, *Medieval Technology.*

singularities attached to eminent historical figures. Rather, they were now produced and sold by artisans and merchants as commodities, as representational or spectacular artifacts, or as precious toys. The famous clock-makers and goldsmiths in Augsburg and Nürnberg, for example, constructed elaborate ships, animated nativity scenes, highly ornamented clocks, and animated pictures, often displaying biblical scenes such as the creation. Among the most important pieces of this period was an automatic toy in the form of a golden coach, made for Louis XIV. Toys were also important for entertainment at dinner tables and feasts. These automata were often made in the form of animals, such as spiders, and were meant, among other things, to "tease the women" sitting at the table. Festive meals were actually staged as "mechanical" works of art in this period: food was served in the form of artificial landscapes, buildings, figures, and parks, and it was often court engineers and mechanics who were in charge of staging these elaborate meals.[47]

The Automata of the Eighteenth Century

The dozen or so eighteenth-century automata that we know of were made by natural philosophers, mechanics, cabinetmakers, and clock-makers; they were built for purposes of experimentation and marketing or to fulfill a commission; and they displayed artistic techniques, were fantasy objects, or were philosophical experiments. Taken together, these automata and their makers circumscribe the cultural and economic milieu of automaton-making in the eighteenth century of which the Jaquet-Droz and Roentgen families were a part. This milieu included the automaton-makers' ambitions, the cultural and economic relationships they had to their respective employers, and the status of their automata within their lives.

As I mentioned earlier, Jacques de Vaucanson is maybe the best known of eighteenth-century automaton makers. He built three automata—a flute player, a galoubet player, and a duck—in the 1730s and started showing them on stages in Paris in 1738.[48] The flute player was almost life-size (1.65 m), made of wood, and played twelve melodies.

---

47. Heckmann, *Die andere Schöpfung*, 98–99, 123–49; Sauer, *Marionetten*, 19; Beyer, *Faszinierende Welt der Automaten*, 53; Swoboda, *Der künstliche Mensch*, 69, 73.

48. A galoubet is a shepherd's flute from Provence. The player accompanies himself on a drum. See Vaucanson, *Le Mécanisme du fluteur automate*, 21.

The air came from three bellows run by clockwork inside the body and passed through the lips to produce tones on the flute. The second musician automaton—the galoubet player—required a similar level of skill and mechanical complexity.[49] Vaucanson's duck became his most famous and most popular automaton, not least because its imitation of life extended to its ability to swallow food, digest, and defecate. Vaucanson's automata provoked the attention of influential people such as Voltaire, Frederick the Great, and the French general minister to Louis XV, the Cardinal de Fleury (a grandson of Louis XIV), who appointed Vaucanson in 1741 to the post of inspector of silk mills. Vaucanson was also eventually appointed "associated mechanician" at the Académie des Sciences.[50]

Several aspects of Vaucanson's work are relevant in relation to other eighteenth-century automata, their makers, and the historiography of automaton-making. First, Vaucanson's ambition to produce automata that imitated living beings in such detail was groundbreaking and trendsetting. His automatic musicians worked like human musicians, and his duck amounted to a mechanical experiment in animal physiology. Vaucanson even studied human flute players to design his mechanical one. The mechanical sophistication and the verisimilitude of his flute player was unprecedented at the time. Second, Vaucanson promoted his automata to natural philosophers by organizing shows in the Académie des Sciences and by publishing a manual on the technical details of the flute player's internal mechanisms as well as the physical theory underlying them.[51] Doing so, he followed contemporary protocol on improving one's standing as a natural philosopher and gaining access to the distinguished Académie.[52] Third, after a few years, Vaucanson took up a prestigious government position in the silk

49. According to the legend, Vaucanson had to build three hundred flutes before he found the one that served his purposes. Swoboda, *Der künstliche Mensch,* 92–93.

50. Vaucanson was supported by court patronage and his appointment at the silk mills, and he won the competition for the inspector position against Denis Diderot. He became a member of the Académie des Sciences in 1747. Ibid., 92; Riskin, "Defecating Duck," 601; Schaffer, "Enlightened Automata," 126.

51. The manual *Le Mécanisme du flûteur automate* was very quickly translated into English and German. It explains the physics of the sound production in the flute and outlines acoustical experiments (about the influence of air vibrations on the pitch and the frequencies of tones) conducted with the flute player. Jessica Riskin suggests that Vaucanson's flute player tests were part of an experimental approach that marked the beginning of what is now called artificial life. Riskin, "Defecating Duck," 615.

52. Hahn, *Anatomy of a Scientific Institution.* I owe special thanks to J. B. Shank here, who shared with me his insights into the institutional structures in the Académie in the eighteenth century and the ways people such as Vaucanson used them.

industry and stopped pursuing the business of automata. He was an ambitious and skilled mechanic whose work provided opportunity for upward social mobility. He promoted his automata systematically, and they became the epitome of such creations and the model for later interpretations of all other eighteenth-century automata.

The "Miracle Writing Machine" (Allesschreibende Wundermaschine) was built by the German mechanic Friedrich von Knaus (1724–1789), who presented it in 1760 to the empress Maria Theresa in Vienna. This automaton is not an android automaton, strictly speaking, since the writing mechanism is not located in the body of the writing figure, but it is one of the first examples of a human figure to display the fundamental cultural technique of writing.[53] There is not a great deal of archival material on Friedrich Knaus's life, but the details that are known indicate that he was a fairly typical mechanic at a European court in the eighteenth century. His father, Ludwig Knaus, became a clockmaker in 1737 at the court of Count Ludwig VIII in Darmstadt and brought thirteen-year-old Friedrich with him. Ludwig Knaus was successful and eventually became court mechanic and valet in 1749.[54] In 1739 Count Ludwig ordered Friedrich Knaus to build a precious clock as a gift for Maria Theresa. The clock's display was a theater of adoration, employing baroque conventions of representing power.[55] After this success, Friedrich Knaus traveled beyond Darmstadt and in 1753 presented a writing machine to Louis XV in Versailles. The machine was later sold to Prince Karl Alexander of Lorraine, brother of the Emperor Franz and governor of the Austrian Netherlands. On this occasion, Karl Alexander gave Friedrich Knaus a position as mechanic at his court and, three years later, took him with him to Vienna, where Friedrich became court mechanic and later inspector of the court's Imperial and Royal Physical Cabinet. It may have been in this period that Friedrich Knaus ennobled himself and added the "von" to his name.[56]

In Vienna, Friedrich Knaus pursued his plan to produce an even

53. For insightful elaboration on handwriting as a cultural technique, see Funk, "Tiere und Zähne."

54. See the folder "Friedrich von Knaus, Allesschreibende Wundermaschine" in the archive of the Technisches Museum, Vienna; and Beyer, *Faszinierende Welt der Automaten*, 44.

55. With this clock, Count Ludwig aimed to express not only his admiration for the imperial couple but also his gratitude, since they had helped him financially in the past. Heckmann, *Die andere Schöpfung*, 236.

56. His titles were mostly his own inventions. The trouble he got himself into at the end of his life was partly because he actually held hardly any of the titles he had bestowed on himself. Ibid., 237; Swoboda, *Der künstliche Mensch*, 115; Beyer, *Faszinierende Welt der Automaten*, 57.

more elaborate writing machine and in 1760 presented the culmination of his efforts. The Miracle Writing Machine is two meters high, the clockwork is situated in a spherical container meant to represent the earth, and the writing figure is a small goddess, holding a quill in her hand and sitting on top of the sphere on a platform shaped like a cloud. Behind the goddess is a board with a piece of paper on which she is writing. The text to be written is programmed on a pegged barrel, but there is also a register that makes it possible to program the machine manually (this would enable a mechanic to make the machine look even more "miraculous" because it would be able to write what is being "dictated" to it). Each peg corresponds to a letter, and the pegs can be arranged on the barrel according to the desired text. The pegs conduct the motion via a set of levers and axles to cams that have the motion of the writing hand encoded on their profiles. After each letter, the writing board shifts to the left, and after writing a few letters, the goddess dips her quill into the ink and continues to write. A book entitled *Selbstschreibende Wundermaschine, auch mehr Kunst- und Meisterstücke* was published anonymously in 1780 in Vienna to boost this machine's fame. It turned out that Friedrich von Knaus himself asked a monk named "Father Marianus" to write the encomium.[57]

The Kaufmann family, the makers of the "trumpeter automaton" of 1812, ran a firm that was dedicated to building musical mechanical toys and instruments and selling them to kings and princes and other well-to-do-clients. They specialized in building musical automata and mechanical musical instruments that were as close to the original as possible in their acoustical reproductions. Johann Gottfried Kaufmann, the father, was born in 1751, the son of a hosier near Chemnitz (Saxony). Johann Gottfried's own parents were very poor, and he initially underwent an apprenticeship in his father's craft. When his parents died in 1770, he went to Dresden, where he undertook another apprenticeship, this time with a mechanic who repaired clocks and other mechanical commodities.[58] His master died in 1772, even before Johann Gottfried had finished his apprenticeship, but he continued his master's workshop on behalf of his master's widow. Johann Gottfried had always been fascinated by music, and he began to combine clockworks

---

57. Heckmann, *Die andere Schöpfung*, 238; Swoboda, *Der künstliche Mensch*, 113–15; Beyer, *Faszinierende Welt der Automaten*, 56–57.
58. Engelmann, "Aus der Geschichte," 797. Rebecca Wolf uncovers crucial details on the relationship between hosiery-making and mechanical arts and automation, such as the parallel skills and tools in the two crafts. Wolf, *Kaufmanns Trompeterautomat*.

with musical instruments, such as a harp and a flute.[59] As his musical clocks became increasingly complicated and elaborate, they received attention from wealthy and influential people. The elector Friedrich August III of Saxony, for example, bought a clock with harp and flute tunes for his wife in 1787.[60] In 1785 Johann Gottfried's son Friedrich was born. He became an apprentice with a clock-maker in Dresden and traveled from 1803 to 1805 through Germany, France, Switzerland, and Austria. He spent considerable time in Vienna and studied music there, as well as mechanics and acoustics. When he returned to his father, his technical and musical training were a major asset to his father's workshop, and he became his father's closest assistant. Together they built their first major mechanical musical work with trumpets and timpani, a huge artifact called Belloneon. The first of these massive instruments was purchased by the Prussian king Fredrick William III in 1805.[61] Another instrument they built together was the so-called harmonichord in 1810; it was not an automatic instrument but an imposing keyboard instrument. Friedrich became a master at playing this instrument, and Carl Maria von Weber composed an Adagio and Rondo with orchestra for it.[62] Friedrich eventually was offered an appointment as harmonichord player from the Count of Darmstadt. However, when he also received an offer of a lifelong salary from the king of Saxony, he decided to stay in Dresden. In the later nineteenth century, Friedrich Kaufmann and the following generations of the family established a so-called acoustical cabinet, which was a widely known and popular attraction.[63]

The Kaufmanns built an android automaton that still exists today (although it is no longer functional) and is conventionally counted among eighteenth-century android automata. The figure, built from 1810 to 1812, displays a trumpet player. It is life-size (169.5 cm) and the mechanism is located in the head and the upper body.[64] The re-

59. By the year 1774 he had already built his first musical clock, a grandfather clock with a harp chime. The biographer Engelmann claims that Johann Gottfried was intrigued by the newly fashionable Swiss musical clocks. Engelmann, "Aus der Geschichte," 797.

60. Schardin, "Musikuhren und Musikautomaten."

61. A picture of the Belloneon is in Engelmann, "Aus der Geschichte," 799. See also Beyer, *Faszinierende Welt der Automaten*, 72.

62. The instrument was a keyboard in the form of a vertical fortepiano. The strings were hit not with a hammer but with a cylinder covered with leather and treated with rosin. Engelmann, "Aus der Geschichte," 843.

63. Beyer, *Faszinierende Welt der Automaten*, 72; Wolf, *Kaufmanns Trompeterautomat*, 29.

64. The trumpeter is kept at the Deutsches Museum under inventory number 4423. The entry for the trumpeter gives this height. Rebecca Wolf has done groundbreaking work in uncovering

markable technical achievement of this automaton was its artificial lar-
ynx and artificial tongue: the automaton could play any tune a human
trumpeter could, and it could even play "double-tunes," two tunes of
the exact same strength and volume.[65] The mechanism was driven by a
cylinder with a spring and a shaft, and the musical program was stored
on a pegged barrel.[66] During a performance, the barrel moved in its
axial direction. For one complete musical piece, the barrel made about
six revolutions, which corresponded to a time period of thirty to forty
seconds. The "reading" of the peg pattern on the barrel occurred by
means of a set of seven levers. The sound was produced through so-
called tongues, valvelike devices that produced sound when air flowed
through them. Three bellows provided the air for the tongues. The
bellows' motion was coordinated by the spring cylinder and the shaft,
transmitted via a set of cogs.[67]

The automaton-makers Vaucanson, Knaus, and Kaufmann exem-
plify, together with the Jaquet-Droz and Roentgen families, the world
of automaton-making in the eighteenth century. Despite many parallels, individual artisans' relationships to their automata varied a great
deal. Neither father and son Jaquet-Droz nor Roentgen and Kinzing,
for example, published a book, manual, or other text on their music-
making women automata, as Knaus and Vaucanson did. The automata
by the Jaquet-Droz family and by Roentgen and Kinzing were products of their specific workshops, and their makers had different priorities and different social and entrepreneurial agendas than Vaucanson,
Knaus, and Kaufmann. Both the Jaquet-Droz family and David Roentgen had contacts with courts, like Knaus and Vaucanson. However,
while Roentgen was already an established deliverer of luxury furniture to courts when he built his automaton, the Jaquet-Droz family's
automata came into being after Pierre's contact with the international
court society: a pension from the Spanish king, granted after the king
had seen a display of a variety of precious clocks, enabled Pierre and his
son to produce their three automata. Their business activities and production were in general on a lower scale than those of the Kaufmann
and the Roentgen-Kinzing manufacture, and neither father nor son

details about the trumpeter's mechanism and its functioning. Wolf, *Kaufmanns Trompeterautomat*, 93–104.

65. Ibid., 97–100; Engelmann, "Aus der Geschichte," 799.

66. On pegged barrels and their history, see chapter 4. The trumpeter's mechanism could actually be seen through a tilted mirror that was attached in the interior of the trumpeter's body. Engelmann, "Aus der Geschichte," 799.

67. Wolf, *Kaufmanns Trompeterautomat*, 95–97.

Jaquet-Droz was employed at a court. Friedrich von Knaus was the son of a court clock-maker and himself became a court clock-maker, distinguishing himself professionally and rising in status from a ducal court to the imperial court in Vienna. His book on his writing machines was published for purposes of self-promotion, but the strategy ultimately did not work out for him: he was impoverished and abandoned when he died in 1789.[68] In contrast, the Kaufmanns already had a successful family business, specializing in mechanical musical instruments and serving both aristocratic and bourgeois consumers, by the time they built their automaton, and they managed to sustain their economic success for two more generations until the late nineteenth century.

A few more influential artifacts are worth mentioning in this context, even though none of them are android automata. Between 1778 and 1791, Wolfgang von Kempelen built a series of four speaking machines, of which the last one is still extant in the Deutsches Museum in Munich.[69] Kempelen was a civil servant and a royal and imperial councillor in Austria-Hungary.[70] He became most famous for building the famous hoax automaton that displayed a chess player. But at the same time that he was building the chess player, the years before 1769, he was also building musical instruments to explore whether they could reproduce the sounds of the human voice.[71] Kempelen analyzed the mechanism of the human voice and speaking apparatus in depth and in 1791 published the results of these studies in a comprehensive treatise.[72] The same treatise also describes the speaking machine in depth, with the explicit suggestion that readers study and improve the machine's principle. The machine consisted of a set of bellows, acting as lungs, and a rubber mouth, connected through a resonator box. In the box, an ivory reed served as the glottis.[73] The Maillardet family, who were clockmakers from La Chaux-de-Fonds and apprentices and co-workers of

68. Heckmann, *Die andere Schöpfung*, 238; Beyer, *Faszinierende Welt der Automaten*, 57.

69. The anatomy of speaking had been a concern for a while in the later eighteenth century. The Academy of Sciences in St. Petersburg, for example, ran a prize competition in 1779 to build a machine that could pronounce clear vowels. Swoboda, *Der künstliche Mensch*, 98.

70. He was, among other things, the director of the Hungarian salt monopoly and in charge of the police apparatus in Transylvania. Ibid., 101.

71. Ibid., 103. Riskin comments that, like Vaucanson, Kempelen built both genuine and fraudulent automata. Riskin, "Defecating Duck," 620. This distinction has important epistemological valences for Riskin's general argument about automata.

72. Kempelen, *Mechanismus der menschlichen Sprache* (1791). See also Heckmann, *Die andere Schöpfung*, 253.

73. For descriptions, see, e.g., Riskin, "Defecating Duck," 618–19. Apparently it was much easier for the machine to produce French, Italian, and Latin words than German words. See also Hankins and Silvermann, *Instruments and the Imagination*, chap. 8.

father and son Jaquet-Droz, developed a series of mechanical spectacles in the late eighteenth and early nineteenth centuries, among them two magicians that still exist. These machines each display a figure clothed as a magician, and the spectator can ask the magician questions by submitting preprinted cards to the machine. The machine then dispenses an answer on another card.[74] Another kind of mechanical device that is often taken to be an important variation on the automata theme, in particular in its role in the history of data processing and computing, is the mechanical calculation machine. Among the most important mechanics and mathematicians in this context are Philipp Matthäus Hahn and Gottfried Wilhelm Leibniz.[75] Some of these calculators originated in artisan contexts similar to the ones considered here.[76] However, since they replicate a human capacity but display no human figure, I do not consider them in depth.

74. Beyer, *Faszinierende Welt der Automaten*, 60–61; Chapuis and Gelis, *Le monde des automates*, 161–64. Beyer mentions one of the second-generation Maillardet mechanics who very successfully traveled around with his automata. When he became ill and could no longer travel, his business soon went into decline. This suggests once more how important it was for automaton-makers to be committed to the project of traveling with them and staging them in many places. If this did not happen, automata could soon be forgotten. See also the documentation of automata I discuss in chapters 2 and 3.

75. Heckmann, *Die andere Schöpfung*, 91; Weber, "Physik und Technologie," 114. Engelmann, *Leben und Wirken*. On calculating machines, see also the work by Matt Jones on Leibniz's cooperation with artisans: "Reckoning Matter: Calculating Machines, Innovation, and Thinking about Thinking from Pascal to Babbage," chap. 2, "Artisans and their Philosophers," book manuscript in Voskuhl's collection.

76. Heinz Zemanek makes this context very clear. See Zemanek, "Die Automaten," 245.

TWO

The Harpsichord-Playing Android; or, Clock-Making in Switzerland

The harpsichord-playing woman automaton came into being in an unusually productive and well-developed artisan environment. The Jaquet-Droz family was well established in La Chaux-de-Fonds and had connections, talent, education, and opportunities. Histories often explain their success in automaton-building either through the individual ingenuity of father and son and their teachers or through their interest in simulating life and promoting a mechanistic worldview.[1] I complement these histories by foregrounding the broader economic and cultural conditions that were operating in La Chaux-de-Fonds, Switzerland, and on the European continent in the eighteenth century and that had considerable impact on the family's ambitions and opportunities. I use these conditions to explain how and why father and son Jaquet-Droz built the harpsichord-playing woman automaton and what the automaton's purpose and status were in their lives. I argue

1. Roland Carrera, F. Faessler, and other Swiss historians describe Pierre Jaquet-Droz as a pioneer of Swiss clock-making and a student of Johann and Daniel Bernoulli in Basel. Carrera, Loiseau, and Roux, *Androiden*; Faessler, "250e anniversaire"; *Beiträge der Schweiz zur Technik*. Wilhelm Schmidt-Biggemann interprets the Jaquet-Droz automata as products of the contemporary mechanistic philosophy. Schmidt-Biggemann, *Maschine und Teufel*, 101–4. Jessica Riskin positions the Jaquet-Droz automata firmly in the eighteenth-century philosophical program of simulating life. Riskin, "Defecating Duck," 604–6, 625, 631.

that the automaton was a product much more of culture, politics, and economics than it was of epistemology or natural philosophy.

In this chapter I take a journey to La Chaux-de-Fonds, a small, rural village in western Switzerland that housed clock-makers, local dignitaries, natural philosophers, and patrons. The village's social makeup provided a range of opportunities for Pierre and Henri-Louis Jaquet-Droz. I start with Pierre Jaquet-Droz's youth and university studies and then explore his beginning career as a young clock-maker, integrating this with larger features of Europe's early industrialization in the eighteenth century. I trace the channels through which he was introduced to the European court society as a young man and investigate how he and his son designed and built their android automata in the early 1770s. I also inquire into the activities of the Jaquet-Droz family during the late 1770s and 1780s, after they finished their work on androids, to uncover their self-understanding as artisans, artists, businessmen, and citizens. The specific place they accorded their automata among their various business enterprises is far from obvious, especially since soon after father and son finished their automata, the androids practically disappeared from their lives and the two artisans returned to their core clock-making trade, founding firms for the international selling of timepieces.

There has been a great deal of discussion about the Jaquet-Droz automata's significance in the general cultural landscape of the European Enlightenment, as well as their effect on contemporary spectators. Among these discussions one finds assumptions of the kind I discussed in chapter 1: that the Jaquet-Droz automata embodied the key principles and practices of the "Age of Reason," that they captured people's imagination and made them think about the mechanical constitution of the human body and soul, and that the automata were part of a widely shared excitement about mechanical androids that was characteristic of the eighteenth century as a whole. The documents known to us about the Jaquet-Droz automata do not consistently support these assumptions. In the second part of this chapter, I look at such documents, as well as at how the automata were staged and how observers commented on them from 1774 into the 1790s. I discuss manuscript and printed texts about the Jaquet-Droz automata and draw conclusions that differentiate the kinds of interest that the automata provoked in their time among audiences and the general public.

My findings break down into three parts. First, individuals who saw the automata and wrote about them in letters or diaries were less interested in the automata's mechanical sophistication or their significance as mechanical replicas of human bodies than they were in social aspects of the automata's exhibitions, such as networking with the Jaquet-Droz family and with influential people who were in the audience. Second, all descriptions of the three automata that were printed in periodicals, calendars, and newspapers between 1774 and 1791 were transcriptions of an original text written by father and son Jaquet-Droz and printed in a brochure that was presumably handed out during automaton exhibitions. Descriptions of the automata in these print media are therefore practically identical to each other, and they do not tell us about individual people's experiences of the automata or reactions of groups of people during automaton showings. Third, longer and more detailed texts written by editors of travelogues or calendars, among them Johann Bernoulli and Georg Christoph Lichtenberg, about the Jaquet-Droz family's life and their three automata were also copied from the original brochure that father and son Jaquet-Droz wrote. These texts are, on the whole, more specific and more deliberate, but the automaton descriptions within them are, nevertheless, not individually authored texts that could give us insight into people's responses to the androids. The texts give us insight, instead, into economic and cultural conditions of mass print media at the time. The authors who wrote longer texts on the Jaquet-Droz family did engage some aspects of Swiss clock-making and industry, but the texts show no interest in the automata's unusual mechanical sophistication or metaphysical implications of the ways in which they blur the boundary between humans and machines.

At the end of chapter 3, I merge the conclusions that I draw in this current chapter with analogous findings presented in that chapter on the making of, and commentary on, David Roentgen's harpsichord-playing automaton, joining them into a larger commentary on automata, the artisan industry, and automaton commentary.

Pierre Jaquet-Droz's Youth and University Studies

Pierre Jaquet-Droz, born on 28 July 1721 in La Chaux-de-Fonds, was the descendant of an old and established family in the Neuchâtel mountains. The profession of Pierre's father, Abram Jaquet-Droz, cannot be established with certainty. Although a family tree that Pierre put together just before his death records that his grandfather was a justice and his

great-grandfather a councillor of the honorable community in Le Locle (a neighboring village), it does not mention his father's profession.[2]
Abram Jaquet-Droz married Madelaine Droz, daughter of Daniel Droz,
from an estate called Sur le Pont near La Chaux-de-Fonds. Through this
marriage, Sur le Pont (which consisted of two houses) passed into his
possession and later into his son Pierre's.[3] It is likely that Abram cultivated the land that came with the estate, and the local economy at the
time makes it plausible to assume that he was also a clock-maker: being
an *horloger-paysan* (a peasant clock-maker) was a common occupation
in the Neuchâtel mountains at the time. The dynamics of early industrialization in Switzerland entailed that traditional, agricultural means
of subsistence coexisted with industrial, modern ways of production,
to their mutual benefit.[4] This economy was to have a profound impact
on Pierre Jaquet-Droz's life as well.

Pierre spent his childhood and the first part of his life as a young artisan on the Sur le Pont estate. Unlike many other clock-makers in the
area, he received a substantial education in his youth, first in his native
village school and by the local pastor, then at the faculty of philosophy
at the University of Basel.[5] His education was meant to prepare him for
university studies in theology and for a future occupation in the clergy.
This choice of profession was not unusual at the time for the son of a
well-to-do family. Clerical professions provided one of only two possibilities for upward mobility for nonaristocratic and nonpatrician social
groups, since the only university subjects available to them were theology and medicine.[6] There is no documentation of the actual course of
Pierre Jaquet-Droz's studies at the University of Basel, but matriculation

2. The oldest printed biographies, dating from the 1850s, say nothing about this matter. *Bio-graphie universelle*, s.v. "Droz, Pierre Jacquet," 11:329; *Nouvelle biographie générale*, s.v. "Droz, Pierre-Jacquet," 14:812.

3. Sur le Pont is a rural area northeast of La Chaux-de-Fonds. Perregaux and Perrot, *Les Jaquet-Droz et Leschot*, 42–43.

4. There is an entry in the journal of a man named Abram-Louis Sandoz from 1747 that indi-cates that he went to visit father and son Jaquet-Droz and purchased pendulum cases from them (he says "les Jaquet-Droz") which can only have been father Abram and son Pierre, since Henri-Louis had not yet been born. Chapuis, *Histoire de la pendulerie neuchâteloise*, 109–10. On the phe-nomenon of the *horloger-paysan* and its relation to early industrialism, see Hauser, *Schweizerische Wirtschafts- und Sozialgeschichte*; Bergier, *Die Wirtschaftsgeschichte der Schweiz*.

5. *Biographie universelle* (1854), 329; and *Nouvelle biographie générale* (1857), 812. See also Faessler, "250e anniversaire," 13–14; Chapuis, *Histoire de la pendulerie neuchâteloise*, 109; Perre-gaux and Perrot, *Les Jaquet-Droz et Leschot*, 31.

6. Wernle, *Der schweizerische Protestantismus*, 1:30. The clerical profession was also part of the political establishment, as political power in the principality of Neuchâtel was exercised by four citizenries and the venerable classes of the pastors (*vénérable classe des pasteurs*). See Dixon and Schorn-Schütte, *Protestant Clergy*.

records show "Pierre Droz" as a student for the year 1738, when he was age seventeen.[7] His studies there lasted about two years.[8]

A key question about Pierre Jaquet-Droz's university education is whether, during his stay at the University of Basel, he encountered Johann Bernoulli I and Daniel Bernoulli I, who were both professors there at the time.[9] No sources are available to tell us about this, but speculation is relevant because it shapes our understanding of whether Pierre Jaquet-Droz's automaton work was a product of contemporary scientific and natural philosophical culture or of local, traditional artisan culture, or of both in a proportion to be weighed. An important factor that speaks for a connection between Pierre and the Bernoullis is obviously the exceptional ingenuity and intimate familiarity with physical and mechanical principles that his mechanical work discloses. It set Pierre Jaquet-Droz apart from other small-town clock-makers of his time. We also know that the Bernoullis were involved with automaton work in the first half of the eighteenth century: Basel municipal authorities asked them in several cases between 1709 and 1743 to conduct investigations of self-moving mechanical spectacles owned by itinerant lecturers and artisan journeymen who had asked permission to present their work at Basel marketplaces and fairgrounds. Such permits were usually issued or denied after expert investigation.[10]

It is likely that Pierre Jaquet-Droz received at least basic training in

7. Wackernagel, *Die Matrikel der Universität Basel*, 5:89. Pierre Jaquet-Droz's name appears under no. 448, as "Pierre Droz, de la Chaux de fond, 1 lb." (1 libra, indicating a one-pound matriculation fee). Other sons of distinguished families from La Chaux-de-Fonds, Le Locle, and Neuchâtel, who later became clock-makers, physicians, historians, pastors, or artists, also studied at the University of Basel, and their names are found in the matriculation records.

8. Carrera, Loiseau, and Roux, *Androiden*, 10.

9. Johann I, born in 1667, had become professor of mathematics at Basel University in 1705, where he remained until his death in 1748. His son Daniel, also a distinguished mathematician, was first appointed to the chair for anatomy and botany at Basel University in 1734, then lectured in physiology in 1743, and was eventually appointed to the chair of physics in 1750. Walter, *Soziale Grundlagen*, 223–28; Wolf, *Biographien*, 3:151–202; Fleckenstein, *Johann und Jakob Bernoulli*.

10. Applications by itinerant lecturers are collected in the Staatsarchiv der Stadt Basel, Straf- und Polizeiakten, F4 Schaustellungen, 1667–1865. There are requests from the years 1709, 1716, 1724, 1738, and 1743. Reference to the Bernoullis is in the Manuel du Conseil, which collects the protocols of the meetings of the city's "Small Council" (Kleiner Rath), dated from 7 September 1743. I was pointed to this archive by Perregaux and Perrot, *Les Jaquet-Droz et Leschot*, 37; and Carrera, Loiseau, and Roux, *Androiden*, 16. On curiosities, itinerant lecturers, and natural philosophy, see also Hochadel, *Öffentliche Wissenschaft*; Hankins, *Science and the Enlightenment*, 46–81; the essays collected in Blondel and Bensaude-Vincent, *Science and Spectacle*; and the essays in Evans and Marr, eds., *Curiosity and Wonder*. Daniel Bernoulli was a distinguished mechanic and was also versed in mechanical acoustics, which could have been a theoretical foundation for Pierre Jaquet-Droz's later work in flute and organ clocks. See Huber, *Daniel Bernoulli*, chap. 17 on Daniel Bernoulli's theories of the flute and the organ pipe.

natural philosophy while at the University of Basel and that education played both social and intellectual roles later in his life, because the circles in which he moved as an adult included local natural philosophers, pastors, political dignitaries, and other educated men. However, biographers' speculations on connections between the Bernoullis and Pierre Jaquet-Droz often go rather far and conceive of them in rather narrow terms: they range from the claim that Pierre Jaquet-Droz learned "the scientific method" for his mechanical constructions from the Bernoullis to assumptions that Pierre Jaquet-Droz was Daniel Bernoulli's "assistant in the years 1738 and 1739."[11] Some accounts tend to convey the impression that his presumed association with the Bernoullis was the main factor to explain his and his son's ability to build their automata. Against the background of the life that Pierre Jaquet-Droz led in the forty years following his university education, which I am laying out in this chapter, the influence of the Bernoullis fades. His immersion in clock-making and court worlds proves to have as much influence as his education in natural philosophy.

In 1740, at age nineteen, Pierre Jaquet-Droz left the University of Basel and matriculated at the faculty of theology of the Académie de Neuchâtel as a candidate for a clergy position in the Calvinist Church.[12] That same year, he also spent considerable time at his parental home in La Chaux-de-Fonds. He ended up never taking a position as a clergyman and probably never even engaged in the training.[13] As he spent time at his home, he must have decided to get involved in the local clock-making culture and lay the grounds for his future occupation as a clock-maker and automaton mechanic.[14]

Jaquet-Droz immersed himself in the distinctive and thriving clock-making culture of his hometown, La Chaux-de-Fonds. He became an apprentice and then, by the late 1750s, a distinguished clock-maker with a specialization in so-called Neuchâtel pendulum clocks (*pendulerie*

11. Perregaux and Perrot, *Les Jaquet-Droz et Leschot,* 37; Carrera, Loiseau, and Roux, *Androiden,* 16.

12. Faessler, "250e anniversaire," 14. The biographers Perregaux and Perrot also found Pierre Jaquet-Droz's name in the *Actes de la compagnie des pasteurs* in Neuchâtel for the month of June 1740. Perregaux and Perrot, *Les Jaquet-Droz et Leschot,* 38.

13. The next entry on Pierre Droz in the register of the Compagnie des Pasteurs says "quit the study of theology." Perregaux and Perrot, *Les Jaquet-Droz et Leschot,* 38.

14. Pierre Jaquet-Droz's earliest biographer states that he had to wait to come of age for his clerical training and that that was the reason he spent time at his parental home in the same year that he matriculated in the register of the Compagnie des Pasteurs. At his parental home, the biographer says, he saw one of his sisters occupied in output work for the watch-making industry. *Biographie universelle,* 11:329.

neuchateloise)—medium-size clocks that were characterized by their precision, artistic ornamentation, and combination of time-keeping with other functions such as mechanical play and mechanical music. Over those twenty years, he received training, financial support, patronage, and even a royal pension, primarily from resources in his local environment in the Neuchâtel mountains. Some historians, as I have said, put great emphasis on the Bernoullis' influence on Pierre Jaquet-Droz's life and work. Their stories offer an appealing narrative in which a clockmaker from rural Switzerland received scientific training and mentorship from members of a distinguished eighteenth-century family of natural philosophers, and they enable us to understand Pierre Jaquet-Droz's automata as products of a "modern" scientific-philosophical environment and as epistemic tools in the scientific enterprise to understand, and simulate, the world.[15] The Bernoullis may have helped him approach his clock-making with ideas and principles from natural philosophy that many other clock-makers did not have access to. But in his youth and early adulthood, he relied on a complex compound of resources that he received primarily from his hometown.

Clock-Making in La Chaux-de-Fonds in the Early Eighteenth Century

By the time the young Pierre Jaquet-Droz returned from Basel University, clock-making was well established in the Neuchâtel mountains, with diverse modes of production in La Chaux-de-Fonds and neighboring villages. Already at the end of the seventeenth century, clock industries had started to develop on the basis of existing artisan activity. The Neuchâtelian clock and watch industry was likely preceded by the making of iron and wooden clocks; and while at the end of the Middle Ages the area had not been a prosperous place (lacking infrastructure and important cities), one or two centuries later such conditions were present. Land reforms in the fourteenth and fifteenth centuries helped advance agricultural activity, and industries such as mills, mines, brickworks, and forges developed near rivers and streams. By the end of the seventeenth century, the mining and iron industries went into decline, but they had given rise to other industries in the vicinity, such as

---

15. Carrera, Loiseau, and Roux, *Androiden*, 16. See also Haspels, *Automatic Musical Instruments*, 121–23, 192–95; Bassermann-Jordan, *Uhren*, 297, 331, 421–30.

toolmaking shops and foundries.[16] The emergence of clock-making in the area is inseparably bound up with preexisting artisanship that dealt with metal, wood, and mill-building, specifically that of locksmiths, goldsmiths, and the makers of municipal clock towers. In the fifteenth and sixteenth centuries, such artisans dealt with both large and small equipment and were probably in charge of repair and maintenance of any machinery in the village. Toward the end of the fifteenth century, they started making iron clocks small enough for well-off people to keep in their private houses. Clock-making was subsequently transplanted from village to village and valley to valley in the area, as clockmakers traveled and trained apprentices who themselves migrated away from and then again back to the La Chaux-de-Fonds area.[17]

Protoindustrialism in Post-Reformation Europe

Larger factors operating in early modern Switzerland, such as the Reformation and the earliest stages of industrialization, contributed to the emergence of Neuchâtel clock-making. The figureheads of the Swiss Reformation, Ulrich Zwingli and John Calvin, had founded in the sixteenth century a religious and ultimately political movement that had profoundly revised the role of Christianity in the world of labor and money and had formulated new principles of economic thinking, social order, and ethics.[18] The Reformation's influence was amplified through the convergence of political and religious power in the leading Swiss centers at the time, especially in Geneva, located just south of the Neuchâtel mountains. In Geneva, John Calvin had created a model Protestant state under a theocratic constitution, and under his rule Geneva had developed from a "small, undistinguished town in rural Savoy" into an "international center of religious, cultural, and financial pre-eminence."[19] From the 1540s onward, Geneva's influence extended

16. Some industries dealt mainly with watch-making, others, like those in La Chaux-de-Fonds, with the making of pendulum clocks. Cardinal and Mercier, *Musées d'horlogerie*; Chapuis, *Histoire de la pendulerie neuchâteloise*, 2–3; Perregaux and Perrot, *Les Jaquet-Droz et Leschot*, 22.

17. Chapuis, *Histoire de la pendulerie neuchâteloise*, 6, 74–75, 100–101.

18. Hauser, *Schweizerische Wirtschafts- und Sozialgeschichte*, 79–90. The connection between Protestant ethics and capitalism has been, of course, the subject of thorough and continuous debates ever since Max Weber's thesis on the relationship of a specific "protestant ethic" to the "spirit" of modern capitalism. Poggi, *Calvinism and the Capitalist Spirit*, 13–74.

19. *Wartburg*, Geschichte der Schweiz, *123*; Bergier, Zu den Anfängen des Kapitalismus; Calvin's ethics—that service to God was the only justification of any action—naturally had an impact on mechanical trades such as clock-making and the goldsmith industry.

to the entire western part of Switzerland, including La Chaux-de-Fonds and its surrounding villages, affecting in particular economic conditions of mechanical and trade crafts such as clock-making, goldsmithing, textile-making, and printing.

The Reformation and the Counter-Reformation also precipitated enormous migration. While Switzerland was in general an attractive refuge for persecuted Protestants, Geneva, as a Francophone metropolis and a site of the Reformation, was particularly appealing for French Calvinists (Huguenots) many of whom were skilled and well-connected artisans. Continuously over a period of more than two centuries, Geneva thus attracted a considerable proportion of an economically active and innovative stratum of persecuted people. The immigrants brought with them not only skill and initiative but also commercial expertise, capital, knowledge of international markets, and reliable trade and credit relations. They reactivated crafts and trades such as the silk and cotton trade, introduced clock-making in Geneva, and initiated early capitalist habits in production and trade that were outside of the strictly regulated medieval guilds, such as the putting-out system and home industry. Protestant immigrants and their resourcefulness are also credited with having prepared the general grounds for industrial production in Europe, by promoting ways of organizing labor and material that were conducive to such production.[20]

The historical transformations described by the term Industrial Revolution, and the economic systems preceding the Industrial Revolution, are two additional backdrops for the production of my two women automata.[21] Industrial Revolution connotes, in its broadest definition, the changes, often taking several hundred years, from a rural, handicraft- and agriculture-based economy to an urban, machine-driven, in-

20. Between the sixteenth and the eighteenth centuries, the Counter-Reformation in France, Italy, Spain, and the Netherlands (under Spanish rule) persecuted Protestants and forced them to seek refuge in Switzerland and in the Protestant territories of the Holy Roman Empire. The French Wars of Religion in particular gave rise to repeated surges of violence in the second half of the sixteenth century. The persecution culminated in Louis XIV's revocation of the Edict of Nantes in 1685. Between 1550 and 1720, about three thousand French Calvinists each year came to the city of Geneva alone. Innovation in production of commodities would have been impossible without the refugees' initiative and support. Bodmer, *Der Einfluß*, 13–20, 87–88; Braun, *Das ausgehende Ancien Régime*, 115–17; Wartburg, *Geschichte der Schweiz*, 123; Bergier, *Die Wirtschaftsgeschichte der Schweiz*, 59–60; Winzer, *Emigranten.*, 22–26. See also Hugh Trevor-Roper's response to Max Weber. Trevor-Roper, *Crisis of the Seventeenth Century*; Innes, *Social Concern in Calvin's Geneva*, 1; Mörikofer, *Geschichte der evangelischen Flüchtlinge*; Biucchi, "Industrial Revolution in Switzerland," 641; Hauser, *Schweizerische Wirtschafts- und Sozialgeschichte*, 80.

21. David Landes aptly expresses one of the political and epistemic issues over the nature and status of the Industrial Revolution by stating, "One of the most divisive issues of the revolution is the question of how far back to look for its origin." Landes, *Unbound Prometheus*, xi.

dustrial manufacturing economy. The term Industrial Revolution has been criticized for its inherent teleological assumptions and its connotations of radical and sudden overturn, in the light of historical studies that reveal widespread overlap of traditional and modern modes of subsistence and production in Europe between the thirteenth and the eighteenth centuries.[22] Franklin Mendels coined the term protoindustrialization in 1974 as an alternative historical term for the emergence of the dynamic socioeconomic systems of the eighteenth and nineteenth centuries, inside and outside of Britain; he suggested that this term better described the transformations in the organization of labor, capital, and materials that preceded industrialization, and made it possible.[23] Such transformations changed the balance of labor supply and local subsistence, enabled the accumulation of capital, and led to agricultural surpluses. In the 1980s, Charles Sabel and Jonathan Zeitlin joined efforts to broaden historians' perspectives on Europe's industrialization. They suggested investigating "artisanal, flexible alternatives to mass production" that were widespread in the decades and centuries preceding industrialization.[24] The discussion on the status and impact of guild regulation in the economy of early modern Europe is also ongoing. Among the points of debate are the guilds' influence on general economic development, the pattern of their decline, and the ways in which our twentieth- and twenty-first-century understanding of industry and modernity continuously shapes our interpretations of the history of technology.[25]

In regard to the history of my two women automata, it is important to understand preindustrial economic processes outside of the paradigm of the Industrial Revolution: there is no other way to explain the prosperous artisan contexts in Switzerland and the Rhineland in

22. See, for example, Safley and Rosenband, *Workplace*; Horn, Rosenband, and Smith, *Reconceptualizing the Industrial Revolution*.

23. Mendels, "Proto-Industrialization"; Kriedte, Medick, and Schlumbohm, *Industrialization*. A discussion of how "protoindustrialism" challenges and revises Marx's suggestion of an "era of the manufacture" that supposedly preceded industrialization is found in Wehler, *Deutsche Gesellschaftsgeschichte*, 102–3. Maxine Berg discusses Mendels's term extensively in her critical history of Britain's industrialization. Berg, *Age of Manufactures*, 66–72.

24. Sabel and Zeitlin. "Historical Alternatives." The quote is from their edited volume *World of Possibilities*, 1. See also Ogilvie and Cerman, *European Proto-Industrialization*.

25. Epstein, "Craft Guilds." See also the essays collected in the special issue "Reflections on Joel Mokyr's *The Gifts of Athena*," *History of Science* 45, no. 2 (2007), edited by Maxine Berg, Liliane Hilaire-Pérez, Larry Stewart, Kristine Bruland, and Joel Mokyr; the essays discuss Joel Mokyr's 2002 book *The Gifts of Athena*, which deals with the role of the technological and scientific knowledge that has so quickly expanded over the past two hundred years and profoundly impacted the economic and social history of the world. Mokyr investigates the historical roots of this phenomenon, often referred to as the "knowledge economy."

which they came into being. The economic invigoration and diversification that was occurring in preindustrial, post-Calvinist Geneva in the course of Huguenot artisan immigration influenced conditions in the Neuchâtel mountains profoundly over the two centuries following the Reformation. Clock-making was first introduced into the city of Geneva in the sixteenth century, developed into a hybrid between guild-type restricted production and laissez-faire, outputting-type protoindustrial production, and spread from there northward to the Neuchâtel mountains.[26] Here as well, the immigrants brought with them a variety of tools and techniques, and they encountered the already existing industries and crafts in rural villages such as La Chaux-de-Fonds that I described above.[27]

The clock and watch industries in the Neuchâtel mountains had originated and developed outside of the medieval guild rules to begin with. The situation in the countryside was in general more diverse than in towns, and rural economies tended to profit more from pre- and protoindustrial changes in labor and capital, since they were less influenced by guild regulation. The actual influence of guild regulation varied widely at the time in French-speaking Switzerland. Early modern Switzerland was also not a centralized monarchy but a confederation of sovereign city-states and peasant republics. There was, therefore, no centralized or mercantilized economic policy. And while some of the Swiss towns in the eastern, German-speaking part adopted methods and policies from contemporary mercantilist and physiocratic theory, there was an economically liberal spirit operating in the principality of Neuchâtel and its mountains.[28]

The history of Pierre Jaquet-Droz intersected with key changes ongoing in Switzerland and Europe during the period "before" the Indus-

---

26. Braun, *Das ausgehende Ancien Régime*, 119.

27. The refugee policy in the principality of Neuchâtel was more restricted than that in the city of Geneva throughout the post-Reformation period, since Neuchâtel's reigning dynasty, the princes of Orléans-Longueville, were too strongly connected with the French monarchy to support refugees openly. Nevertheless, already at the end of the seventeenth century, a large number of refugees immigrated to Neuchâtel.

28. Hauser, *Schweizerische Wirtschafts- und Sozialgeschichte*, 151, 161, 97, 182–83; Biucchi, *Industrial Revolution in Switzerland*, 647; Martin Körner, "Town and Country in Switzerland," 249; Pfleghart, *Die schweizerische Uhrenindustrie*, 27–44. See also my discussion of mercantilism in chapter 3. In 1706, after the death of the last ruling member of the Orléans-Longueville dynasty, the principality of Neuchâtel became a politically and legally unusual construction in the eighteenth century. The Prussian kings seized monarchical power over the principality, while the area was at the same time becoming part of the Swiss Confederation. Watt, *Making of Modern Marriage*, 30ff.; Baker, *Model Republic*, 344–45; Nabholz et al., *Geschichte der Schweiz*, 2:102–3; Favre, *Neuenburgs Union*, iv–v, 24–33.

trial Revolution. His success relied on factors such as his skill, social networks, and class status, as well as the modes of production around him. They contributed to the unusual productivity in his workshop that made possible the construction of his automata. He was not part of an industrial revolution; rather, he worked in the midst of manifold change in production and consumption in a period that preceded the steam engine and the factory.

Economic transformation in the Neuchâtel mountains was also driven by climate conditions. In Swiss mountainous areas with a high level of rainfall, such as Pierre Jaquet-Droz's hometown, the main agricultural activity was stock-breeding, which was less labor-intensive than farmland cultivation. Small-scale farmers and farm laborers, as well as their wives and children, therefore turned to industrial occupations, in particular mechanical homework and clock-making, to supplement their livelihood. In more arable farming areas in Switzerland, there was much less production of such commodities by rural industries.[29] By the end of the eighteenth century, Switzerland had become one of the most industrialized places on the European continent, despite accounts in contemporary travel reports that continued to portray the country as rural and idyllic, firmly rooted in the structures of a traditional agricultural society.[30] Switzerland held a special place in the European imagination; its starkly contrasting scenes of pastoral beauty and proto-industrial activity dramatized the relationship between tradition and modernity. These contrasts also characterized the Neuchâtel mountain region in which Pierre Jaquet-Droz found the resources and incentives to build android automata.

The Local Economy in La Chaux-de-Fonds

Between 1720 and 1790, Pierre Jaquet-Droz's first craft as a young artisan—pendulum-making—grew into its most productive period. La Chaux-de-Fonds remained its center until about 1810 and made the

---

29. The home industry prospered in forest lands, hilly regions, and moors, where the yield from the soil proved inadequate for the inhabitants' subsistence. Biucchi, "Industrial Revolution in Switzerland," 628; Kellenbenz, "Rural Industries in the West," 67–74; Levy, *Social Structure of Switzerland*, 25; Bergier, *Die Wirtschaftsgeschichte der Schweiz*, 98.

30. I discuss some aspects of the cliché of Switzerland as an idyllic agricultural nation in my investigation of travel literature on Switzerland later in this chapter. See also Hentschel, *Mythos Schweiz*, 61–81, 99–103.

trade famous abroad. Numerous other mechanical industries clustered around pendulum-making, featuring a range of crafts, corporations, working conditions, and methods of dividing labor.[31]

Many of La Chaux-de-Fonds's families were involved in the town's emergence as a center of clock-making. They tended to be related to one another through intermarriage, professional cooperation, credit agreements, or local political office. They had a great deal of influence in Pierre Jaquet-Droz's life, and he became part of their network after he returned from Basel University. The first generation of pendulum masters and their families (named Brandt-dit-Grieurin and Ducommun-dit-Boudry) established themselves in the seventeenth century; Pierre Jaquet-Droz became a member of the second generation through friendship and intermarriage.[32] These families distinguished themselves through technical innovation in clock-making, for example, but also through advanced commercial activities such as developing business relations abroad. It is plausible to believe that they taught Pierre Jaquet-Droz such practices.[33] A particularly distinguished man was Josué Robert, a master clock-maker, counsel, deputy judge, and chair of charity organizations. He essentially founded the industry of making pendulum clocks in La Chaux-de-Fonds.[34]

There were close family and economic relations between the Roberts and the Jaquet-Droz households: Pierre Jaquet-Droz was close to Josué Robert's two sons David and Louis-Benjamin (David married Pierre's sister) and may have had some type of informal master-apprentice relationship with Robert.[35] In 1750 Pierre Jaquet-Droz married the daughter of Abraham-Louis Sandoz-Gendre, a local cabinet and pendulum merchant who provided Pierre Jaquet-Droz with cases for his pendulum clocks. The Brandt-dit-Gruérin, Robert, Jaquet-Droz, and Sandoz families formed a robust network based on friendship, familial relations,

---

31. Diderot and d'Alembert's *Encyclopédie* states, according to Alfred Chapuis, that about fifteen professions were attached to the industry of pendulum-making. Chapuis, *Histoire de la pendulerie neuchâteloise*, 99–100.

32. Chapuis, *Histoire de la pendulerie neuchâteloise*, 93–97; Perregaux and Perrot, *Les Jaquet-Droz et Leschot*, 26.

33 Jeanneret, *Biographie Neuchâteloises*, 1:500–506. See also Faessler, "250e anniversaire," 9.

34. Chapuis, *Histoire de la pendulerie neuchâteloise*, 102–3; Perregaux and Perrot, *Les Jaquet-Droz et Leschot*, 26–28.

35. There are three names of apprentices of Robert between 1730 and 1739 but no indication that Pierre Jaquet-Droz, who only returned from Basel University in 1739, was his apprentice. But Robert may have guided him through his first works. Chapuis, *Histoire de la pendulerie neuchâteloise*, 104; Perregaux and Perrot, *Les Jaquet-Droz et Leschot*, 31; Faessler, "250e anniversaire," 16; Carrera, Loiseau, and Roux, *Androiden*, 10.

and professional cooperation.[36] Throughout his career Pierre benefited from this reservoir of experience, skill, and social and cultural capital.

Around 1750, twelve years after Pierre Jaquet-Droz returned from university, more than 120 clock-makers (masters and companions) were practicing their craft in La Chaux-de-Fonds.[37] They constituted, by a wide margin, the majority of the artisans in the area. In comparison, there were only four foundry workers and nine goldsmiths. The clock-makers distinguished themselves not only in the production of precious and ornate artifacts, but also in innovations in the technical modules of clock and watch mechanisms. Even though in many historical accounts father and son Jaquet-Droz eclipse almost all of their distinguished contemporaries, their success was inseparable from the intellectual fecundity of this environment. The artisans and artists in the Neuchâtel mountains formed an economically and socially tight-knit group, and they borrowed from each other's assets and art. It is difficult to meaningfully isolate any individual achievement within these two generations of artisans.[38]

Although pendulum clock–making was blossoming economically and culturally in the Neuchâtel mountains at this time, it was not a systematically or institutionally taught craft.[39] The first theoretical works on the art of clock-making started appearing in the late seventeenth century, mostly in London and Paris, the European clock-making centers, and the pace of publication picked up momentum around the 1740s. These works were typically textbooks for the systematic study and practice of clock-making and clock repair. Antoine Thiout's treatise from 1741 stands out, because it summarized most of the works known

36. The mother of Josué Robert was a Brandt-dit-Gruérin, which is indicated in Robert's family tree. In a family pact from January 1747, Josué Robert divided up goods to the children from his first and second marriages, and this pact, too, indicates close relations between the Roberts and the Jaquet-Droz families. Perregaux and Perrot, *Les Jaquet-Droz et Leschot*, 43–44.

37. There were 61 makers of "small clocks" and 68 makers of pendulum clocks. The crafts were so differentiated that the artisans' names came up in separate categories in census tables from 1750. Chapuis, *Histoire de la pendulerie neuchâteloise*, 106–7.

38. Ibid., 109. A man named Abram-Louis Perrelet, for example, invented the self-winding watch, and a man named Daniel Gagnebin, who became a fatherly friend of Pierre Jaquet-Droz later in his life, invented a machine to apply studs and ridges onto cylinders with great precision for pendulum (and musical) clocks. Ostervald, *Descriptions des montagnes*, 12; Chapuis, *Histoire de la pendulerie neuchâteloise*, 106. For the novel ways in which Jaquet-Droz used the "endless screw" in general applications, see Haspels, *Automatic Musical Instruments*, 23, 30, 94–95.

39. General mechanical engineering, though, had been systematized since the Renaissance. For an influential and comprehensive work from the eighteenth century, see Leupold, *Theatrum machinarum generale*. In the wake of Newton's *Principia*, mathematical mechanics became a paradigm of all inquiry into the natural world. However, this was not part of artisanal training during this period in rural areas such as the Neuchâtel mountains. Carrera, Loiseau, and Roux, *Androiden*, 10.

at the time. It provided critical service to both pendulum clock–makers and watchmakers and was widely read in the Neuchâtel mountains. It remained the most important reference work until Le Paute's and Berthoud's were published in the next generation.[40] In general, however, clock-making was taught and practiced differently in the Neuchâtel mountains than in the European metropolises; it also had different goals and sought different market segments.[41]

It is important to also mention in this context a popular legend about the origins of clock-making in the Neuchâtel mountains, one that explains the industry's beginnings in quite a different way. The legend tells of a late-1670s encounter between a traveling English horse trader named Peter, who had a watch that needed repair, and a young locksmith named Daniel JeanRichard from Bressel, a small village a few kilometers from Le Locle and La Chaux-de-Fonds. JeanRichard was born around 1665 and must have been a young teenager at the time of this encounter.[42] He had probably been exposed to mechanical artifacts earlier in his life, but this might have been his first opportunity to see, handle, and examine a watch. He managed to get the watch working again, and this success is said to have sparked his subsequent occupation of clock-making. Six months later, he finished his first watch.[43] His work attracted the attention of his community, and they entrusted him with repairing their watches. A few years later, Daniel JeanRichard started pursuing clock-making in nearby La Sagne, at the age of fifteen, and after 1700 he established himself in Le Locle. There, he trained many apprentices, among them his three sons, and the industry thus began to flourish in the Neuchâtel mountains.

In many histories of Swiss clock-making, this encounter between Daniel JeanRichard and the Englishman is taken to be the origin of

40. Among the most prominent works were Derham, *The artificial clock-maker* (1696); Sully, *Regle artificielle du tems* (1714); Le Roy, *Avis contenant les vrais moyens* (1719); Thiout, *Traité de l'horlogerie* (1741); Le Paute, *Traité d'horlogerie* (1755); Berthoud, *Essai sur l'horlogerie* (1763); Cumming, *Elements of clock and watch-work* (1766). See also Milham, *Time and Timekeepers*, chap. 10; Walter, *Soziale Grundlagen*, 283; Landes, *Revolution in Time*, 143–54.

41. Neuchâtel clock-makers started in the late seventeenth century to build clocks geared toward a market segment with lower incomes. Neuchâtel watches were cheaper than those from Geneva, and Genevans liked to "look down upon them as shoddy work." Landes, *Revolution in Time*, 279.

42. Historians disagree on the date of Daniel JeanRichard's birth. David Landes cites the year 1665, Aymon de Mestral 1672, and other historians cite years in between these two dates. De Mestral explains this by referring to the relatively large number of JeanRichards in the area and lost baptism registers. Landes, *Revolution in Time*, 259; Mestral, *Daniel JeanRichard*, 5. The earliest report to tell the story of Daniel JeanRichard is presumably Ostervald's *Descriptions des montagnes* (1766).

43. Favre, "Daniel JeanRichard," 150.

clock-making and watchmaking in the Neuchâtel mountains.[44] The eminent historian of Neuchâtel clock-making Alfred Chapuis commented extensively on this matter in 1917. Against the background of his comprehensive and thorough historical research, on which I have relied throughout this chapter, it is unsurprising that, while conceding that JeanRichard was a talented and active artisan, Chapuis finds that other clock-making families initiated the trade. He holds that Neuchâtel clock-making has older and more diverse roots than are often imagined in twentieth-century accounts.[45] The problems here are similar to the ones I encountered in the historiography of automaton-making: there is substantial decontextualization and dehistoricization, as well as a cultivation of narratives that center around individual mechanically minded figures and pay little attention to social context.[46]

An interesting complementary interpretation of the legend surrounding Daniel JeanRichard has been given by the economic historian Albert Hauser, who assumes that JeanRichard's decisive contribution to Neuchâtel clock-making was that he initiated and pursued the division of labor in the clock-making industry.[47] JeanRichard purchased components of clocks from Geneva and other places and hired apprentices to assemble the clocks. He also used machinery to cut cogwheels, as clock-makers in Geneva did, and thus initiated methods to standardize production and introduce interchangeable parts.[48] His workshop was a model for many others, although his did not expand a great deal and was soon surpassed by others. JeanRichard probably used the protoindustry of lace-making as a model and applied the home and outputting production methods of that industry to the needs of the up-and-coming clock-making in-

---

44. The story even appears in the *Encyclopædia Britannica*, under the entry "Switzerland." It also appears in Flaad, *Untersuchungen zur Kulturgeographie*, 109–11; and David Landes devotes an entire chapter to the founding myth of Neuchâtel clock-making. Landes, *Revolution in Time*, esp. 257–58. Ulrich Im Hof is more cautious, saying that JeanRichard was retrospectively elevated to the status of the legendary founder of Jurassic clock-making. *Handbuch der Schweizer Geschichte*, 2:718.

45. Chapuis suggests that clock-making grew out of existing artisan environments and not the isolated efforts of any individual person. Chapuis, *Histoire de la pendulerie neuchâteloise*, 90.

46. Perregaux and Perrot's account, *Les Jaquet-Droz et Leschot* (1916) is, once again, an exception to this.

47. Hauser, *Schweizerische Wirtschafts- und Sozialgeschichte*, 92.

48. When he broadened his mechanical business and started building watches himself, he must have relied on machine tools for making individual parts. The legend relates that he heard from passing travelers that wheel-cutting machines were in use in Geneva. From their descriptions, it is said, he constructed a wheel-cutting machine for himself. Mestral critically and cautiously rehearses this tale, pointing out persistent problems with historical evidence. Mestral, *Daniel JeanRichard*.

dustry.[49] Hauser's argument interprets the legendary founding of Daniel JeanRichard's business as part of the existing, extensive network of clock-making outwork industry in the Neuchâtel mountains.

The consequences of the new production methods made themselves manifest in the social history of clock-maker villages such as La Chaux-de-Fonds. Entrepreneurial activity in rural areas often grew out of former agricultural activity, and new milieus and classes were emerging in those areas, with specific expertise and skills and new forms of social mobility. They differed from other rural social groups in their ambitions and self-understanding, as well as in their position vis-à-vis the sociopolitical strata of the Holy Roman Empire and the late ancien régime. The result was rural communities that resembled small-town bourgeois culture much more than traditional rural, agricultural societies.[50] Pierre Jaquet-Droz's life and work overlapped with these social changes, as his hometown La Chaux-de-Fonds integrated clock-making artisanship, capitalist commercial methods, and cultural and intellectual elites. The changes in the modes of production and social stratification in Switzerland from the sixteenth to the eighteenth century are surprisingly absent in the literature on the Jaquet-Droz family and their automata.

Pierre Jaquet-Droz's Life and Work as a Young Artisan

Pierre Jaquet-Droz immersed himself in La Chaux-de-Fonds's distinct economy—prolific clock-making industry in combination with agriculture—in the period between his university studies at Basel and his matriculation at the faculty of theology of the Académie de Neuchâtel around 1740. It is during this time that he must have given up his

49. Lace-making was a well-established craft in the principality of Neuchâtel in the seventeenth century. Around the middle of the eighteenth century, there were about twenty-eight hundred lacemakers in the principality. Hauser, *Schweizerische Wirtschafts- und Sozialgeschichte*, 156; Flaad, *Untersuchungen zur Kulturgeographie*, 11, 109–11.

50. The expanding industries in the area (silk, cotton, lace-making, and watchmaking) were decentralized in the rural areas and did not bring about an urban agglomeration. However, in some regions up to two-thirds of the rural population were no longer peasants. They became involved in "ways of living and thinking which had ceased to be those of an agricultural society." Biucchi, "Industrial Revolution in Switzerland," 639. Körner, "Town and Country in Switzerland." At the same time, the outputting system also brought with it increasing dependency and poverty, since tools or raw material were often received on a credit basis from the contractor. Braun, *Das ausgehende Ancien Régime*, 113.

career as a clergyman.[51] He spent most of the years between 1740 and 1747 on the Sur le Pont estate. He might have been a clock-maker apprentice then, and his first master might have been Josué Robert, or an employee of Josué Robert's who lived on Sur le Pont.[52] However, rather little is known about his apprenticeship.[53]

As Pierre Jaquet-Droz was coming of age in this period, he became part of a tightly-knit group connected through professional and family relations. Many of La Chaux-de-Fonds's cultural, political, and intellectual elite were regular guests at Sur le Pont, including merchants, clock-makers, artisans, natural philosophers, natural historians, and physicians.[54] Among the most influential were the Gagnebin brothers, Abraham and Daniel, who were about fifteen years older than Pierre Jaquet-Droz and were well-educated, well-connected physicians and natural philosophers who corresponded with prominent philosophers such as Albrecht von Haller and Jean-Jacques Rousseau. They were supportive companions to Pierre Jaquet-Droz.[55]

During the years of his apprenticeship, Pierre Jaquet-Droz produced a variety of mechanical objects, including music boxes, heavily ornamented watches, and pendulum clocks, which were becoming increasingly fashionable in the mid-eighteenth century.[56] His work soon

51. Accounts of Pierre Jaquet-Droz's youth are based on anonymous reports from the early nineteenth century, which were reproduced in Neuchâtel chronicles and biographical dictionaries. *Biographie universelle*; Jeanneret, *Etrennes Neuchâteloises*; Jeanneret, *Biographie Neuchâteloises* (Jeanneret relied in this biographical dictionary on Ostervald's treatise on Neuchâtel of 1765, several contemporary travel reports discussed later in this chapter [by Meiners and Bernoulli], and his own *Etrennes Neuchâteloises*, the standard source); Bachelin, *L'horlogerie neuchateloise*; *Nouvelle biographie générale*. The 1970s biographer Carrera also emphasizes the contribution of Jeanneret as a chronicler of the Neuchâtel mountains. Carrera, Loiseau, and Roux, *Androiden*, 11.

52. Perregaux and Perrot quote once again the abbot Jeanneret's *Etrennes Neuchâteloises* on this. Apparently this employee of Josué Robert's was not the best. They say "le premier maître de Pierre fut un mauvais ouvrier de Josué Robert." Perregaux and Perrot, *Les Jaquet-Droz et Leschot*, 43.

53. Perregaux and Perrot say about his apprenticeship, "tout est inconnu." Perregaux and Perrot, *Les Jaquet-Droz et Leschot*, 42.

54. On intermarriage between these families, and the resulting robust networks, see Perregaux and Perrot, *Les Jaquet-Droz et Leschot*, 43–44; Faessler, "250e anniversaire," 14; Carrera, Loiseau, and Roux, *Androiden*, 11.

55. See Sinner de Ballaigues, *Voyage historique*, 1:212, on Daniel Gagnebin's natural philosophy. Perregaux and Perrot speculate extensively on the close intellectual bonds between Jaquet-Droz and Daniel Gagnebin, saying that they may have read books and discussed ideas together. Perregaux and Perrot, *Les Jaquet-Droz et Leschot*, 89. When Jean-Jacques Rousseau visited Abraham Gagnebin in 1765 in La Ferrière to work on botanical classification, he saw the Gagnebins' natural history cabinet, whose catalog contained thousands of entries of natural objects. Jacquat, "Abraham Gagnebin"; Wolf, *Biographien*, 3:227–40; Walter, *Soziale Grundlagen*, 282–83; Sinner de Ballaigues, *Voyage historique*, 209–10.

56. These are mostly clocks with carillons and elaborate cases. See images in Cardinal and Mercier, *Musées d'horlogerie*, 31–51.

became more complex, attracting attention and drawing people to La Chaux-de-Fonds. He became more ambitious and started to travel more and expand his professional horizon. Between the later 1740s and the mid-1750s, he traveled regularly to Paris, the European clock-making center, to visit cabinetmakers and other artisans and to purchase cases for his clocks.[57] He was also in correspondence with Ferdinand Berthoud, a famous clock-maker and the author of the most influential clock-making textbook of the time; in May 1753 Berthoud wrote a long letter to Jaquet-Droz. It is possible that one of Pierre Jaquet-Droz's visits to Paris was on Berthoud's invitation, and it is likely that they met.[58]

During this period, in 1750, Pierre Jaquet-Droz married Marie-Anne Sandoz, the daughter of the local merchant Abraham-Louis Sandoz-Gendre. Sandoz-Gendre was an ambitious citizen and a key influence on Pierre Jaquet-Droz and his professional decisions. It was Sandoz-Gendre who initiated and planned Jaquet-Droz's crucial visit to the Spanish court in the late 1750s, and he also accompanied his son-in-law on this trip. Pierre Jaquet-Droz and his wife founded a household on the Sur le Pont estate, but soon the house became too small, and the family and workshop moved to a larger dwelling, closer to the center of La Chaux-de-Fonds.[59] The couple's first child, daughter Julie, was born in 1751, son Henri-Louis in 1752, and finally another daughter, Charlotte, in 1755. Marie-Anne never recovered from giving birth to her last child, and she died in December of 1755. The child, Charlotte, died soon after, in February 1756.[60]

In this same period, influential people started to notice Pierre and his work. Milord Maréchal, also known as Lord Keith, the governor (the representative of the Prussian king) of the principality of Neuchâ-

---

57. In September 1753, Abram-Louis Sandoz wrote in his journal that Pierre was in Paris to acquire "des nouvelles positives des deux fils Abram et Louis Perret." Chapuis, *Histoire de la pendulerie neuchâteloise*, 111–12. It was just at this time that he constructed a series of particularly elaborate pendulum clocks with complicated movements that became important objects for his journey to Spain in the late 1750s. On these pendulum clocks, see Beyer, *Faszinierende Welt der Automaten*, 68; Cardinal and Mercier, *Musées d'horlogerie*, 37–44. See also Perregaux and Perrot, *Les Jaquet-Droz et Leschot*, 45; Carrera, Loiseau, and Roux, *Androiden*, 12; Faessler, "250e anniversaire," 16.

58. On Berthoud, see Jeanneret, *Etrennes Neuchâteloises*, 32–43. We do not know whether Pierre met other clock-makers or automaton-makers, such as Vaucanson. Chapuis raises this question explicitly in *Histoire de la pendulerie neuchâteloise*, 112.

59. This was a place called Jet d'Eau. The building was where the old post office had been. Perregaux and Perrot, *Les Jaquet-Droz et Leschot*, 89–90; Faessler, "250e anniversaire," 17; Carrera, Loiseau, and Roux, *Androiden*, 15.

60. Pierre Jaquet-Droz never remarried. His sister and his parents-in-law took charge of bringing up his two children. Perregaux and Perrot, *Les Jaquet-Droz et Leschot*, 52–53; Faessler, "250e anniversaire," 17; Carrera, Loiseau, and Roux, *Androiden*, 12.

tel at the time, became interested in Jaquet-Droz's elaborate and unusual mechanical works; he was another key figure in this decisive period in Jaquet-Droz's life. Lord Keith was a descendant of an aristocratic Scottish family and a gifted and extremely well-connected man. After leaving Scotland in the 1710s, he went first to Spain and then to Switzerland. When Jaquet-Droz and his work were introduced to him, he advised Jaquet-Droz to make people outside of Switzerland aware of his clocks.[61] Lord Keith suggested going to Spain—where he still had influential contacts—to exhibit clocks and to present them at the court of King Ferdinand VI, a well-educated monarch who promoted arts and mechanics.[62]

The encounter with Lord Keith was fortunate for Jaquet-Droz, since even though he had built a series of fancy and innovative objects and had received admiration and respect for them, he still needed to sell his works and make them known to larger circles—not easy to do in the rural area where he lived. Even the better-off families had difficulty affording the luxurious commodities that he produced. Pierre Jaquet-Droz had to look abroad for potential buyers.[63]

Journey to the Spanish Court

Pierre Jaquet-Droz's journey to Spain was a considerable effort, given both the delicacy of the mechanical objects he was transporting and the ambitious project of meeting the Spanish king and queen at court. His father-in-law, Abraham-Louis Sandoz-Gendre, set up the itinerary for the journey and saw to the travel arrangements. Jaquet-Droz organized local contacts and letters of introduction to the Spanish king. Madrid was not a completely unknown place for Neuchâtelians: both Josué Robert and Jaquet-Droz had business relations there, as they had sold pendulum clocks to Madrid residents before. There were also other Neuchâtelians residing in Madrid at the time, with whom Pierre Jaquet-Droz came to have regular interactions.[64]

In April 1758 Jaquet-Droz, his father-in-law, and a worker named

61. Faessler, "250e anniversaire," 18–20.
62. The "good" King Ferdinand VI had helped found the Royal Academy of Fine Arts in San Fernando and employed distinguished musicians at his court. Schulz, "Spaces of Enlightenment," 189.
63. Perregaux and Perrot, *Les Jaquet-Droz et Leschot*, 61.
64. Among those were a Huguenin, a Rognon, a Perret, and a Ducummon-dit-Boury. Faessler, "250e anniversaire," 20; Perregaux and Perrot, *Les Jaquet-Droz et Leschot*, 68.

Jacques Gevril embarked on their journey, carrying with them a specially constructed container designed for the transport of six extremely
delicate pendulum clocks. Arriving in Madrid after forty-nine days of
travel, they were welcomed in the house of Sieur Jacinto Jovert, a friend
of Lord Keith's. Because of King Ferdinand's ill health, Jaquet-Droz had
to wait five months before he was received; he spent this period moving in the Neuchâtelian circles in Madrid, making his name known
there as well as—through the grapevine—back in Switzerland.[65]

In September 1758 Pierre Jaquet-Droz finally gained an audience
with the king to present his pendulum clocks and mechanical works.
The king allegedly made him play them more than one hundred times.
Among the objects he displayed were a clock with a striking mechanism
and a flute that featured a cage containing a singing bird that moved its
wings. Another was a pendulum clock that was designed as a mechanical bell-striker; its figure responded to inquiries about numbers from
the audience by beating on a drum (the clock was called Le nègre). The
mechanism that seemingly could "understand" and respond to inquiries about numbers worked by means of magnets. Another pendulum
clock was called The Stork (La cigogne), probably part of a mechanical
water or fountain arrangement. The Stork was apparently very amus-
ing to see, as the ambassadors of Portugal, Denmark, and Holland each
demanded one for their own lords. The most sophisticated clock was
called The Shepherd (Le berger). It was a pendulum clock that featured
a shepherd sitting on top playing his flute. This clock reportedly reaped
the most success and remains today in the royal palace in Madrid.[66]

That all the pieces Pierre Jaquet-Droz displayed were sold for good
prices demonstrates that his work was compatible at the time with the
still-flourishing court society and its tastes and needs: he delivered
machinery and props that people in high positions liked and needed
for their purposes. Ornamented pendulum clocks accompanied by
music were a genre that he could work within, vary, and push to its
limits, well beyond the demands of mere time-keeping, to accomplish
this success. He was thus learning about the uses and possibilities of

65. "Journal de A. L. Sandoz," Nb36, Nb37, Nb38, Nb39; "Journal pour Abram Louis Sandoz;
Justicier de la Chaux-de-Fonds; 1757–59"; "Journal du Voyage d'Espagne de Jaquet-Droz, 1758–
1759," D 1853; "P. Jaquet-Droz; Lettres d'Espagne, 1758–1759," Nb 60; all in the archives of the
Bibliothèque de la Ville La Chaux-de-Fonds. See also Musée neuchateloise 3 (1866): 77, 104; and
Tissot, Voyage de Pierre Jaquet-Droz.

66. Supposedly, Pierre Jaquet-Droz received two thousand "pistoles d'or" for his clocks. Perregaux and Perrot, Les Jaquet-Droz et Leschot, 53–56; Chapuis, Histoire de la pendulerie neuchâteloise,
112; Faessler, "250e anniversaire," 20.

spectacular machinery, music, and clock-making mastery within the context of courts, stages, and highbrow audiences during his apprenticeship and in Spain. His long stay in Madrid also meant that he could make his name known beyond the king and queen and socialize with Neuchâtelians in Madrid and their clients and peers. These factors proved crucial in the design and purpose of his three automata.

In January 1759, Jaquet-Droz and his father-in-law embarked on their journey back to the Neuchâtel mountains, where they arrived in late March. He was now a prominent and celebrated pendulum clock-maker far beyond the borders of the Neuchâtel mountains. His workshop expanded and became a flourishing business. All industry in Neuchâtel, not only the pendulum-clock industry, was boosted by Pierre Jaquet-Droz's success, because the commissions and orders that came in after the journey to Spain also went to the workshops of his numerous competitors.[67] Crucially, the trip provided the financial conditions under which he and his son could spend several years conceiving of, designing, and building three mechanical androids.[68]

The Three Automata

As a young adult, Pierre Jaquet-Droz had taken into his household a young man of humble background, Jean-Frédéric Leschot. Jaquet-Droz treated Leschot like an older sibling of his son Henri-Louis, allowing the two young men to learn and live together. They were initiated into the clock-making trade together until around 1767, when fifteen-year-old Henri-Louis was sent to the university. Leschot became one of the three principals of the first Jaquet-Droz workshop and later, after the automata were built, a crucial figure in the founding of the various branches of the Jaquet-Droz main firm, the *maison*.[69]

---

67. The trip to Spain was not the instigator of the prosperity of the pendulum-making industry in La Chaux-de-Fonds, as some believe. It was an important event, but the larger success was the result of sustained efforts of several generations of artisans. Chapuis, *Histoire de la pendulerie neuchâteloise*, 114.

68. It was the court grapevine, among other things, that spread Pierre Jaquet-Droz's fame among rich families on the European continent, an effect of his networking in Madrid. Perregaux and Perrot, *Les Jaquet-Droz et Leschot*, 89. Some sources, such as *Nouvelle biographie générale*, 13:813 (and others copy from this), state explicitly that Philipp V granted a pension to Pierre Jaquet-Droz. This is inaccurate, since Philipp V was Ferdinand's father and died in 1746.

69. Perregaux and Perrot estimate that Leschot's move into the Jaquet-Droz household occurred before the voyage to Spain, around 1756. Perregaux and Perrot, *Les Jaquet-Droz et Leschot*, 90. In 1769 already, Leschot had shares in the profits of the Jaquet-Droz firm, and in 1782 he

Pierre Jaquet-Droz's son Henri-Louis was born in 1752 and received his early education from his aunt, his maternal grandparents, his father, and the local pastor. In 1767 Pierre Jaquet-Droz sent him to Nancy to study under the abbé de Servan, a historian, engineer, and mathematician.[70] Henri-Louis studied physics, mathematics, music, and drawing, and this training enabled him to bring new directions to his father's work.[71] In 1769 Henri-Louis returned to the Neuchâtel mountains, where his father and Leschot had continued building precious and sophisticated pendulum clocks. Around this time, the three men must have started working on the android automata, with Henri-Louis putting into practice the skills and expertise he had acquired in Nancy.

By early 1774, four pieces of exceptional craftsmanship were completed: an automaton theater called The Grotto (La grotte), a one-meter by two-meter animated picture, and three androids.[72] The androids depicted a small boy who wrote texts that were programmed on a disc of wedges; another boy, who sketched four drawings programmed on a cylinder of cams; and the harpsichord-playing woman, who played five melodies, supposedly composed by Henri-Louis. The musician's performance was programmed on a drum furnished with studs and ridges, which followed the principle of the strike mechanisms that Pierre had used for his pendulum clocks. I discuss this mechanical design in detail in chapter 4. Father and son Jaquet-Droz exhibited their mechanical masterpieces for a few months in La Chaux-de-Fonds and then took them on tour to cities in mainland Europe, to London, and finally, in the 1780s, as far as we know, to Geneva, where they had founded branches of their clock-making firm.

---

figures as an associate. See Papiers Leschot et Jaquet-Droz 1716–1957, no. 77, Archives d'Etat, Genève; Jaquet-Droz Papers D 166, D 476, D 3840; and Leschot Papers D 158, D 1053, Musée International d'Horlogerie; Aeschlimann, "Pierre Jaquet-Droz et Jean-Frédéric Leschot."

70. Lanier, L'abbé Michel Servan.

71. During this time, Henri-Louis may also have seen or heard about an automaton theater called Rocher in the park of the Château de Lunéville in Nancy, which was installed in 1742. It may have served as a model for an automaton theater named The Grotto (La grotte) that he and his father built together with their androids. Marc Vanden Berghe, "Henri-Louis Jaquet-Droz," 151; Perregaux and Perrot, Les Jaquet-Droz et Leschot, 91.

72. Automaton theaters and animated pictures—tableaux mécaniques—were among the fashionable mechanical treasures in the eighteenth century. See Heckmann, Die andere Schöpfung, 248–49; Brieger, Das Genrebild, 148. Henri-Louis Jaquet-Droz's La grotte vanished in the later eighteenth century, and the remaining descriptions mostly refer to an engraving dating from the mid-1770s: Perregaux and Perrot, Les Jaquet-Droz et Leschot, 19.

The Jaquet-Droz Company and Its Branches

Sometime in 1773, Henri-Louis Jaquet-Droz left La Chaux-de-Fonds for London to set up, together with Leschot, a branch of the Jaquet-Droz firm for production and trade. The automata must have been finished or nearly finished by then. The impetus for the London branch was international demand (mostly from the Middle East and the Far East) for automatic pendulum clocks and other exquisite mechanical objects. In founding a foreign branch, Pierre and Henri-Louis were following an established business model for eighteenth-century clock-making.[73] Henri-Louis was in charge of the branch in London, living there for about two months a year and spending the rest of the year in La Chaux-de-Fonds and on the road. Pierre traveled to London and stayed with his son when his presence was needed.[74] In 1783 Henri-Louis founded a new London association, together with the clock-maker brothers Henry and Jean-Marie Maillardet, who were also originally from La Chaux-de-Fonds and may have been trained as young men in the Jaquet-Droz workshop.[75] Henri-Louis became quite notable during this time as a widely traveling businessman and a member of several distinguished societies.

In La Chaux-de-Fonds, the old production house existed until well into the 1780s with both father and son Jaquet-Droz having shares in it.[76] There is an eyewitness report of the workshop in an October 1782 diary entry, many years after the three automata were finished, by a councillor named François de Diesbach. Diesbach traveled through La Chaux-de-Fonds, was impressed with its wealth and architecture, and "before dinner," he writes, had the chance to visit "Monsieur Jaquet-Droz," who showed him pendulum clocks and organ clockworks. Pierre Jaquet-Droz told him that his son was in London and that the autom-

---

73. The brothers Huguenin and the brothers Ducommun-dit-Verron, for example, fellow La Chaux-de-Fonds clock-makers, had houses in Bordeaux and London. On the business association between Henri-Louis Jaquet-Droz and Leschot, see Notes diverses (mécanique, etc)—En tout ou partie de la main de Jean-Frédéric Leschot (1764–1824), manuscript collection "History," vol. 5, no. 954, in the archives of the Bibliothèque de Genève. See also Perregaux and Perrot, *Les Jaquet-Droz et Leschot*, 94; Chapuis, *Histoire de la pendulerie neuchâteloise*, 114. On London clock-making and trade with China, see Chapuis, *Relations de l'horlogerie suisse avec la Chine*.

74. Perregaux and Perrot, *Les Jaquet-Droz et Leschot*, 112–14.

75. For correspondence and business transactions between Henri-Louis Jaquet-Droz, Leschot, and the Maillardets, see the papers in nos. 955 and 956 of the manuscript collection "History," as well as no. 961 of "Theology, Literature, Sciences," in the archives of the Bibliothèque de Genève.

76. Perregaux and Perrot, *Les Jaquet-Droz et Leschot*, 114.

ata were in Paris.[77] The automata, at that stage, were thus with neither of the two original makers, and in the 1780s Pierre Jaquet-Droz was producing clocks for foreign markets and not additional automata.

Henri-Louis Jaquet-Droz was the head of all his firms, solely in charge of the artistic as well as the commercial parts.[78] He did not have a strong physical constitution, however; he suffered from tuberculosis. Production and business negotiations obliged him to make arduous journeys, and he did not respond well to the rough climate of the Neuchâtel mountains or the humid climate in London. Given his responsibilities in Paris, London, and his home country, he was constantly overworked. Pierre Jaquet-Droz found himself equally taxed, dividing his life between Paris, London, and Switzerland.[79] In 1784 Henri-Louis was no longer able to tolerate London's climate, and he took the business that he had established there with Leschot in 1773 to Geneva. His father followed him to Geneva a few years later, after he had continued his workshop in La Chaux-de-Fonds. From then on, Henri-Louis's own production house was in Geneva as well as that of the association Jaquet-Droz & Leschot, while the one founded as Jaquet-Droz & Maillardet remained in London. In each of these centers, commodities such as watches, tobacco cases, manicure sets, and perfume bottles (with pearls, enamels, and precious stones) were designed, built, assembled, and sold, and many of these objects contained small, hidden, very sophisticated mechanisms. Also produced were clocks with musical mechanisms and astronomical instruments for scientific purposes.[80]

Henri-Louis Jaquet-Droz and Leschot settled well in Geneva, becoming members of important circles and formal associations and receiving citizenship in 1785. Genevans were interested in them not least to reinforce Geneva's claim as the clock-making capital of Switzerland. In 1786 Henri-Louis was made a member of the prestigious Société des Arts, and he came to greatly influence Genevan clock-making, art, and industry through projects and inventions—ranging from toolmaking to urban planning—that he administered with Leschot. In reunion

77. Perregaux and Perrot had access to this diary through a descendant, Max de Diesbach, who communicated to them an extract from François's journal. Ibid., 115.

78. Ibid., 117. For correspondence between Henri-Louis and his associates, see papers in no. 961 of the manuscript collection "Theology, Literature, Sciences" in the archives of the Bibliothèque de Genève.

79. Perregaux and Perrot, Les Jaquet-Droz et Leschot, 112–17.

80. See Leschot's papers (Papiers Leschot), Ms. fr. 4498–4500, manuscript collection "French Manuscripts," in the archives of the Bibliothèque de Genève. See also Perregaux and Perrot, Les Jaquet-Droz et Leschot, 113–14.

with Pierre, they even planned new android musicians, but there is no evidence that they ever built them.[81]

Looking at how the Jaquet-Droz family ran their businesses after they finished their three original android automata helps us understand the status that these objects had in their lives, as well as the artisans' priorities and self-understanding as artists, businessmen, and citizens. The automata highlight one unarguably successful episode in their lives, but the Jaquet-Droz family's sense of their own legacy revolved much more around other commodities and objects in their repertoire. As we have seen, in the decade or so after they completed the automata, they invested a great deal of time into founding and running business ventures that had nothing to do with automata. These endeavors took priority for both Pierre and Henri-Louis, and they followed contemporary customs and expectations of what skilled and ambitious clock-makers did.[82] As the workload and economic risk of their business ventures increased, their interest in touring and exhibiting their automata disappeared. Indeed, the three automata were almost certainly sold, since they do not appear in any of Henri-Louis Jaquet-Droz's or Leschot's accounting books, inventories, or wills.[83]

The Jaquet-Droz business in Geneva was in its prime in the years 1786 and 1787, and it had a highly visible presence. Although its prosperity appeared to be stable, it turned out to be of short duration. A few setbacks and losses occurred in the later 1780s when customers did not pay their bills. It may have been that the firm was forced to sell the three automata in this context. In 1789 the French Revolution made matters worse. Trade with England became more complicated, and clock-making faced the challenge of modifying its heretofore luxury products for a broader stratum of clients. Because of his poor health, Henri-Louis Jaquet-Droz moved from Geneva to nearby Chambésy on the lake, but his respiratory problems remained. He moved on to Provence and then to Naples, but to no avail: he died on 15 November 1791, less than a year after his father's death, leaving his associate Leschot as their only business successor. The Jaquet-Droz businesses were

---

81. Papiers Leschot; Comptabilité de Jean-Frédéric Leschot et Henry-Louis Jaquet-Droz, Ms. suppl. nos. 955, 956–60, Bibliothèque de Genève.

82. The same is true for the Roentgens, as I discuss in chapter 3.

83. See Leschot's and Henri-Louis Jaquet-Droz's wills in the Archives privées no. 77, Papiers Leschot et Jaquet-Droz 1716–1957, in the Archives d'Etat, Genève (Leschot's is 77.02.02; Henri-Louis Jaquet-Droz's is 77.07). Henri-Louis's inventory is listed under Jur. Civ. F, no. 825 (1791) in the Archives d'Etat, Genève. Their accounting books are nos. 955 and 956 of the manuscript collection "History" and no. 961 of the collection "Theology, Literature, Sciences," in the archives of the Bibliothèque de Genève.

liquidated through the trust of Ami-Isaac Dassier, a friend of Henri-Louis's.[84] Some reports say that upon Henri-Louis Jaquet-Droz's death the automata were sold to Spanish impresarios.[85]

In spite of the automata's fame, it is rather difficult to keep track of their whereabouts after the 1780s. We know that, right after they were completed, father and son Jaquet-Droz exhibited them locally in La Chaux-de-Fonds for a short while, then showed them in Paris and London, and eventually took them to Geneva, where the artisans had settled. Some historians of the Jaquet-Droz family believe that until the 1870s, the automata were "on tour," appearing on various stages in continental Europe, but there is little evidence to support this belief.[86]

The Harpsichord Player's Reception

Although the harpsichord player and the Jaquet-Droz family's other automata did push the envelope of the mechanical arts of the time, they were not a categorically new phenomenon for contemporary audiences, or even spectacular beyond the spectacle culture of the early modern period. Texts on the Jaquet-Droz automata written during the years that the automata were on display are rather reticent about the automata themselves, their spectacular craftsmanship, and their metaphysical implications as "machine-men"—themes that took on such relevance in the nineteenth and twentieth centuries. Sources at our disposal comprise diaries and letters from La Chaux-de-Fonds locals or travelers who came through and printed reports in contemporary periodicals, almanacs, travelogues, and calendars. Citations and mentions of the Jaquet-Droz automata in such documents appeared for the twenty or so years after the automata were finished and first exhibited, after which period references began to ebb.

Manuscript and printed references to the automata give us insight into the circumstances of their exhibition, but both are often silent about the automata's actual mechanical function and their effect as artificial humans. Writers of letters or diary entries were, instead, inter-

---

84. Papiers Leschot; Comptabilité de Jean-Frédéric Leschot et Henry-Louis Jaquet-Droz; Papiers Leschot et Jaquet-Droz 1716–1957. The company encountered financial problems also because the famous clock firm of James Cox never paid its debts after James Cox died. Landes, *Revolution in Time*, 105; Le Corbeiller, "James Cox and His Curious Toys."

85. Schindler, "Die legendären Automaten."

86. Examples for this unevidenced claim are Schindler, "Die legendären Automaten"; and Altick, *Shows of London*, 350.

ested in the automaton shows as social events and in their own social relations to the Jaquet-Droz family as artisans and notables, while the printed texts were all copied or translated verbatim from a brochure that the Jaquet-Droz firm produced, probably in 1774, to be handed out at automaton shows.[87] This brochure merely provided plain descriptions of the android automata and the automaton theater, and the texts copied from it thus all convey a similar impression and tell us little about people's reactions, awe, wonder, or thoughts on the role of automata in the problem of unstable human-machine boundaries.

Observers' Reports on the Automata

Among the first eyewitness reports of the automata is a June 1774 entry, about six sentences long, in the diary of David Sandoz, a La Chaux-de-Fonds area notable. Sandoz said that he went to La Chaux-de-Fonds and "saw the 'curiosities' of the Jaquet--Droz family."[88] The particular display that Sandoz saw must have involved a combination of the automaton theater La grotte with at least two of the three androids: a whole mechanical scenery. His entry says that in one of the curiosities, "there is a man who walks to the mill with a donkey and who returns from there with a sack on the donkey, and a dog that barks." It also features "young women who dance to the tune of a harpsichord played by a young woman of the size of a six-year-old child; she moves her head while she is playing." Sandoz also saw "a boy who draws a portrait of King Louis XV with a black pencil, etc." His entry closes by noting that there was "also a cow with its calf, moo tones, goats that climb a rock."[89] Sandoz does not seem to perceive any one of the mechani-

87. Perregaux and Perrot speculate that the text was written by a foreigner and that the Jaquet-Droz firm did not have the opportunity to revise the text, for their name is misspelled in it ("Jacquet Droz" rather than "Jaquet-Droz"). The brochure is reproduced in Perregaux and Perrot, *Les Jaquet-Droz et Leschot*, 102–5. Oliver Hochadel investigates the relationships between brochures, publicity, and spectacles for the related case of eighteenth-century itinerant lecturers who performed electrical experiments. Hochadel, *Öffentliche Wissenschaft*, 194–99.
88. The entry is from 29 June 1774. This must have been one of the earliest exhibitions of the Jaquet-Droz mechanical theater. "Curiosities" is Sandoz's own term: "Avons vu les curiosités du sieur Jaquet." Transcriptions of his diary, from 1770 to 1778, are printed in Favre, "Journal de David Sandoz," 228. See also Perregaux and Perrot, *Les Jaquet-Droz et Leschot*, 101. All translations in this book are my own unless otherwise stated.
89. Favre, "Journal de David Sandoz," 228. There is no mention of the writing automaton. Pierre and Henri-Louis Jaquet-Droz may have varied the cast, or the writer may have been under maintenance. It seems unlikely that Sandoz would not mention an android at all.

cal theater's elements as foregrounded, and the two androids do not seem to have been staged at the center. His note does not display any particular excitement either about the mechanical theater or about the androids or their metaphysical implications.

Shortly afterward, in July 1774, another La Chaux-de-Fonds notable, Isaac Droz, reports in a letter to the governor of the principality of Neuchâtel that "the automata with which the Jaquet-Droz family had occupied themselves in the last few years are finished."[90] Droz's report continues with a rather detailed paragraph about the exhibition of the Jaquet-Droz's mechanical works. It says the automata are now accessible to the public and that the street leading to their house is occupied every day by coaches and wagons. Rain does not discourage people from coming, and showings are staged all day, starting around six in the morning and ending at seven or eight at night. Pierre and Henri-Louis Jaquet-Droz, with two workers, take turns in presenting their automata. Isaac Droz also mentions that among those in attendance are honorable persons from neighboring areas, high-level administrators from other cantons with their ladies, and also an incognito French ambassador with other gentlemen. Droz included in his letter three drawings "drawn by these beautiful children"—which is a little imprecise, given that only one automaton was a draftsman. In fact, Droz's letter pays little attention to the three androids or the automaton theater, not indicating exactly which pieces were on display and certainly not providing any details or any statement about the effect the display had on him. The letter seems much more focused on the crowds and the notables, which is understandable since it was written to the governor.

On 21 January 1775, about six months after David Sandoz's diary entry and Isaac Droz's letter, Abraham-Louis Droz, another prominent citizen of La Chaux-de-Fonds, wrote to his brother Pierre-Frédéric. The letter contains about two paragraphs on matters having to do with the Jaquet-Droz family. Droz starts out by reminding his brother that Pierre Jaquet-Droz of La Chaux-de-Fonds made a very rare and curious "piece" called La grotte or La bergerie, in which one could see many automata

90. Perrin, "Un solliciteur," 70. The letter is from 9 July 1774. In the archives of the library of La Chaux-de-Fonds (Bibliothèque de la Ville La Chaux-de-Fonds), there is a small booklet of 130 pages that contains (among other things) copies of letters that the engraver and jeweler Isaac Droz wrote between 1769 and 1787 to the governors de Lentulus and de Béville. Among other topics, they deal with the establishment of a clock-making industry in Berlin ("l'établissement d'une fabrique d'horlogerie à Berlin"). Perrin, "Un solliciteur," 61–65. See also Perregaux and Perrot, Les Jaquet-Droz et Leschot, 101.

who played musical instruments.[91] It goes on to say that, in addition, "one saw a child of two years who held a brush in his hand and drew like a skilled person of this art." "The writing automaton," the letter points out, "wrote everything that one dictated" without a person's touching him "directly or indirectly."[92] Droz's description of the writer automaton's skills is interesting, since the expression "neither directly nor indirectly" is a direct quote from the brochure that the Jaquet-Droz firm printed about their three androids and the automaton theater. There is a whole history, which I tell at the end of this chapter, about how this brochure's text eventually became copied many times in various print media in the 1770s and 1780s. It becomes apparent here, in the context of Abraham-Louis Droz's communication, that the brochure enabled individual letter writers who had witnessed automaton showings to use the brochure's preproduced phrasing to describe the automaton's motions, instead of writing their own reports. This might be one of the reasons that we find so few individualized accounts of the effect that the Jaquet-Droz automata had on spectators. In Abraham-Louis Droz's letter, too, the phrasing about an automaton's writing what is dictated to him is almost identical to the phrasing in the brochure.[93] His letter also mentions, like Isaac Droz's, the distinguished personalities who saw the automata, and he closes this paragraph by saying that "the son of said lay judge Jaquet-Droz" left for Paris "with said piece" to show it to the new king Louis XVI (crowned in 1774) and "also to go to London."[94] In the second paragraph, Abraham-Louis tells his brother about his desire to join the Jaquet-Droz family as a clock-maker and perhaps go to London with them, as he was hoping to spend some time there practicing his profession.[95]

A letter of 31 March 1776 from a Suzanne-Louise Nicolet to her brother, a pastor-in-training and tutor in Utrecht, mentions the Jaquet-

91. This letter is printed in *Musée neuchâtelois* 24 (1887): 257–58, as part of a series of articles on Pierre-Frédéric Droz, clock-maker, traveler, and metallurgist in the eighteenth century. The quote is on page 257.

92. Ibid.

93. Abraham-Louis Droz writes: "L'automata écrivain écrit tout ce qu'on lui dicte sans que personne le touche soit directement soit indirectement." *Musée neuchâtelois* 24 (1887): 257. The brochure says, "Cet automate écrivain . . . écrit . . . tout ce qu'on juge à propos de lui dicter, sans que personne le touche ni directement, ni indirectement." Perregaux and Perrot, *Les Jaquet-Droz et Leschot*, 103.

94. He consistently talks about "one piece": "une pièce, "la susdite pièce," etc. This might be because it is a whole scenery. *Musée neuchatelois* 24 (1887): 257–58.

95. He sends with the letter a set of watches that he made and invites his brother to travel together with him and meet at his father's house, should his brother's plans coincide with his own. *Musée neuchatelois* 24 (1887): 258.

Droz firm's brochure and gives a few more insights into the automata's whereabouts immediately after they were completed. The Nicolets were friends of the Jaquet-Droz family, and she wrote this letter a little more than a year after Abraham-Louis Droz wrote his. It is about ten sentences long and starts by saying that concerts of La Chaux-de-Fonds's Thursday Society were discontinued "since M. Jaquet-Droz's departure" and that he was in London with his pieces, "where he works."[96] She also tells her brother that she has tried several times to "send him the description of all their precious masterpieces which [father and son Jaquet-Droz] themselves printed."[97] She offers to send him a sample piece written by the writing automaton and a drawing by the draftsman automaton, saying that she herself saw the automaton write what was dictated to him without anyone appearing to touch him and that one could see his eyes follow his writing. She also says she saw the "other draftsman" (this is a little imprecise) follow his work on paper with his eyes, seeming to examine it. She makes no mention of the harpsichord player, which suggests, once again, that the exhibitions did not always include all three. Her language resembles that of the brochure but not in an unusual way, as she uses common words and phrases. In November of 1776 she writes again, stating that "M. Jaquet-Droz" had not returned and probably would never return, and noting that "his father" also left a while ago in order to join him in London ("M. Jaquet-Droz" is thus always Henri-Louis here) to help him make new precious pieces.[98]

Finally, the correspondence of Julie de Lespinasse contains a mention of the Jaquet-Droz automata in Paris. De Lespinasse ran a distinguished salon in Paris in the 1750s and 1760s and was famous for her passionate letters, which filled several volumes. Through her intellectual engagements she was acquainted with Diderot and d'Alembert, among others. In an undated letter, she writes, "Before dinner, I will go to rue de Cléry and see the automata which are astonishing, as they say."[99] There is reason to assume that this letter dates from mid-February 1775.[100] Later in 1775, the automata went to London.[101]

96. Perregaux and Perrot, *Les Jaquet-Droz et Leschot*, 102.
97. Ibid.
98. Ibid.
99. Lespinasse, *Correspondance*, 282.
100. A report in the *Journal de Politique* of January 1775, discussed later in this chapter, says that the automata were exhibited at l'Hôtel Lubert, rue de Cléry (see also Perregaux and Perrot, *Les Jaquet-Droz et Leschot*, 107). This is in the tenth arrondissement. She may well have seen them there.
101. See Abraham-Louis Droz's letter to his brother, of January 1775. *Musée neuchatelois* 24 (1887): 257–58.

Judging by these reports, there seems to have been a whole set of mechanical toys and musical devices on display when the Jaquet-Droz family showed their work, and it seems that the selection varied with individual showings. The reports also suggest that people did not become overly excited or carried away by the automata. Instead, people emphasize the important role that the stagings played as part of the social life of the Neuchâtel mountains: networking, receiving distinguished personalities, and rubbing shoulders with famous artists and artisans.

The First Printed Text about the Harpsichord Player

The Jaquet-Droz company and their harpsichord player (together with the other automata) must have left La Chaux-de-Fonds in the second half of the year 1774, since in January 1775 they were already in Paris. The androids did not return until the 1780s.[102]

Short, descriptive texts about the harpsichord player (and the other automata) were printed in eighteenth-century periodicals from 1774 until the early 1790s. I trace these earliest texts about the harpsichord player to understand how they were composed, copied, and integrated into other contexts in print media, thus also identifying general publishing conventions for journals, calendars, and travelogues at the time. I explore what other texts and contexts the harpsichord player was made part of, between what other reports she was "sandwiched," and who found her or her makers important.

The first written description of the harpsichord player appeared in the brochure that the Jaquet-Droz firm produced around 1774. The document describes the performance of their automata and their automaton theater, and copies were probably given out in the towns where the automata were exhibited.[103] There is no original of the brochure extant, but transcriptions and one engraving have survived. The document first describes the writing automaton, then the draftsman, then the harpsichord player, and finally the automaton theater, each in about one paragraph.[104] The description of the harpsichord player states that

102. See, once again, Abraham-Louis Droz's letter of January 1775 and Suzanne-Louise Nicolet's letter of late March 1776, in Perregaux and Perrot, *Les Jaquet-Droz et Leschot*, 102. These trips raise the question of whether the automata were a big sensation in a capital like London or Paris, or whether they were merely one among many curiosities and sensations.

103. No earlier written text about the Jaquet-Droz automata is known. See Perregaux and Perrot, *Les Jaquet-Droz et Leschot*, 103.

104. The description of the automaton theater is a bit longer. Ibid.

the figure "represents a young girl between ten and twelve years old, sitting on a stool, and playing a harpsichord." The text continues, saying that "the automaton's body—the head, the eyes, the arms, the hands, and the fingers—conduct various movements which seem natural. The automaton herself plays on her harpsichord various melodies in two or three voices, with great precision. Since her head can move in all directions, as can the eyes, she casts her glances equally often to her hands, to the music, and to the audience; her pliable body leans forward occasionally to have a closer look at the music; her chest drops and rises alternately, in order to indicate breathing."[105] This text highlights both the harpsichord player's music-playing and her bodily motions, and it is worthwhile to note that it weighs these two dimensions equally: it talks as much about the automaton's moving head, eyes, and breathing as it does about the actual music-making and the melodies. The text makes clear that the automaton does "more" than just make music. The automaton's breathing, furthermore, is singled out prominently at the end. By mentioning the functions of the bodily motions (looking at music and audience) and their effects (appearing natural and engaging to the listeners), the text evokes a whole musical-performative scene, rather than just emphasizing the mechanical reproduction of music. I return in chapter 4 to the mechanical means of this performance and its cultural correlates in eighteenth-century music-making.

Between 1775 and 1790, nine texts were published containing references to the Jaquet-Droz family and their harpsichord player. The references ranged from short remarks about Pierre and Henri-Louis Jaquet-Droz to longer accounts of their automaton, of clock-making, or of Switzerland's industry and geography. These texts appeared in the weekly magazines *Journal de politique et de littérature* (published in Brussels) and *Journal de Genève*, in the pocket calendars *Göttinger Taschenkalender* and *Almanach de Gotha*, and in five travel reports authored, respectively, by a Swiss dignitary, a Swiss natural philosopher, a German professor, a British tutor, and a British clergyman. Six of these texts mention the Jaquet-Droz family or their automata briefly, while the other three provide more detailed descriptions of the harpsichord player's performance.

It is remarkable that the descriptions of the harpsichord player in

---

105. Ibid. The French original uses the pronoun "elle" in the text and the article *cette* for *automate*, even though the grammatical gender of the paragraph's subject (*l'automate*) is masculine. I use the English pronouns "she" and "her" throughout to indicate this choice of pronouns and articles in the French original.

such texts are copied or translated verbatim from the Jaquet-Droz firm's 1774 brochure and thus are practically identical to one another.[106] They reflect little engagement on the part of the authors with the automaton itself. The articles share this with the references to the women automata in the letters and diaries I discussed above (and also with a letter about the dulcimer player that I discuss in chapter 3). I argued that those letters and diaries lacked careful engagement not only with the automata, their specific performance, and the superb craftsmanship that they embodied, but also with larger issues, such as metaphysical or ethical consequences that might emerge from an encounter with mechanical humans. I explained the letters' and diaries' indifference by suggesting that not only was automaton-making a historically specific practice, but so was the reporting of responses to automata. The responses to the harpsichord player (and the other automata) functioned within specific social networks and served social functions in a manner similar to that of other spectacles at the time.

In the case of printed responses in eighteenth-century mass media, I explain, in the following, the lack of specific engagement with the automata by considering the economic exigencies and literary conventions of the publishing markets of the time. Contemporary conventions of filling periodicals with short reprinted texts from other sources took priority over commissioning original commentary on the automaton's epistemological, cultural, or ethical implications—commentary that later accounts lead us to believe were a part of contemporary texts on automata. At the end of this chapter, I discuss how even prominent eighteenth-century natural philosophers, Georg Christoph Lichtenberg and Johann Bernoulli, when they composed texts about the harpsichord player for the calendar or travel report that they edited, were content to copy the automaton's description from elsewhere. They apparently saw no need or desire to elaborate on the automaton's meaning or implications.

The Harpsichord Player in the Media Industry of the Eighteenth Century

The production and dissemination of texts about the harpsichord player occurred in the context of profound changes in public communication in Europe in the second half of the eighteenth century: fast

---

106. Some of the later ones may have been copied from one another, not from the original.

expansion of publishing and literacy, commercialization of literature, and newly emerging relationships between reading, sociability, and political participation affected the ways that people wrote, printed, read, and discussed texts. The new cultural and economic conditions of literary production, along with the changing roles of authors, editors, and publishers, greatly influenced the content and form of texts in periodicals, almanacs, and travelogues. Texts about automata were microcosms of this very transformation.

There was a rapid increase in the production of books, newspapers, and periodicals between 1770 and 1800. Together with this, the number of readers also increased significantly, for many reasons, including the proliferation of public libraries, reading cabinets, and reading circles and the extension of the public education system. During the last third of the eighteenth century—right at the time that the first texts about the harpsichord player were printed—the literacy rate, the number of bookstores, and the number of authors doubled.[107]

Reading became a foundation of newly developing social activities (much as music-making did, as I discuss in chapter 4), and it was practiced in reading circles, salons, and public libraries. It generated new foundations for, and relationships between, literature and sociability.[108] Periodicals played a significant role in creating this reading public; many different kinds were edited and published for different social groups. The educational and moralizing style of these periodicals was itself a product of the Enlightenment, and they in turn popularized the thoughts and principles of the Enlightenment. They created new reading groups, explored new themes and topics, and spawned a market in which writers, editors, booksellers, and tutors could make a living.[109]

The texts about the two women automata were part of this newly emerging economy and cannot be taken in isolation as objective accounts of the dulcimer player and harpsichord player. The texts discussed in the remainder of this chapter, in fact, do not mention what their authors thought about the automata. Rather, my analysis explains how the texts about automata came into being and that automata and texts about them were the results of two different modes of production—artisan production and literary production. The texts tell

107. For the German-speaking lands, see, for example, Wehler, *Deutsche Gesellschaftsgeschichte*, 305–10; Wilke, *Literarische Zeitschriften*; Bahr, "Aufklärung," 91.

108. Goldsmith and Goodman, *Going Public*; Seibert, *Der literarische Salon*. See also note 47 in chapter 4.

109. *Deutsche Literaturgeschichte*, 121–23. See also my discussion of "moral weeklies" in chapter 4, page 150.

us how textual worlds, intellectual worlds, and worlds of artifacts came together to make up the universe of the two women automata.

Short Mentions of the Harpsichord Player

Texts that mention the Jaquet-Droz firm or their automata were published in diverse media and were part of a variety of reading experiences. The first periodical mention of the harpsichord player appeared in the *Journal de politique et de littérature* in January 1775, seven months after the android's first presentation to the public. The individual issues of this journal were divided into two parts, a political journal and a literary gazette.[110] Political articles covered news from Russia, Italy, Germany, and Great Britain, and the literary gazette dealt with spectacles, natural history, and literary novelties. Curiously, the account of the harpsichord player appears in the political part of the journal.[111] A one-page account describes the Jaquet-Droz firm's three androids and automaton theater La grotte, devoting one paragraph to the writing automaton, one to the draftsman, and a few words to the harpsichord player. The descriptions of the writing automaton and the draftsman are almost identical to those in the brochure (with occasional omissions and variations), and the one-sentence description of the harpsichord player is identical to the first sentence of the brochure description.

Two reports, from 2 and 9 February 1775, in the *Mémoires secrets pour servir à l'histoire de la République des lettres en France* may have been written after the same set of exhibitions in Paris.[112] The first mentions that "Mister Jaques Droz, a young man of 22 years of the Principality of Neuchâtel in Switzerland," had attracted curious people in the previous few days with several automatic figures, of which one in particular "made Parisian artists despondent."[113] The report then presents a short description of the writing automaton that is copied from the original brochure by the Jaquet-Droz firm. The description of the automata ends here (with no mention of the harpsichord player), and the

110. The journal was issued weekly between 1774 and 1778 in Brussels. It resulted from the fusion of two earlier magazines, the *Journal de Politique* and the *Gazette de Littérature* and became the *Mercure de France* after 1783.

111. It is the last article of the political part of that week's issue, printed just before the literary part. *Journal de politique et de littérature*, 1775, 99. The mix of topics and categories was typical for contemporary periodicals.

112. Bachaumont, *Mémoires secrets*, 7:273, 282. I owe this reference to Perregaux and Perrot, *Les Jaquet-Droz et Leschot*, 108.

113. Bachaumont, *Mémoires secrets*, 7:273

author proceeds to tell a story about how Jacques de Vaucanson himself assisted in the staging of the Jaquet-Droz automata. According to this report, Vaucanson was astonished at the rapid and precise execution of the writing machine, without any apparent communication with its maker. The report states that Jaquet-Droz supposedly offered to explain the mechanism, but "the academicien" refused and wanted to resolve the problem himself.[114] A week later, the report of 9 February 1775 states that the Prince of Soubise (a minister of Louis XVI) introduced to the queen "the new mechanic who astonishes all of Paris." This artist, the report says, was notified two days in advance of the honor he was to have, in which time he managed to program his draftsman to draw a portrait of the king and queen.[115]

Similar brief mentions of father and son Jaquet-Droz or their harpsichord player appeared in reports from eighteenth-century travelers to Switzerland.[116] One example is the work of William Coxe, a well-educated English clergyman, tutor, and travel companion to noblemen and gentlemen. He toured most of the European continent in the second half of the eighteenth century and also wrote historical works, memoirs, and editions of correspondence. He published *Travels in Switzerland*, an edition of letters that he composed during a trip to Switzerland in 1776, shortly after the Jaquet-Droz family presented their automata in La Chaux-de-Fonds. A letter from September 1776 describes his expedition to Le Locle and La Chaux-de-Fonds, whose economic situation he explores in detail, mentioning Pierre Jaquet-Droz briefly by saying that he was "now in Paris." This reference to the Jaquet-Droz family only in passing is in line with other writers' observations that were published in the months after the automata were first presented.[117]

Another example is the *Voyage historique et littéraire dans la Suisse occidentale*, published in 1781 by Jean-Richard Sinner de Ballaigues. Sinner was bailiff of the Erlach Castle in the canton of Bern and, from

114. Ibid.

115. Ibid., 7:282. Perregaux and Perrot surmise that he had already built in La Chaux-de-Fonds the cams necessary to have the automaton draw the king's and queen's portraits, in preparation for the journey to Paris.

116. A comprehensive overview of travel activity in Switzerland and the principality of Neuchâtel in the late eighteenth century is found in Beer, *Travelers in Switzerland*; Wäber, *Landes- und Reisebeschreibungen*, 34–52, 196–200.

117. Statements about Pierre Jaquet-Droz's current place of residence were ubiquitous in works like this. Often, however, they were vague conjectures or information picked up from questionable sources. Apart from a brief visit to show the automata in 1775 (which was soon after they were completed), Pierre Jaquet-Droz never took up residence in Paris. The timing matches my findings in the letters and diary entries discussed earlier in this chapter.

1748 to 1776, director of the university library at Bern.[118] He intended his book to be less a travel report than a series of observations on the history, geography, and arts of western Switzerland. Two chapters deal with a journey to the Neuchâtel mountains, reporting on the villages La Chaux-de-Fonds and Le Locle, the local industry, and the beginnings of clock-making in the area.[119] Sinner mentions Pierre Jaquet-Droz, together with his automata and other machines that were "known in all of Europe."[120]

Manuals for travel in Switzerland also made mention of the Jaquet-Droz firm and their automata. One of the most widely circulated from the 1780s, by the editor and professor Christoph Meiners from Göttingen, was a handbook to help young men prepare their voyages. It was organized as a collection of letters from various Swiss locales, and several letters from September of 1788 by Meiners focused on the area around La Chaux-de-Fonds and Le Locle.[121] These letters recount in detail the area's political and economic characteristics and report on local clock-making workshops (*magasins d'horlogerie*). Meiners mentions father and son Jaquet-Droz, saying they "moved away from La Chaux-de-Fonds. The son settled in Geneva which I only learnt after having left Geneva."[122]

Finally, among the briefer mentions, the physician Johann Gottfried Ebel, born in Silesia and a citizen of Zurich after 1801, wrote a manual for travels to Switzerland that was first published in 1790, with many editions to follow. This first proper travel guide for Switzerland, the best and most comprehensive handbook until well into the third decade of the nineteenth century, offers practical advice on how to prepare for and conduct a journey to Switzerland, including suggestions for travel, hiking routes, maps, equipment, and budget calculations. The second part of the work describes Switzerland in encyclopedic style: entries are alphabetically ordered and provide information on route-planning, sightseeing, and hostels. It is the entry on La Chaux-de-Fonds that

118. His catalog of manuscripts was received with enthusiasm in the contemporary learned world. Burri, *Johann Rudolf Sinner von Ballaigues.*

119. Sinner de Ballaigues, *Voyage historique,* 201–29. The story he provides is the often-rehearsed legend about Daniel JeanRichard.

120. Ibid., 212. This phrase is repeated often in other works on the Jaquet-Droz family in the following years.

121. Meiners, *Briefe über die Schweiz,* 239–58.

122. Ibid., 258. The letter of 10 September 1788 is from La Chaux-de-Fonds. The detail about Geneva is consistent with reports of Henri-Louis Jaquet-Droz moving to Geneva and founding a business there.

mentions briefly the Jaquet-Droz family and their automata.[123] These brief references show that the name Jaquet-Droz remained known during these decades and was closely connected with La Chaux-de-Fonds.

Longer Texts on the Harpsichord Player

Texts that offer more detail about the Jaquet-Droz firm and their mechanical works appeared in similar media (periodicals, pocket calendars, and travel reports) and in similar economies of text production, but they differed in composition and authorship. All of the detailed accounts copied the description of the harpsichord player more or less verbatim from the brochure text, without any more meaningful engagement, but they integrated that description into longer texts, framing it with other material and adding new contexts to it. For two of these texts, we can reliably identify as authors two distinguished eighteenth-century natural philosophers, Johann Bernoulli and Georg Christoph Lichtenberg. I analyze these longer texts not for what they have to say about the automaton, but rather for how they integrate the prefabricated text with other contexts. From this I infer what kind of meaning these authors ascribe to the harpsichord-playing automaton. Their treatment of the harpsichord player is ultimately evasive and reticent, too, but the treatment is, in each case, part of a larger text that the authors composed and that effectively functioned as an "environment" for the descriptions of the harpsichord-playing automaton.[124]

Detailed texts (in two cases, with their authors determinable) about the harpsichord player appeared in a pocket calendar, in a travel report, and in an almanac. First, the 1780 issue of the *Taschenbuch zum Nutzen und Vergnügen* (a popular pocket calendar better known as the *Göttinger Taschenkalender*, edited by the eminent Enlightenment natural philosopher Georg Christoph Lichtenberg), carried a seven-page article on the Jaquet-Droz family, their hometown, and their mechanical works.[125] This was about five years after the automata were first presented. Second, natural philosopher Johann Bernoulli, in his eighteen-volume

123. Ebel, *Anleitung*, 200.

124. Mario Biagioli and Peter Galison discuss the legacy of Foucault's and other French philosophers' revisitations of authorship and textuality in relation specifically to scientific authorship. Biagioli and Galison, *Scientific Authorship*, 3–5.

125. Lichtenberg, *Taschenbuch*, 66–73. The calendar's title translates roughly as "Pocketbook for practical use and enjoyment."

travel compendium *Sammlung kurzer Reisebeschreibungen*, of 1781 to 1787, translated and expanded a travel report about western Switzerland that was originally written by a Swiss man in 1764. Bernoulli devoted about twelve pages in this report to the Jaquet-Droz family and their works.[126] Third, there appeared in the 1789 edition of the widely read *Almanach de Gotha*, a publication similar to the *Göttinger Taschenkalender*, a four-page description of the three automata and the automaton theater La grotte. In its original format, this almanac was published annually from 1764 to 1889.[127]

Portable almanacs were among the most important and widely circulated media of the quickly growing literary market of the late eighteenth century. They bring into focus the proliferation and commercialization of literature at the time. From the mid-seventeenth century, they were increasingly carried, and their format was adapted for this purpose: they became small, handy, affordable booklets, and luxury versions were often extravagantly decorated. Publishers and booksellers created and sold specialized almanacs for music, for theater, for wine experts and smokers, for women, for children, and for the various estates and professional groups, such as officers, clergymen, doctors, actors, musicians, and book traders, but also for servants and prostitutes.[128] Almanacs were key media in popularizing the ideas and visions of the Enlightenment. They tended to have larger readerships and wider distributions than other media, and they provided cultural and intellectual hinges between the various social groups and estates of Old Europe.[129]

Two historians of calendar culture, Maria Lanckorońska and Arthur Rümann, illustrate how this culture enabled and reflected the social and cultural transformations of the time. They take an eighteenth-century poem by the prominent Enlightenment poet Johann Christoph Gellert

---

126. The collection's title translates as "Travel reports and other news that serve the advancement of knowledge of men and countries." Bernoulli, *Sammlung* (1781–87).

127. *Almanach de Gotha*, 90–92. The "Gotha"—as it is affectionately nicknamed—is to this day the main reference work for genealogies of the European higher nobility. On its history, see the issues of the years 1816, 1863, and 1913; also see Lanckorońska and Rümann, *Geschichte der deutschen Taschenbücher*, 13.

128. The social, cultural, and literary aspects of calendar culture in this period are well researched. See, for example, Mix, *Die deutschen Musenalmanache*; Baumgärtel, *Die Almanache*; Gladt, *Almanache und Taschenbücher*; Bunzel, "Almanache und Taschenbücher"; and Klussmann and Mix, *Literarische Leitmedien*.

129. They crossed the abundant political and religious boundaries in the Holy Roman Empire and the ancien régime and were read in the whole German-speaking territory as well as the French-speaking territory. They thus were important instruments in the creation and differentiation of literary publics.

about the Old Reich's estate society and add two lines about calendar culture.[130] The original poem by Gellert describes the rules of social and political order of the estate society in early modern Germany.

Enjoy the things that God granted to you
Renounce with joy what you don't have.
    Each estate has its own peace
    Each estate has its own burden.[131]

Lanckorońska and Rümann's twist goes as follows:

Enjoy the things that God granted to you
Renounce with joy what you don't have.
    Each estate has its own peace
    Each estate has its own burden.
    And each estate was given
    Its own pocket-sized book.[132]

The Göttingen pocket calendar was published annually under slightly varying names from the early 1770s into the first decade of the nineteenth century. It was founded originally by Johann Christian Dieterich, a prominent eighteenth-century publisher, who was able to attract numerous important authors.[133] He first recruited J. C. P. Erxleben, professor of physics and veterinary medicine at the University of Göttingen to be the calendar's editor. When Erxleben died in 1777, Dieterich offered the editorship to Georg Christoph Lichtenberg, a professor of experimental physics at the University of Göttingen and an eminent philosopher of the Enlightenment, who edited it from 1777 until his own death in 1799. In the course of Lichtenberg's editorship, the calendar acquired enormous prominence, and it and he became in a way "one and the same entity" in the German-speaking literary landscape.[134] The calendar soon reached a circulation of eight thousand—

130. Lanckorońska and Rümann, *Geschichte der deutschen Taschenbücher*, 11.
131. This is the fourth stanza from Gellert's poem *Zufriedenheit mit seinem Zustande* (Contentment with one's state).
132. "Genieße, was dir Gott beschieden,/Entbehre gern, was du nicht hast,/Ein jeder Stand hat seinen Frieden,/Ein jeder Stand hat seine Last./Und jedem Stande war beschieden,/Daß ihn ein Taschenbuch erfasst." Lanckorońska and Rümann, *Geschichte der deutschen Taschenbücher*, 11.
133. Dieterich also founded the Gotha almanac in 1766.
134. Mautner, *Lichtenberg*, 199.

"virtually the whole of educated Germany."[135] Lichtenberg's salary for this position was free lodging in Dieterich's house.[136]

Lichtenberg was a leading advocate of Enlightenment principles such as education and the wide dissemination of knowledge, and his calendar was an instrument for this purpose. He divided it into three parts: an editorial (which he wrote himself every year, and some of his editorials became his most prominent texts); a section for news, novelties, and gossip; and a section for new inventions and mechanical and other peculiarities.[137] Lichtenberg himself was probably the author of at least half, and perhaps most, of the texts in the calendar.[138] He was also responsible for collecting the curiosities, the novelties, and the gossip that were so characteristic of short articles in calendars.[139] The collaboration of ambitious publishers like Dieterich with editor-philosophers like Lichtenberg to produce pedagogical and entertaining mass media was a typical Enlightenment phenomenon. It was common for an editor to be also the main contributor to and overseer of the periodical as a whole, and such "literary entrepreneurs" became increasingly influential economically and culturally in the late eighteenth century.[140]

Lichtenberg's article on the Jaquet-Droz family and their mechanical works in the 1780 issue of his Göttingen calendar appears under the category of inventions and physical wonders, bearing the title "P. Jaquet Droz and H. E. Jaquet Droz Father and Son."[141] Lichtenberg frames and introduces the work of the Jaquet-Droz family with an account of the Neuchâtel mountains in late-eighteenth-century Switzerland.

135. *Dictionary of Literary Biography*, 94:184.

136. This arrangement was common at the time. On the details of Lichtenberg's involvement in the Göttingen pocket calendar, see Baasner, *Georg Christoph Lichtenberg*, 25–26; Mautner, *Lichtenberg*, 198–207; Lichtenberg, *Briefe an Dieterich*.

137. The first editorial in the 1778 calendar was the essay "Über Physiognomik," which made Lichtenberg as well as the calendar famous overnight. Sautermeister, *Georg Christoph Lichtenberg*, 83–110; Mautner, *Lichtenberg*, 199–201.

138. Mautner, *Lichtenberg*, 199, says "most" of the contributions; Baasner, *Georg Christoph Lichtenberg*, 26, says "almost half."

139. Mautner holds that this activity of Lichtenberg's coincided with his passion for collecting things, but Lichtenberg also saw it as a distraction from his other works. Mautner, *Lichtenberg*, 199.

140. Kant once compared the figure of the ambitious, commercially successful publisher with the "director of a factory." Wehler, *Deutsche Gesellschaftsgeschichte*, 316. A fictitious character in a novel by Friedrich Nicolai (Sebaldus Nothanker) was a publisher who had ten to twelve authors sitting at a long table in his house, and each of them had to write his or her quota to receive the daily wage. Quoted in *Deutsche Literaturgeschichte*, 128.

141. This text was written by Lichtenberg himself. Mautner, *Lichtenberg*, 200. "H. E." Jaquet-Droz is a really interesting and unusual error. The only other place where the first name of the younger Jaquet-Droz is taken to be "H. E." rather than Henri-Louis is in Busch's *Versuch eines Handbuchs der Erfindungen*, 1:189, published in 1790.

He employs the cliché of Switzerland's supposedly idyllic mountains and valleys to start off, asking the reader: "What would a widely traveled, wise man answer if you asked him: What ingenuity do you expect from a people whose winters last almost seven months, a people that knows hardly anything about the tender days of spring and autumn, and, in the remaining five months, is scorched by the sun? Would he assume to find under such a sky a clockmakers' republic people that nature herself seems to have formed into mechanics, and altogether one of the most dexterous and industrious people in Europe?" Lichtenberg answers his rhetorical question by claiming that the wise man would have to assume thusly. He tells the reader that he is speaking of the Neuchâtel mountains between Switzerland and Burgundy, home of thousands of artists and of "the famous Berthoud" and "the two Misters Droz," whose work he wants to discuss. These latter two, he explains, are from a village called La Chaux-de-Fonds, which contains "together with the neighboring village Le Locle 721 clockmakers, 98 goldsmiths, and 1323 lace-making women, not even counting the painters and other mechanical artisans." Lichtenberg further illustrates the economic situation in la Chaux-de-Fonds by noting that every year "15,000 golden and silver pocket watches are made" (not counting the pendulum clocks), as well as all kinds of clock-making tools, used and "admired even by London and Paris artists." On this note, Lichtenberg makes fun of the increasingly "globalized" world of trade and commerce in the eighteenth century, saying that "there is no doubt" that the occasional pocket watch from Le Locle and La Chaux de Fonds is kidnapped to London ("like Madeira wine to the West Indies"), in order to be offered, with a great increase in imagined value and real price, to the gallant connoisseur as a piece of precious English work. "But let us move on," he writes, "to the works of the Jaquet-Droz family." Lichtenberg says that "these two artists" have built complex clocks in the past, of which "the two most noteworthy" are in the possession of the king of Spain. He then quotes "unanimous testimony of the experts," according to which father and son Jaquet-Droz built artworks that "surpass the famous works of Vaucanson."[142]

After this introduction, Lichtenberg gives details of the Jaquet-Droz family's three androids and automaton theater. He takes the text from the 1774 brochure, translates it into German, cuts out an occasional phrase, and paraphrases a few passages. Still, his text stays fairly close

---

142. Lichtenberg, *Taschenbuch* (1780), 66–68.

to the French original. The order in which details of the automaton are surveyed remains exactly the same, and Lichtenberg adds nothing to the Jaquet-Droz firm's original brochure text.

Several characteristics of Lichtenberg's text resonate with his calendar's pedagogical mission; with contemporary preoccupations about natural history, anthropology, and cosmology; and with contemporary conventions of textual production. One of Lichtenberg's goals for the calendar, ironically, was diligence and care in the selection and writing of original articles, rather than mere copying of texts from other sources, as was common practice with other contemporary calendars.[143] Given the high demand for curiosities, novelties, and gossip on the part of the calendar's audience, this care was not always possible. Lichtenberg mined contemporary popular works just as much, asked friends to help him out with texts, and sought inspiration by other means.[144]

Although Lichtenberg was one of the leading commentators of his time on matters of anthropology and natural philosophy, his main concern in these passages is not to elaborate on the automata's relevance to questions about the constitution of human selfhood. Introducing spectacular android automata to a relatively large audience, he prioritizes anecdotes, statistics, and a "milieu study" of Switzerland over reflections on "machine-men." The text on the Jaquet-Droz automata is sandwiched, furthermore, between other stories about curiosities and inventions and entertaining or educational short treatises on historical and geographical topics, such as Japan, the history of the bell, the slave trade, common errors (*populäre Irrtümer*), religious superstition, fashion, and women's rights in Russia. Many of these neighboring articles are copied from other sources, too.[145] The *Taschenbuch* as a whole is integrated into a calendar, which provides notations, as was common practice at the time, of meteorological and astronomical data, holidays, birthdays and genealogies of European aristocratic dynasties, engravings, and tables of common measures of length, weight, and volume.[146]

The author of the second longer text is Johann Bernoulli III, another leading natural philosopher of the eighteenth century. He was the son

---

143. In a letter to Friedrich Nicolai, Lichtenberg himself boasted that most of the articles in his calendar were "new and not merely, as is common for some of those calendars, copied from rather ordinary sources." Cited in Mautner, *Lichtenberg*, 199.

144. Mautner explains details of this process. Ibid., 200–201, 280.

145. Examples would be an article from the *Taschenbuch* from 1782 on Japan that was copied from the *Transactions of the Royal Society* and stories about curiosities from the *Journal de Paris*, a French daily newspaper that was published from 1777 to 1792. See Lichtenberg, *Taschenbuch* (1782), 1, 15.

146. See also Caetano da Rosa, "Androiden."

of Johann II and the grandson of Johann I, who were both professors of mathematics at Basel. Johann III also published on mathematical and astronomical topics, but his most influential works were his writings on natural history (anthropology and geography) and his eighteen-volume collection of travel reports.[147] For the seventeenth volume of this series, published in 1783, Bernoulli translated and expanded a travel report about western Switzerland that was written by a Swiss savant named Samuel-Frédéric Ostervald and originally published in 1764.[148] Bernoulli devotes two full chapters (out of a total of fourteen) to the area around La Chaux-de-Fonds and Le Locle. The Jaquet-Droz family and their works figure in a lengthy passage covering about twelve pages. This description of the Jaquet-Droz automata is one of Bernoulli's additions, since Ostervald's original report was published seven years before the automata were finished.[149]

Bernoulli uses conventions similar to Lichtenberg's to frame the description of the Jaquet-Droz automata and also makes similar choices as to his text's form and content. It is quite possible that Lichtenberg and Bernoulli used similar sources for material to preface their descriptions of the Jaquet-Droz automata.[150] Bernoulli's text is considerably longer than Lichtenberg's, dealing in much greater depth with the villages La Chaux-de-Fonds and Le Locle; however, the passage on the Jaquet-Droz automata remains the same length, as it is also copied from the brochure.

Like Lichtenberg, Bernoulli starts out by describing the geographical and natural historical characteristics of the village Le Locle and presents statistics about the inhabitants' occupations and métiers (listing 331

147. Bernoulli, *Sammlung* (1781–87).

148. Ostervald's original work is entitled *Descriptions des montagnes et des vallées qui font partie de la principauté de Neufchâtel et Valangin*. He also authored the article "Neuchâtel" in d'Alembert and Diderot's *Encyclopédie*.

149. Bernoulli himself tells the reader in the preface to this volume that his translation contains "far more than the French original." The authorship of individual texts in Bernoulli's collection, however, is not unambiguous. In the preface to this "First Supplementary Volume," Bernoulli alludes to "translators and contributors" who were late in delivering their work. Therefore, there may have been more than one writer working on this volume. Bernoulli, *Sammlung* (1783), vi, viii.

150. There are several examples of correspondences between the uses of statistics by these two and other authors. The figures in Bernoulli's text, for example, bear some resemblance to those in Sinner de Ballaigues, *Voyage historique*, 220, where Sinner states that, according to a census in 1766, there were 3,000 inhabitants in Le Locle and 300 clock-makers, while Bernoulli indicates about 3,094 inhabitants and 331 clock-makers (p. 137). Another interesting correspondence is between Bernoulli and Coxe, both of whom say that the annual production of watches in the area was 40,000. Bernoulli, *Sammlung* (1781–87), 142; Coxe, *Travels in Switzerland*, 108. Sinner de Ballaigues and Lichtenberg, in contrast, state the annual production at 15,000 watches. Sinner de Ballaigues, *Voyage historique*, 220; Lichtenberg, *Taschenbuch* (1780), 68.

clock-makers, 761 lace-makers, 78 goldsmiths, and 15 hosiery-makers). He also provides the total number of watches ("not counting pendulum clocks") made in Le Locle and La Chaux-de-Fonds and spends considerable time and space illustrating the variety and prosperity of the village's clock-making industry, rehearsing at length the legend about Daniel JeanRichard and the beginnings of clock-making in the area.[151] Bernoulli also explains the stages and divisions of labor in the clock-making industry around La Chaux-de-Fonds and Le Locle. When he talks about the Jaquet-Droz family and their mechanical works, he first describes several of the spectacular clocks that Pierre Jaquet-Droz built in the 1760s and also recounts Jaquet-Droz's visit to the king of Spain. He then describes the three androids and the automaton theater. Bernoulli's descriptions are, like Lichtenberg's, literal and almost complete translations (with a couple of errors) of the corresponding passages in the 1774 brochure by the Jaquet-Droz family. Bernoulli's text is slightly longer than Lichtenberg's, because he omits fewer passages, but like Lichtenberg, Bernoulli adds no details about the automata to the brochure's text.

The third and final longer text I examine is a three-page account of the Jaquet-Droz automata in the *Almanach de Gotha*, which was published about six years after Bernoulli's 1783 travel report. The reading experience that the Gotha almanac provides is similar to that of the Göttingen calendar.[152] The heading under which the automata appear is "Chefs d'œuvre" (master works or masterpieces), a typical rubric for news in calendars. The first part of the Gotha almanac contains a calendar, a list of holidays, a long section on genealogies of European aristocratic dynasties, and geographical and astronomical data; the second part, after the calendar, collects curiosities, treatises to educate and entertain, and short stories. The text on the Jaquet-Droz automata appears in the second part and is a description with no introduction or any other explanatory context. For this reason, and because this text is one of only two French texts that I consider here (along with the previously discussed *Journal de politique*, whose content was identical to that of the brochure), it is interesting to compare its details to the original in the brochure, since a period of fifteen years elapsed between their respective composition.

The account of the harpsichord player in the *Almanach de Gotha* amounts to a close paraphrase of the text in the 1774 brochure. The syntax is slightly altered, and on many occasions synonyms are used. The almanac, for example, uses the French word "sein," rather than the brochure's "gorge," when describing the motions of the harpsichord player's chest.[153] Another example is the peculiar wording for the instrument the harpsichord player is playing. While the 1774 brochure uses the phrase *clavessin organisé*, the text in the Gotha almanac changes this to *clavecin organique*.[154] But for the most part, the order in which the three automata are presented and the sequence of details about each automaton within the individual paragraphs remain exactly the same. To see how similar the two texts are, we can compare the first sentences of the paragraph that describes the harpsichord player. The 1774 brochure says:

La troisième figure représente une jeune fille de 10 à 12 ans, assise sur un tabouret et qui touche un clavessin organisé. Cette automate, dont le corps, la tête, les yeux, les bras, les mains, et les doigts ont divers mouvement qui paraissent naturels, exécute elle-même sur son clavessin divers airs de musique en deux ou trois parties, avec beaucoup de précision.

The *Almanach de Gotha*, fifteen years and a great deal of transcribing and reprinting later, says:

La 3eme pièce représente une fille de 12 ans assise sur un tabouret devant un clavecin organique. Cette musicienne automate imite on ne peut plus naturellement les mouvemens du corps, des yeux, des bras & des doigts, d'une personne qui joue; & touche sur le clavecin plusieurs pièces de musique à deux & à trois parties, avec toute l'exactitude imaginable.

Descriptions of the Jaquet-Droz harpsichord-playing automaton were printed in a variety of media and written by a variety of authors for a variety of purposes over a period of one and a half decades. It is remarkable that, over this range of time, space, and function, they were so consistently and faithfully copied from one another. We can safely assume that none of the texts' authors saw the automata in action; their accounts are not individual and deliberate accounts of impressions and do not engage the relevance of androids to the relationship between humans and machines. Rather, the texts' form and content, and their sim-

153. "Sein" refers more directly to the chest of a female body, the "bosom."
154. The term "clavecin" would nowadays translate as "harpsichord."

ilarity, are rooted in the contemporary literature-producing machinery of which both the texts and the media were products. The compilation and composition of the texts followed this machinery's conventions, and the texts' function has as much to do with this machinery as with the documentation of the staging of android spectacles.[155]

The texts that I selected here more or less exhaust one of the most authoritative bibliographies on the Jaquet-Droz automata, in Faessler, Guye, and Droz's *Pierre Jaquet-Droz et son temps*, and I even included some texts here that were overlooked there. It is, of course, not clear how many other accounts were produced in the 1770s to 1790s. Another wave of interest, and thus of publications, occurred in the later nineteenth century (see my chapter 6). As far as the two and a half decades after the harpsichord player's production are concerned, however, accounts like the ones presented here are the most typical.[156]

Conclusion

My findings regarding printed texts about the harpsichord-playing automaton, taken together with the eyewitness reports, suggest that there was no unambiguous, collective excitement about this automaton—or other automata—in the eighteenth century that would transcend social boundaries and constitute a mass phenomenon comparable to the mass phenomena in the nineteenth and twentieth centuries. And yet, the persistent idea that the automata of the eighteenth century were widely known and appreciated has served important roles in narrating the history of automata and artificial humans and the history of industrial modernity. Ideas of this kind deploy the eighteenth century to suggest parallels between the Enlightenment and the culture of artificial humans of the twentieth century, assuming that we can understand our own relationship nowadays to artificial humans (and to technology in general) by reference to the "Age of Reason." Among the

155. One of the most comprehensive works in the history of science on the historical specificity of text production and intertextuality is Frasca-Spada and Jardine, *Books and the Sciences*. Another fascinating example is Anke te Heesen's cultural history of the materiality and practice, and of the political and artistic significance, of "cutting and pasting" in the period around 1900. Heesen, "News, Papers, Scissors."

156. Faessler, Guye, and Droz, *Pierre Jaquet-Droz*, 164–65. Oliver Hochadel discusses the problem of primary source material on spectacles in the eighteenth century. He mentions in particular that, at least for the case of itinerant lecturers, reports on spectacles were rarely written by observers; normally, they were written by the itinerant lecturers or impresarios themselves. See Hochadel, *Öffentliche Wissenschaft*, 194–95.

most significant problems with this idea is that the kind of "publics" that were emerging in the eighteenth century were not mass societies of the kind that have, since the late nineteenth century, so profoundly shaped our own current societies.

Pierre and Henri-Louis Jaquet-Droz were "clockmaker-peasants" (*horloger-paysans*) who received unusual education, who were fortunate enough to be introduced into the contemporary court culture of spectacle, and whose commodities fit well into the economic landscape of protoindustrialism and overseas trade. Those are my main explanatory parameters to understand why and how they built the harpsichord-playing android automaton. I followed references to the harpsichord player through a variety of texts and traced the contours of her travels, meanings, and functions. Spectators' reports that I found in letters and diaries did not suggest pronounced interest in the automaton's performance. The texts' casual and reticent nature reflected, rather, the social function of the showing of spectacles at the time. Mentions of and commentary on the harpsichord player were often mediated through other texts and followed unexpected routes. These texts were products of specific writing industries and practices that shaped and constituted their form and content. Thus they are not documents of, or windows into, responses from a generally conceived "public" to the automata. I find a similar situation in the history of texts on the dulcimer player in chapter 3. The relationship between eighteenth-century automata and contemporary reactions is a complex compound of histories of artisanship and of reading and writing texts. Our own twenty-first-century understanding of automata in the Enlightenment and the human-machine boundary operates, not least of all, on key assumptions that stem from the texts about automata between 1780 and 1810.

THREE

The Dulcimer-Playing Android; or, Furniture-Making in the Rhineland

Like the Jaquet-Droz family's harpsichord player, the dulcimer player was built in an unusually successful and prolific artisan environment: the furniture manufacture of David Roentgen and the clock-making workshop of Peter Kinzing, both in the small principality of Neuwied on the Rhine, about sixty miles south of Cologne. Roentgen and Kinzing's collaboration did not begin with the production of the dulcimer player. From the early 1770s onward, they designed and built fancy commodities that combined furniture with musical works and clockworks for a wide range of wealthy customers. Such commodities included cabinets and desks with mechanical interiors, delicate clocks in elaborate cases, and furniture with integrated musical instruments (harpsichords, for example).[1] Their dulcimer-

---

1. Fabian, *Kinzing und Roentgen*, 59–60. Throughout this chapter, I rely on the detailed and comprehensive historical works by Dietrich Fabian, J. M. Greber, and Michael Stürmer on the Roentgen family and their furniture manufacture. Fabian and Greber spent decades making the family and their lives in the Pietist brotherhood accessible to other historians. While Fabian's and Greber's and primary interests lie in the economic and aesthetic history of early modern furniture-making, I integrate a detailed understanding of the Roentgens' lives with an analysis of the construction of mechanical androids in the European Enlightenment. I rely on Stürmer's work in particular to understand the economic and technological conditions of "old" artisanship in the protoindustri-

playing automaton built in 1784 was one such piece of musical furniture among many others.

David Roentgen's business attracted a large and distinguished group of clients, including political and clerical dignitaries from all over the European continent. Between 1742 and 1793, the manufacture that he ran with his father Abraham became the largest and most productive of its kind in the entire Holy Roman Empire. In these five decades, he produced more than two thousand pieces of every conceivable type of furniture and sold them to the major European political and clerical courts. Among his clients were Catherine the Great, Louis XVI, Marie-Antoinette, and Frederick the Great. At the height of his success, between 1780 and 1790, David Roentgen was the most distinguished and sought-after furniture-maker in Europe.

The success of Roentgen's furniture manufacture had its roots in the old European society of the late eighteenth century. This was a world in which the convergence of traditional court society, protoindustrialism, increasing overseas trade, Enlightened economic policy, post-Reformation spiritual reform, religious persecution and migration, and emerging middle classes with new habits of consumption and fashion-making created a period of profound and sustained transition. The same phenomena manifested themselves in David Roentgen's life and formed the backdrop for the production of the dulcimer-playing automaton: in his small principality of Neuwied (a "model state" as some political economists at the time called it), he received political and economic privileges from the reigning prince in the form of free land, tax relief, and guild exemption; he was apprenticed in protoindustrial modes of production; and he was a member of a Pietist sect, the Moravian Brothers (Herrnhuter Brüdergemeine), which allowed him access to a high level of education, mobile capital, a pool of skilled labor, an elaborate work ethic, and distinguished trade and business contacts among the nobility and the otherwise wealthy.

Like the Jaquet-Droz family in Switzerland, the Roentgens in the Rhineland lived and worked in mixed forms of what we now label "traditional" and "modern" forms of production, consumption, religious practice, and state intervention. These mixed forms, typical for the late eighteenth century, were often distinct from more "obviously" traditional,

---

alizing world of the second half of the eighteenth century and to understand the specific economic consequences of the Roentgens' affiliation with the Moravian Brothers. Fabian, *Abraham und David Roentgen*; Greber, *Abraham und David Roentgen*; Stürmer, "Die Roentgen-Manufaktur"; Stürmer, *Handwerk*; Stürmer, *Scherben*; Stürmer, *Luxus*.

early modern, or modern versions. The distinctness in regard to mixes of the "traditional" and the "modern" of eighteenth-century Europe is relevant here not only in general regard to the production and consumption of goods, but also in specific regard to the production and consumption of machine-men and machine-women.[2]

The roots of the Roentgen furniture manufacture lay in the London apprenticeship of David's father Abraham Roentgen. Abraham traveled broadly as a young carpenter, eventually joining the Moravian Brothers; he spent the rest of his life living and working in their closely knit spiritual and economic communities. David initially became involved in his father's firm as an apprentice, then took on more responsibilities in his early twenties, and within twenty years turned his father's small village joinery into an internationally renowned business that served clients in France, Sweden, Russia, and the German lands. The dulcimer-playing automaton was a product of this firm: I contextualize its design and production as part of David's efforts in the years between 1779 and 1784 to establish and maintain client relationships with the French king and queen.

We know little about exactly when, how, or by whom the dulcimer player was built or whose idea or invention it was. We do, however, have information on events surrounding its creation, including details of the organization of the Roentgen manufacture (and David Roentgen's movements) right at the time that the automaton was made. The dulcimer player was built in the early 1780s and was presented at the French court early in 1785. In that same period, between 1779 and 1783, the manufacture underwent its final and largest expansion. David himself was less and less present in his manufacture during this time. Instead of building furniture, he was traveling and acquiring clients, corresponding, and tending to what had become an international business. In exactly the years that the dulcimer player was built "back home," he was moving his manufacture to a top-ranking position among European furniture-makers, supplying all the important courts with his goods.

2. As I discussed in chapters 1 and 2, a number of approaches have been proposed to understand transitions in Europe from early modern to modern society. Such proposals identify characteristics of the "early modern" and the "modern" period in different places: the self, institutions, public spaces, and spaces of commodity production. I also want to point out Hans-Ulrich Wehler, who has called the late eighteenth century an "unprecedented universal-historical break in the West" (*beispiellose universalgeschichtliche Zäsur im Westen*), and Isabel Hull, who has called the same period "the moment of fundamental transformation toward the modern." Wehler, *Deutsche Gesellschaftsgeschichte*, 12; Hull, *Sexuality, State, and Civil Society*, 6.

David Roentgen's career, including the making of the dulcimer player, displays a tension between the local production of furniture and an international network of elite clients. Indeed, the dulcimer player was made in a manufacture from which its principal, the automaton's supposed creator, was mainly absent maintaining his complex economic relations with an international court culture that needed and appreciated spectacle objects such as automata.

That the dulcimer player was a local product of an international business puts in perspective the reasons and purpose for its construction. We can trace the economic, cultural, and political dimensions of the Roentgen business at the time, David's travel habits and responsibilities in his workshop (and those of his deputies) in the 1770s and early 1780s, and his legal and spiritual conflicts with the Moravian Brotherhood, and on this basis we can conceive of the dulcimer player as the product of a unique business model in the unique environment of the Holy Roman Empire. Her design also follows a specific, recognizable cultural scenario of the time—keyboard-playing women—which allows us to state much more specifically for this case the eighteenth-century dynamics of replicating in a mechanical android body techniques that were part of larger social and cultural schemes of subject-formation.

European Furniture-Making before the Roentgens

The development of David Roentgen's workshop into Europe's foremost furniture manufacture took place under the influence of trends from London and Paris, the leading sites for furniture-making at the time, and also as a spin-off of earlier currents of furniture culture in the sixteenth and seventeenth centuries. Before 1700 furniture design prospered as part of the making of natural history cabinets (*Wunderkammern*), which were produced, among other places, in Augsburg and Dresden.[3] High-priced domestic objects, including furniture, retained their luxury character in the eighteenth century, but they increasingly became available to larger segments of society, as middle-class households became more affluent and developed habits of conspicuous consumption.[4]

---

3. Impey and MacGregor, *Origins of Museums*; Meadow, "Merchants and Marvels."
4. On the ramifications of furniture and the "politics of everyday life" in French absolutism, see Auslander, *Taste and Power*, 35–110 (quote on page 1). On interrelations between furniture and

During the course of the second half of the seventeenth century, Paris became the capital of furniture-making, as the court of Louis XIV established workshops in a gallery of the Louvre housing a colony of artisans and artists. Paris remained the center of furniture-making through the eighteenth century, stimulating and developing the leading furniture styles and tastes. German-born artisans residing in Paris eventually helped David Roentgen make business contacts in the 1770s. Among them were Johann Franz Oeben, who had come to Paris around 1745, and his son Simon, as well as the engraver Johann Georg Wille, who had migrated as a young man in 1736 from Hesse to Paris.[5] Throughout his lifetime, David Roentgen followed Parisian events closely to expand his artistic and economic horizons. He conquered the Parisian market successfully later in his life, once he had established himself in the German-speaking lands in the 1770s. London as well was home to skilled and well-known cabinet makers from the late seventeenth century onward.[6] Abraham Roentgen learned the trade in London in the 1730s as a young traveling apprentice.

Furniture-making became a more widespread, systematized, and codified trade in the first half of the eighteenth century. English and French artisans and artists published large volumes with patterns that provided European artisans with models for clock-case forms, carved ornaments, inlaid work, and metal mountings. These manuals established and disseminated patterns, tastes, models, and techniques.[7] Over the years, Abraham and later David Roentgen developed characteristic styles and designs under this influence, to which they eventually added, as their particular trademark, elaborate mechanical interiors and gadgets for furniture. This was unique to their work: neither the Parisian nor the London furniture-makers provided such a myriad of mechanical interiors for their products.

---

other aspects of eighteenth-century France (such as gender, consumption, orientalism, and selfhood), see the collection of essays in Goodman and Norberg, *Furnishing the Eighteenth Century.*

5. *Allgemeine deutsche Biographie,* 24:85, 43:257–60; *Neue deutsche Biographie,* 19:420. See also Fabian, *Kinzing und Roentgen,* 81, and Greber, *Abraham und David Roentgen,* 12.

6. On the relationship between French and English furniture-makers and their competition as leaders on the European market, see Bertuch's and Kraus's commentary ("Einleitung") in the first issue of their periodical *Journal der Moden,* 29–30.

7. Chippendale, *Gentleman and cabinet-maker's director* (1754), of which a second edition appeared in 1755 and a third in 1762; Sheraton, *Cabinet-Maker and Upholsterer's Drawing-Book* (1793); and Hepplewhite, *The cabinet-maker and upholsterer's guide* (1789). See also Ince and Mayhew, *Universal System of Houshold Furniture* (1760). The standard work in French was the three-volume Roubo, *L'art du menuisier ébénist* (1774). Attempts at codifying knowledge and skill in trades such as carpentry, baking, and tailoring were also prominent in D'Alembert and Diderot's *Encyclopédie.*

Abraham Roentgen's Workshop in Times of Pietism and Mercantilism

Abraham Roentgen, the founder of the Roentgen manufacture, was born in 1711 in Mülheim, near Cologne.[8] He became an apprentice in his father's joinery and then embarked on his travels at age twenty. He worked with masters in The Hague, Rotterdam, and Amsterdam and moved to London in the 1730s. Finding work there, he soon attracted clients, in particular because of his specialization in marquetry and the installation of mechanical appliances in furniture.[9]

As he worked with these masters in the 1730s, Abraham learned craft techniques, organizational models, and artistic styles that became fundamental for his workshop and for training his son David. He founded his own workshop in Neuwied in 1750 and ran it after the models that he had seen among the big London cabinetmakers in the 1730s. He used their methods of dividing labor, managing supplier networks, and establishing far-reaching, international trade relations. These innovations gave him economic and artistic advantages over his competitors on the European continent, many of whom were working within the traditional guild economies of France's ancien régime and the Holy Roman Empire's estate society.[10] Along with greater division of labor, specialization, and internationalization, London also featured a larger variety of employment relationships between individual artisans and firms. The novelist Sophie de La Roche wrote a widely read 1780s travelogue about a London "joiners' factory" run by Georg Seddon, which was among the largest of its kind in the second half of the eighteenth century. She reported that about four hundred artisans worked there and that the production was organized by the division of labor. This workshop was the kind Abraham Roentgen worked for when he was a young artisan in London.[11]

8. See UA R.22.36.81, in the central archives of the Herrnhut Brotherhood, Unitätsarchiv; this is a manuscript biography of Abraham Roentgen.

9. Ludwig Roentgen, *Das erste Buch*, 7–8. Ludwig Roentgen, David's brother, wrote this autobiography in 1811. The first chapters deal with his grandparents, his parents, and his childhood in the institutions of the Moravian Brothers. See also Greber, *Abraham und David Roentgen*, 30.

10. For Michael Stürmer, the Roentgen manufacture's success was a product, even a symptom, of a period when artisanship in Old Europe was in a transition between medieval guild order and industrial production. Stürmer, "Die Roentgen-Manufaktur," 26; Stürmer, *Handwerk*, 212–75; Stürmer, *Luxus*, 15.

11. La Roche, *Tagebuch einer Reise*, 362–65. La Roche's travel reports became well-known classics and were widely talked about. She provided broad social and intellectual panoramas of the areas she visited. Meighörner, *Sophie von La Roche*; Greber, *Abraham und David Roentgen*, 31.

The size and organization of manufactures and workshops in England and on the European continent were subjects of interest and inquiry for a wide range of people in the eighteenth century. Scholars of government and economics, writers, and industrialists embarked on journeys and produced texts about the relations of manufactures and production to state and society. Among these works were multivolume encyclopedias, geographical reports, theoretical treatises, textbooks, and epistolary novels. Such works provided detailed insights into the wide regional variety of manufactures, as well as theoretical reflection on the status and meaning of industry, trade, and luxury for the common good.[12] Artisan worlds such as the Roentgens' figured as exemplars in this literature, and the small princely states of Büdingen and Neuwied, where they lived and worked, were popular objects of empirical and theoretical inquiry.[13]

In early modern Europe, mercantilism was the leading school of thought on the relationships between economics, industry, state, and moral order, and it was the lens through which travelers and writers interpreted manufactures such as the Roentgens'. Mercantilism was the economic equivalent to political absolutism: the state was supposed to stimulate and guide economic growth by developing agriculture, manufacturing, and mining and by restraining imports and encouraging exports. The resulting growth was meant to serve primarily the political and military power of the absolutist state and to establish and stabilize its power vis-à-vis rivaling states. Mercantilism, like absolutism, went hand in hand with nation-state building.[14]

In the counties where the Roentgens settled with their religious community and built up industry, the goal of state-building held sway

12. We learn about the complicated interrelations between economic conduct and religious affiliation, about regional differences in natural and institutional resources, and about the relationship between economic prosperity and city planning. Justi, *Vollständige Abhandlung* (1758–61); Dietmann and Haymann, *Neue europäische Staats- und Reisegeographie* (1750–70); Krünitz, *Oeconomisch-technologische Encyklopädie* (1773–1858); Beckmann, *Beyträge zur Oekonomie* (1779–91); Jung-Stilling, *Versuch eines Lehrbuchs* (1785); Bernoulli, *Sammlung* (1781–87).

13. Andre Wakefield makes an important argument about a strategic function that the sciences of economics and the state served, saying that they were meant to establish "patronage for their authors and good publicity for the German principalities" and that the science of cameralism served as the "public face" of the principality's government. Quoted from the dust jacket of Wakefield, *Disordered Police State*.

14. Apart from many shared principles, mercantilism also displayed enormous variation in early modern Europe. Heckscher, *Mercantilism*, 1:46–110. See also Tribe, *Strategies of Economic Order*, chap. 2; and Pfeisinger, *Arbeitsdisziplinierung*, 36–39. Growing resistance to mercantilist policies developed in the eighteenth century and was expressed in the principles of physiocracy. Groenewegen, *Eighteenth-Century Economics*, 222–46.

as well. Princely rulers sought to populate and cultivate their lands after the devastating Thirty Years' War, and they were interested in religious groups that had strong traditions of artisanship, including such persecuted Protestants as the Moravians, the Huguenots, and the Mennonites. Princes invited them to immigrate to stimulate the local economy and rewarded with princely privileges those who founded new industries and manufactures. Again, the Roentgens' religious community was an exemplary case of such economic policy. Privileges they received included freedom of religion and free houses, but also exemption from guilds and military service, tax relief, and exemption from tariffs and customs duties.[15]

Abraham Roentgen's religious conversion occurred during his time as a traveling apprentice in London, and it influenced his and his son David's work, business, and lives for decades to come. In 1738 Abraham heard Count Nikolaus von Zinzendorf, the founder and bishop of the Moravian Church, preach in London.[16] Zinzendorf was a Lutheran-Pietist theologian who established his first congregation in 1722 when, in a small estate in Berthelsdorf in Saxony that he had inherited from his grandmother, he provided asylum to religious refugees from Moravia.[17] Zinzendorf established the first settlement of his new community in Herrnhut, a village just outside of Berthelsdorf, which also gave the community its name. In 1736 Zinzendorf was expelled from Saxony and moved to the area of the Wetterau, about thirty miles northeast of Frankfurt. There he founded two more settlements, Marienborn in 1736 and Herrnhaag in 1738. In the years following the founding of these settlements, Zinzendorf traveled to the Baltic Sea, England, North America, and the West Indies on missionary work.[18]

The Moravian Brothers—as well as the entire Pietist movement—were a phenomenon of a period that Ernst Troeltsch once characterized as the end of German Protestantism's first significant era as a state religion. Since the Reformation, German Protestantism had developed in symbiosis with the small princely territories of the Holy Roman Empire. While the Protestant movement and ideology had shaken up the

---

15. For the ways in which "early modern" privileges and "modern" patents related to each other, see Biagioli, "From Print to Patents."

16. Roentgen, *Das erste Buch*, 10–13; Stürmer, *Luxus*, 21.

17. The refugees were descendants of the original Bohemian-Moravian sect of the United Brethren. Zinzendorf's community was thus a renewal of a sect whose roots went back to the fifteenth century, to the reformer Jan Hus. See Günther, "Die Wurzeln der Brüderunität."

18. Meyer, "Zinzendorf und Herrnhut"; Beyreuther, *Nikolaus Ludwig von Zinzendorf*, 177–250.

Roman Catholic supremacy and made possible these territories' political and religious sovereignty, the princely territories pursued political agendas that appropriated and instrumentalized Protestantism in turn. Pietism was the most significant revival movement within Protestantism after the Reformation. It came into being around 1600 during a widely perceived crisis of spirituality, and its heyday continued until the end of the eighteenth century. It engendered a vast network of economic and theological influence, extending from England, the Netherlands, and Germany to Scandinavia, eastern Europe, and North America.[19]

Within just two decades, the Moravian Brothers developed into an economic elite spread out over the European continent, forming a network of closely knit communities and settlements that featured skilled labor, mobile capital, trade and credit contacts, and high standards of education. The movement integrated this emphasis on work, trade, and industry with an intense life of faith and piety, asceticism, and moral control. At the same time, it created remarkably modern economic and legal institutions (such as social security, credit and capital investment, and accounting) that existed entirely outside of the traditional estate society. The Brotherhood established a formal constitution for its community as well as three separate governing bodies to run its financial, legal, and spiritual affairs.[20] While the Herrnhut Brothers were often sought after by the rulers of the territories in the Holy Roman Empire, they were just as often persecuted and expelled when a new ruler or government brought about a change in the state's religious affiliation. Because of the constant threat of religious expulsion, the Brotherhood developed a system of investing and reinvesting capital that kept it mobile and unattached to property.[21]

After Abraham Roentgen had converted to the Moravian Church in London in January of 1738, he moved closer to its community in London and attended its services regularly.[22] He formally requested

19. Troeltsch, *Aufsätze*, 488–89; Brecht and Deppermann, *Geschichte des Pietismus*.

20. The Brotherhood's economic and legal governing bodies were the *Oeconomicum* and the *Civile*. For the financial affairs and accountancy in the settlement of Neuwied, see UVC R5 I K 1, and for the civil code of Neuwied's community, see UVC R6 I K 1–13, in the Archiv des Unitäts-vorsteher-Collegs, Herrnhut. The Brotherhood's *Oeconomicum* meticulously organized and recorded all of its financial transaction, real estate transactions, inventories, and private loans to members. The Brotherhood's habits of investing wealth and capital were similarly "outside" of the estate society, in ways that were customary nowhere else in the surrounding court culture. Peter Kriedte points out how relevant Max Weber's essay on sects (versions written in 1906 and 1920) is in this regard. Kriedte, *Taufgesinnte*, 37–40.

21. Stürmer, *Luxus*, 22–23.

22. Roentgen, *Das erste Buch*, 14.

membership in the spring of 1738.[23] In the summer of that year, he returned to Germany and traveled first to the Brotherhood's community in Marienborn, near Frankfurt, where the territory's ruling count, Ernst Kasimir I of Ysenburg-Büdingen had rented a castle to Zinzendorf.[24] Abraham was soon admitted to this community. In April 1739, he married Susanne Maria Bausch, a Pietist woman born in Frankfurt in 1717, who had become a member of the community just two months earlier.[25] In the same year, the couple moved from the castle to a new settlement of the Brotherhood a few miles from Büdingen, in Herrnhaag. Four years later, in August, their son David was born. Thus, by the time Abraham Roentgen started his first business in Germany in the 1740s, he was completely integrated into this remarkable religious community and was supported by its economic infrastructure. The education, the work ethic, and the favorable trade conditions it provided were crucial elements in the rise of his and his son's business over the following decades.[26]

Count Zinzendorf conceived of his brotherhood's settlements as artisan communities. The original settlement in Herrnhut in Saxony became a model that later settlements, including the ones in Herrnhaag and Neuwied (where Abraham moved in 1750), followed in regard to architectural and social design.[27] The community in Herrnhaag was under the continued protection of the count of Ysenburg-Büdingen, who granted the artisans numerous privileges, hoping that these measures would spawn economic growth. As a result, the community, organized around communal buildings, workshops, and family housing, grew quickly.[28]

23. The protocols of the Brotherhood's Council of Elders (*Ältestenrat*) first mentioned his name in October of 1738. See R.2.A. no. 2, 2a, and R.2.A. no. 2, 2b, Archiv der Brüder-Unität. See also Greber, *Abraham und David Roentgen*, 31.

24. The rented castle was Schloss Marienborn, which Count Karl-August von Ysenburg had built in 1673. The large conferences (synods) of the Moravian Church were held in Marienborn in the 1760s. UA R. 2. Synoden. In 1442 Büdingen had become a county through an edict by the emperor, and Count Casimir I had issued a tolerance edict in 1712 that invited many religious emigrants to settle there. See Heuson, *Büdingen gestern und heute*.

25. See the records of this marriage in Herrnhaag's "church book" (Herrnhaagisches Kirchenbuch), UA R. 8. 35a.

26. Fabian, *Kinzing und Roentgen*, 26–45.

27. On the model character of Herrnhut settlements and colonies and their architecture and city planning, see Richter, "Aus der Baugeschichte," 29.

28. On this settlement, see the Herrnhaagische Kirchenbücher, UA R. 8. 35; Meyer, "Zinzendorf und Herrnhut," 36. According to the community model of the Brotherhood, all unmarried people lived in assigned buildings, called "choirs," divided into sections for men, women, widows, and widowers. The principle of the "choirs" is described in an anonymous travel report to Neuwied from 1769 in Bernoulli, *Sammlung* (1781–87), 16:256.

Because of this growth and productivity, the Moravian Brothers were an exemplary subject of the empirical travel studies conducted under mercantilist agendas in the eighteenth century. Visitors and dignitaries frequently visited the Brotherhood's colonies and were interested in the products of their artisan trades, including Roentgen's.[29] The Moravians' social, spiritual, and economic system resonated profoundly with a wide range of Enlightened ideas about economics, social order, work, city planning, and religious toleration.[30]

When Abraham arrived in Herrnhaag in 1739, he started working on a freelance basis. His first clients were individual members of the Brotherhood and the Brotherhood's government itself.[31] Herrnhaag was an advantageous location for an ambitious cabinetmaker, since it was situated at the junction of several principalities and duchies whose rulers needed luxury furniture for their residences.[32] Roentgen had worked for wealthy and noble clients earlier in London, and he soon developed business relationships with the princes of the territories in the area.[33] At the end of the 1740s, Abraham Roentgen's workshop was

29. Neuwied was included in travel reports from the 1750s onward, even before the Roentgen manufacture became famous. Later travel reports typically mention David Roentgen's name in the context of Neuwied artisan work. See Nemnich, *Tagebuch* (1809), 2–23; Lang, *Reise auf dem Rhein* (1789–90); Bach, *Goethes Rheinreise* (1923); Günther, "Die Wurzeln der Brüderunität," 28. The Prussian king Friedrich Wilhelm II visited Neuwied in 1792 and toured the buildings and institutions of the Brotherhood. He took lunch at David Roentgen's house, whom he knew because he had bought furniture from him when he was the crown prince. See UA R. 7. G. b. no. 1. c, Diarien, Berichte, Memorabilien, 1782–1793. The document is also transcribed in Huth, *Abraham und David Roentgen*, 59. See also Fabian, *Abraham und David Roentgen*, 374; Greber, *Abraham und David Roentgen*, 25.

30. The eighteenth-century traveler Joseph Gregor Lang even used the metaphor "state machine" to describe Neuwied in the 1780s, describing how well its "cogs" of liberty and religious tolerance fit into each other. Lang, *Reise auf dem Rhein*, 262. On ubiquitous early modern metaphors of the state as machine in the Enlightenment, see also Stollberg-Rilinger, *Der Staat als Maschine*. Many other travelers remarked on the success of the religious-tolerance policy in Count Alexander's territory. Bernoulli, *Sammlung* (1781–87), 16:261; "Comments on a journey through Germany and the Netherlands in the years 1779, 80, and 81," in Bernoulli, *Sammlung* (1781–87), 14:265–344, esp. 331–32; Bertola de' Giorgi, *Malerische Rhein-Reise*, 172–74.

31. Other trades flourished in the Brotherhood, such as shoemaking, tailoring, and weaving, and there were also bakers, goldsmiths, and carpenters. See Haus- und Tischzinsbücher in UA, R 7. G. a. 2, 53; Krieg, "Das Brüderhaus in Neuwied," 27.

32. Virtually all Brotherhood settlements were built at economically advantageous locations, most of them even directly on established trade routes. The Brotherhood's goal was not only economic self-sufficiency and independence from political rulers, but also the ability to finance its intended worldwide missionizing. See Richter, "Aus der Baugeschichte," 36–37; Kriedte, *Taufgesinnte*, 39–40.

33. He probably also used the Frankfurt fair to find new clients in this period, given how close it was to Herrnhaag. The Frankfurt fair was not only a commercial event but also a place where aristocratic people went for leisure and entertainment. See Fabian, *Abraham und David Roentgen*, 11; Gondorf, "Der Kunsttischler Abraham Roentgen," 15.

prospering and faced no competition from the guild-bound carpenters in the larger Frankfurt area. During this period, the size of his workshop probably reached that of a small village carpenter's workshop, and he had already established the specialization in elaborate mechanical furniture interiors that remained his (and his son's) signature product over the next decades.[34]

Abraham Roentgen's modest prosperity in the 1740s soon ended, however, the victim of increasingly fierce conflicts between the Brotherhood and the county's governing bodies. When Count Ernst Kasimir died, the new count, Gustav Friedrich, expelled the Brotherhood from Herrnhaag in February 1750: they were given three years to leave.[35] The Herrnhaag community of about one hundred members practically dissolved, with the majority moving to Holland, England, and North America, as well as Silesia and Thuringia. Abraham Roentgen moved as part of a small group to the county of Neuwied, the residential city of the count of Wied-Neuwied, about fifty-five miles south of Cologne, where Count Johann Friedrich Alexander had heard of the Brotherhood's imminent expulsion.[36]

Count Alexander (as he came to be called) had become Neuwied's ruler in 1737. He had already distinguished himself in his earlier years by running his territory according to principles of the Early Enlightenment, promoting education, philanthropy, religious tolerance, and theaters and abolishing press censorship. The Thirty Years' War had left his principality devastated and deserted, and he was committed to repopulating and recultivating it.[37] He promoted economic expansion and aimed to attract entrepreneurs who would establish manufactures and factories. Among his initiatives were lotteries in which houses could be won and a promotional brochure for his county.[38] He granted religious

---

34. My account of these aspects of the history of the Roentgens' business relies on details and connections that Greber establishes in *Abraham und David Roentgen*, 35–42.

35. These conflicts revolved around economic as well as theological issues. Meyer, "Zinzendorf und Herrnhut," 46ff. On economic envy and rivalry, and religious persecution, see also Stürmer, *Luxus*, 22.

36. Neuwied is about one hundred miles northwest of Herrnhaag and just ten miles north of Koblenz, an important garrison town, where the Mosel river flows into the Rhine.

37. Tullius, *Die wechselvolle Geschichte*, 65–66; Meinhardt, "Der Werdegang Neuwieds," 69–71, 117–18.

38. With these measures, Alexander closely followed mercantilist ideas on government-funded repopulation. The 1769 travel report about Neuwied, mentioned above, refers to the "lottery houses" in Neuwied. Bernoulli, *Sammlung* (1781–87), 16:263. Traveler Joseph Gregor Lang also mentioned in the 1780s "many manufactures and factories" in Neuwied. Lang, *Reise auf dem Rhein*, 251. See also Meinhardt, "Der Werdegang Neuwieds," 117–40.

freedom, freedom of guilds, tax relief, and generous building permissions within the town of Neuwied.[39] Alexander's mercantilist interests coincided with the Moravian Brothers' need to find a new home. When he heard about their expulsion from the county of Ysenburg-Büdingen, he offered them residence and economic and religious privileges in Neuwied.[40]

In October 1750 about forty refugees migrated to Neuwied, among them Abraham Roentgen and his wife.[41] Their son David, by then seven years old, was already in the Brotherhood's children's school in Niesky, a village in Saxony.[42] Like Herrnhaag, the nascent Neuwied community aimed to build up a settlement modeled on the original in Herrnhut.[43] Initially, the new community faced uncertainty, because financial support from the Brotherhood's headquarters was not immediately granted, and negotiations with Count Alexander over a license to build a settlement and for allocation of the appropriate pieces of land took several years.[44] Even so, Abraham Roentgen started work immediately after his arrival. Neuwied turned out to be an even better and more

39. Alexander, and before him his father, Count Friedrich Wilhelm, and his grandfather, Friedrich III, issued many privileges for various religious and immigrant groups in the late seventeenth and early eighteenth centuries, before and after the Herrnhut Brothers came from Herrnhaag. The privileges regulated exercise of religion, abolition of serfdom, taxing, and building permits. See, in the Fürstlich-Wiedisches Archiv (FWA), FWA 26-10-6; FWA 65-11-13; FWA 80-2-4; FWA 61-6-2; and FWA 65-11-6. I received great help from Dr. Hans-Jürgen Krüger, the Princely Archive's archivist, and his comprehensive knowledge of the archive's holdings. The listed documents are also transcribed, partially or entirely, in the following works by historians: Fabian, *Abraham und David Roentgen*, 310–12; Krüger, "Religiöse Toleranz?" 153; Krieg, "Das alte Herrnhuter," 100; Fabian, *Kinzing und Roentgen*, 369.

40. The extensive negotiations for these privileges are documented in the correspondence between Nikolaus von Zinzendorf and Count Alexander of Neuwied as well as the Brotherhood's records about the Neuwied settlement. UA R. 7. G. a. 1, Briefwechsel Zinzendorfs mit dem Grafen von Wied; UA R. 7. G. a. 3, Akta, das Etablissement in Neuwied betreffend, 1750–1756. In these documents, authorities of the Brotherhood (including Zinzendorf himself) explain the economic and spiritual principles of the Brotherhood to the count and also explain their goals concerning the Brotherhood's settlement in Neuwied.

41. The arrival of the group of people is documented in Neuwied's *Kirchenbuch*, UA R. 7. G. b. 3. a, Kirchenbuch-Extrakte, 1750–1827.

42. Nesky is about 25 miles north of the original settlement in Herrnhut in Saxony, which meant that David was about 300 miles away from the Herrnhaag settlement and about 400 miles away from Neuwied.

43. For drawings, maps, and ground plans of the square (*quarée*) to be built in Neuwied, see the Brotherhood's legal dossiers UVC R. 6. I. K no. 1 (UVC IX.175) and UVC R. 6. I. K no. 3 (UVC IX.176).

44. The complex negotiations concerned, among other things, the limits and concessions of spiritual and worldly authority. See the correspondence between the county of Neuwied's administration and the Brotherhood's authorities in FWA 65-11-4. In August of 1751, the count issued a General Concession for the Moravian Brothers (FWA 65-11-6), which included exemption from military service and from swearing an oath of allegiance to the count. The final concession,

convenient location than Herrnhaag: it was close to the residences of the elector of Trier and other dignitaries, as well as to the fairs in Bonn, Mainz, Trier, and Frankfurt. Work materials could be obtained more easily, since timber from the German lands and abroad, metal fittings, and other resources came directly by ship to Neuwied.[45]

Abraham Roentgen soon established a joinery on the basis of the experience that he had gathered in London and Herrnhaag. He started to standardize the production of his furniture with respect to size, form, and construction, producing a small number of styles, each in a series of three to five pieces. Doing so also entailed a division of labor. Outside of the workshop, Roentgen relied on the works of wood-turners, upholsterers, and locksmiths, while he himself was in charge of administration, acquisition, sales, and the planning and execution of the marquetry designs. It was this organization of the production process that made the workshop so productive and made possible its quick expansion. In the period around 1750, Roentgen successively transformed his workshop into a manufacture.[46]

Abraham Roentgen found new customers among wealthy aristocrats who ran courtly households, and he offered his products at trade fairs, too.[47] This resulted in general in increased demand for more sophisticated furniture, in response to which Abraham broadened his repertoire of products and increased the size of his manufacture. His

---

issued in January of 1756, included thirty-three items. Krüger, "Religiöse Toleranz?" 154–61 (his article also provides partial transcriptions).

45. A traveler, mentioned above, reported in 1796 that the largest part of Neuwied's trade was French and Dutch goods that came by river. Bertola de' Giorgi, *Malerische Rhein-Reise*, 172–74. See also Greber, *Abraham und David Roentgen*, 46.

46. Greber, *Abraham und David Roentgen*, 46–58. The standardization in Abraham Roentgen's workshop was unusual for the time and even for later periods, but it did not go so far that the establishment would have been a factory in the conventional sense. Neither did this happen once David took over in the 1760s, as some have suggested. Carpentry was a trade that functioned in "traditional" ways until well into the nineteenth century: it did not employ (steam-powered) machines to mass produce pieces or use interchangeable parts. There were no machines or factories in the strict sense in carpentry in the eighteenth century. Jacobsson, *Technologisches Wörterbuch* (1781–1802), 1:491, s.v. "Ebenist"; Rössig, *Lehrbuch der Technologie* (1790); Krünitz, *Oeconomisch-technologische Encyklopädie* (1773–1858), 185:241–92; Gülich, *Geschichtliche Darstellung* (1830–45); Karmarsch, *Geschichte der Technologie* (1872), 560. See also Fischer, "Beiträge zur Geschichte der Holzbearbeitungsmaschinen."

47. Count Alexander started buying Abraham's furniture in the 1750s and remained a regular customer until 1791. Through him, Abraham acquired other distinguished clients. See bills and receipts in FWA, Renteirechnungen 1761, S. 215, no. 1279; 1761, S. 223, no. 1388; 1769, S. 300, no. 1022; S. 300, no. 1023. Transcripts of some of these sales records are also found in Fabian, *Abraham und David Roentgen*, 323, 331. On Roentgen's aristocratic clients and the growing network of supporters he developed, see Willscheid, "Der Kundenkreis."

religious affiliation worked in his favor in this regard, as the Brotherhood was well liked within aristocratic circles, not least because of Zinzendorf's own noble background and the fact that aristocratic women in particular found the Moravian Brothers appealing as a spiritual home.[48]

Roentgen's success in adopting English modes of organization and sales was possible because of the many privileges he had been granted in Neuwied, his membership in a distinct religious group, and the weakness of the guilds. What was especially crucial to his success, beyond his wide travel experience as a youth, was the freedom these privileges gave him. He could produce whatever he wanted with however many people he wanted, and he could buy anything from anywhere without paying customs duties—all because of princely privileges. This independence would have been unthinkable if he had been a member of an ordinary guild. As his products became more and more elaborate, the exemption from taxes and customs duty—an exemption that applied to all goods delivered from manufactures to princely courts—became more crucial. The territorial "patchwork" of the Old Reich usually meant customs and fees at every barrier and gate, at considerable cost. Any trade between princely courts throughout the Holy Roman Empire, however, was free of charge, because the princes exempted one another from customs duties. Goods for princes thus had a European market, while the economic horizon of a guild ended at the town wall.[49]

The Workshop in Crisis and David's Coming of Age

Abraham Roentgen's firm expanded quickly during the first twelve years of its existence. In the 1760s, however, he faced an economic crisis that was due to a general recession in the wake of the Seven Years' War. His manufacture required a constant influx of large amounts of capital, as expensive materials—precious metals, expensive woods,

48. See once again the anonymous travel report of 1769 that Johann Bernoulli edited in his collections. The author says he was given a tour by the countess of Neuwied, who showed him the Herrnhut choir for women, and that they visited the Comtessa Reuss von Lobenstein there. Bernoulli, *Sammlung* (1781–87), 16:260. On the topic in general, see Meyer, "Zinzendorf und Herrnhut."

49. Stürmer, "Die Roentgen-Manufaktur," 31, 35. Not even the excessive cost for transportation was an obstacle to the luxury market.

ivory, mother-of-pearl—had to be kept in stock and skilled workers had to be paid wages.[50]

Roentgen was not an independently wealthy man, and he had already taken out loans from his brother-in-law and from the Brotherhood for his move. By the beginning of 1768, unsold furniture worth 2,145 ducats had accumulated in storage (this was equivalent in value to one year's income for three apprentices, or the value of three prestigious town houses).[51] He could no longer run his workshop and support his family on this basis. As a solution to this dilemma, he even contemplated moving his firm to England. He repeatedly asked the Brotherhood council for support, but they were unhappy about his financial insecurity and about the way he ran his business, which was in conflict with the Brotherhood's economic and spiritual principles.[52]

In this time of crisis, Abraham's son David took on major responsibilities in the firm for the first time. He had joined his father's business in 1757 at age fourteen.[53] Born in 1743 in Herrnhaag, he had left his home when he was six to be educated in the Brotherhood's children's houses (*Kinderanstalten*) in Hessen and Saxony. The level of this education was unusually high for a young artisan in that period and far ahead of a general school education. Furthermore, the Brotherhood's education did not instill the rules and beliefs underlying the estate and court society of the time, unlike the education available to most other educated people and their children. Because of the Brotherhood's extensive missionary activities and the French- and English-speaking communities under its roof, there was also a strong cosmopolitan current underlying its educational institutions.[54]

After his schooling, David became a carpenter apprentice at his father's firm in Neuwied.[55] David never worked in any workshop other

---

50. Later in the history of the manufacture, David focused more on producing commissioned furniture. Greber, *Abraham und David Roentgen*, 88.

51. Stürmer, *Luxus*, 31.

52. The situation culminated in the Brotherhood authorities' calling in a loan they had granted to the Roentgens. See the correspondence and loan registry in UVC IX. 23. For a transcript, see Fabian, *Kinzing und Roentgen*, 389; for commentary, see Greber, *Abraham und David Roentgen*, 88.

53. Greber, *Abraham und David Roentgen*, 49.

54. For an autobiographical account of the ways the Brotherhood ran its *Kinderanstalten*, see Roentgen, *Das erste Buch*, 25–42.

55. During this time he lived with his parents, not in the house ("choir") for unmarried males, which would have been the usual arrangement for an unmarried male member of the Brotherhood. On the rules for living arrangements, see Lang, *Reise auf dem Rhein*, 249; Krieg, "Im Chorhaus."

than his father's, and it was from his father that he learned joinery and cabinetmaking. It is likely that his father taught him, both directly and by example, advanced principles of furniture design, business management, and client relations with the noble and distinguished.

Around the year 1767, when he was twenty-four and when the firm was in its deepest economic trouble, David Roentgen took on more responsibilities. To remedy the crisis, he developed a plan to organize a lottery for the unsold furniture sitting in stock at the manufacture. He calculated that the potential revenue from such a venture would be enough to save the business from bankruptcy, while also providing a unique opportunity both to make the Roentgen name known to a much wider circle of people and to display publicly the range of the firm's technical and artistic works. Sometime around the year 1768, he proposed the lottery plan to his father.[56] His father was less enthusiastic about the idea of the lottery, preferring instead to scale down the firm. Abraham had spent difficult years in economically precarious situations and was unwilling to accept more risk or further endanger his relations with his religious community. The idea of the lottery also brought young David into such conflict with the Brotherhood's council that Abraham offered, in August 1768, to send his son away from Neuwied and have him set up a business in another Brotherhood settlement.[57] This offer was entirely in line with the council's goals, because David's ambitious plans were in complete contradiction to their vision of the Brotherhood's life.

Despite this ongoing struggle, in the spring of 1768 David applied to the municipal council of the city of Hamburg in his own name for permission to conduct a furniture lottery there. He was unwilling to abandon this opportunity even when confronted with potential expulsion from the Brotherhood. In the end, he even managed to change his father's mind. As soon as the council of elders heard this, they excluded David from their community. As a consequence, he was not allowed to live in their quarters and was banned from the Eucharist for most of his adult life.[58] To prepare for the lottery, Abraham and David Roentgen

---

56. Gondorf, "Der Kunsttischler Abraham Roentgen," 10–11; Fabian, *Abraham und David Roentgen*, 13; Greber, *Abraham und David Roentgen*, 88.

57. This is a well-researched episode in Abraham and David Roentgen's lives. For accounts that I rely on, see Gondorf, "Der Kunsttischler Abraham Roentgen," 10–11; Fabian, *Abraham und David Roentgen*, 13; Greber, *Abraham und David Roentgen*, 88; Fabian, *Kinzing und Roentgen*, 389.

58. For the struggles between the Roentgens and the Brotherhood, see the protocols of the Brotherhood's Council of the Elders from 1767 to 1769, UA R. 7. G. b. 4. a., Protokolle der Ältesten-Conferenz, 1766–1787; and Huth, *Abraham und David Roentgen*, 7. Examples of the vari-

gathered together one hundred pieces of furniture, whose overall value was 2,145 *Spezies-Dukaten*, with plans to sell 715 tickets for three *Spezies-Dukaten* each. David had a brochure printed with detailed descriptions of all pieces and had it disseminated with a general announcement of the lottery. The majority of the tickets were sold to courts and cities all over the Reich, not locally in Hamburg.[59] The first prize was a desk with a cabinet on top and a hammer piano in the lower part. It combined in characteristically Roentgenian style ornamented luxury furniture with chimes and a musical instrument.

The lottery was a great success, both financially and in terms of marketing: the name Roentgen was now widely known, and the financial strait passed. David was now a principal decision maker in the firm.[60] In the decades between the 1760s and the 1780s, he transformed the workshop into an international manufacture, improved the technology and artisanship, broadened the artistic repertoire of the company, and successively opened up new sources of financing. He achieved a turnover higher than that of the Meissen porcelain manufacture, the reference point for luxury manufacturing in eighteenth-century continental Europe, and he influenced the art of furniture-making in style and technique for generations to come.[61]

At the same time, and for this very reason, his relationship to the Brotherhood, its spiritual authorities, and the Pietist ways of life became more and more strained. Given the clients that Roentgen served and the fame he acquired, his economic success, although it had its roots in the Brotherhood's principles, came into dramatic tension with the Brotherhood's habits and rules, provoking continuous conflicts over his entire lifetime. David was excluded from the Brotherhood for most of his adult life despite his repeated attempts to be readmitted.[62]

---

ous punishments in the Brotherhood, such as temporary exclusion from the Eucharist, are found in Roentgen, *Das erste Buch*, 77.

59. The calculation for the number of tickets and the price was to cover the value of the furniture in stock: 715 x 3 = 2,145. The receipt for Count Alexander's own lottery ticket is in FWA, Renteirechnungen 1769 no. 320 (it was ticket number 1, and the receipt is also transcribed in Fabian, *Abraham und David Roentgen*, 332). See also Greber, *Abraham und David Roentgen*, 88–89.

60. David's increasing involvement can be traced through the signing of bills, receipts, and correspondence for furniture and repairs for Count Alexander's court, where his name appears more frequently over the years. See the count's accounting books and record of receipts in FWA Renteirechnungen and Ausgaben. Here as well, I relied a great deal on the help and support of archivist Dr. Krüger, who guided me through this batch of documents (see also note 34).

61. Stürmer, *Luxus*, 8.

62. David's brother Ludwig was also reprimanded regularly by the Brotherhood's authorities and excluded from the Eucharist. Roentgen, *Das erste Buch*, 77. On the ongoing struggle between the Brotherhood and David Roentgen, see also ibid., 28.

The next steps David took after the firm's success with the lottery in 1768 were exploring markets in eastern Europe and expanding the manufacture's collaboration with other artisans and artists. Since there was already a flourishing cabinetmaking culture in the western part of Europe, particularly in France and England, prospects looked better to the east. David even negotiated with the Prussian king to establish a manufacture in Berlin and Breslau, but these efforts eventually failed.[63] Still, he managed to expand his cooperation with other artisans and artists to broaden the repertoire of his furniture: there were now quite a few components in the Roentgen manufacture whose production exceeded the skills of a cabinetmaker, such as musical instruments, metal works and mechanisms, and technical drawings. In 1771 he employed two painters and also cooperated more regularly with the Kinzing family, which consisted of four generations of clock-makers. He also forged contacts with established artists such as the composer Christoph Willibald Gluck.[64]

As the firm expanded, the space in Abraham's old workshop became too small. Abraham asked the Brotherhood council for permission to extend the existing business spaces but was denied his request. The council argued that since David would not be able to take over the business eventually, because he was not a member of the Brotherhood, it would be better to build a house outside of the Brotherhood's quarters.[65] Similarly, David's repeated attempts to reconnect to the Brotherhood were denied. As a consequence, Abraham made over the business to his son in 1772, in exchange for a pension, although he remained active in the manufacture into the 1780s.[66] Abraham was then age sixty-two, and David was thirty. In 1772 David married Katharina Dorothea Scheurer, the sister of a pastor in the Herrnhut Brethren, from Alsatia.[67] David requested readmission to the Brotherhood together with his wife, but

63. See Huth, *Abraham und David Roentgen*, 50, document 12.

64. The painters were Johannes Juncker, the son of the Frankfurt painter Justus Juncker, and the more senior painter Januarius Zick. *Allgemeine deutsche Biographie*; Adolf Feulner, *Die Zick*, 6–14; *Rheinische Kunstwerke*, Bildtafel 10; Greber, *Abraham und David Roentgen*, 110; Fabian, *Abraham und David Roentgen*, 14.

65. See UA R. 7. G. b. 4. a, Protokolle der Ältesten-Conferenz, 1766–1787, July 1772. A transcript of the corresponding document in the Brotherhood's Neuwied archives (to which I received no access) is found in Fabian, *Abraham und David Roentgen*, 335. The council made the argument (in a council meeting in 1772) that David would be able to own the building only if it was built outside of the Brotherhood's quarters.

66. See David Roentgen's letter to Johannes von Watteville, MS-44, part 1, folder 1, item 1, Roentgen papers, New York Metropolitan Museum of Art. See also Roentgen, *Das erste Buch*, 90.

67. In the same letter of April 1773 to Johannes von Watteville, David Roentgen told about his marriage, his plans to build a house, and his efforts to find loans. MS-44, part 1, folder 1, item 1, Roentgen papers.

the council of elders, after contemplating the request, denied it. Other signs of the bad relationship between him and the council were that he and his wife were not allowed to live with his parents and that the council denied a repeated request for a loan on the father's house.[68]

Now the owner of the business and facing continuing difficulties with the Brotherhood, David established his own legal relationship with the city of Neuwied and the count. He knew how fierce the conflicts could become between his firm and the guilds, whose rules determined prices, limited the number of apprentices to two, and determined the kind of works and products to be built. At the same time, the count granted privileges to local manufactures in Neuwied on a regular basis—including exemption from guild regulations.[69] Thus David requested from his prince the same privileges that his father had enjoyed, and after negotiations in 1774, he received them: exemption from the local carpenters' guild, permission to employ as many workers and apprentices as he wanted, exemption for his employees from taxes, and permission to forbid them to work for anybody else but him. In addition, David himself would have to pay no taxes, was allowed to buy and sell anything from anywhere without paying customs duties, and would receive a plot of land for his house and manufacture free of charge and free of rent for ten years.[70] This set of privileges provided advantageous conditions for David to establish and expand his means of production, allowing him to bring in greater numbers of artisans. He received a building plot on Pfarrstraße in 1774, across the street from his parents' house but outside the Brotherhood's quarters. He built a house large enough to accommodate both his family's and his business's needs, and he planned it so that it could expand as the business grew in the future. Even in the manufacture's most productive years, this building housed all of its business activities (see fig. 3).[71] The

---

68. Greber, *Abraham und David Roentgen*, 112. Some of these problems continued over the following decade. See UA R. 7. G. a. 9, Akta, gräfl. Zusicherungs-Decret für Anbau eines neuen Quarré (1781–82).

69. Count Alexander attracted many "factories" to the county in the second half of the eighteenth century, in addition to those involved with the Brotherhood. As I mention in note 39, Lang's *Reise auf dem Rhein*, 251, describes Neuwied as having "a number of other manufactures and factories."

70. FWA 65-11-6, Privilegia (the document is dated 9 May 1774). The set of privileges was consistent with Neuwied's general policy of attracting immigrants, and not an unusually generous arrangement made specifically for David Roentgen. The document is transcribed, partially or entirely, in Huth, *Abraham und David Roentgen*, 51; Greber, *Abraham und David Roentgen*, 112; Fabian, *Abraham und David Roentgen*, 340.

71. The house still exists today. Fabian, *Abraham und David Roentgen*, 14; Greber, *Abraham und David Roentgen*, 112–14.

FIGURE 3 David Roentgen's workshop in Neuwied, built in 1784. Courtesy of Foto Bollmann, Neuwied, Germany.

construction of the new manufacture, his marriage, and the granting of the privileges consolidated David's life in Neuwied.

The Manufacture's Final Major Expansion

David Roentgen began the final expansion of his business after he received these privileges. By the time he finished expanding, in the mid-1780s, he was the most important furniture purveyor to the French, Swedish, Russian, and Prussian royal courts, and his manufacture had built the dulcimer-playing automaton.

When initiating his workshop and trade expansion, he turned first, in 1774, to Paris, the capital of furniture-making and of European art and culture at the time. He traveled there in August 1774 to investigate new stylistic and artistic currents and to explore the possibility of selling his products at the court of Louis XVI.[72] Upon his return to Neuwied, he had a set of first-rate luxury furniture built in his manufacture, based on new designs and with elaborate internal mechanisms, to take to Paris and sell at the royal court a few years later.[73] The French market for luxury furniture was competitive, and his expansion westward required thorough preparation. The dulcimer-playing automaton eventually became part of this expansion scheme.

Early in 1779, Roentgen went on his second trip to Paris and was granted an audience with Louis XVI and Marie-Antoinette. The royal couple purchased richly decorated cabinets and desks from him and granted him the title "Ebéniste mécanicien du roi et de la reine." His success at the court caused other Parisian dignitaries to become interested in his work and enabled him to establish a local branch of his firm in Paris.[74] Even one of the leading art critics and mediators between French and German culture, Baron Melchior von Grimm, became a client of his.[75]

72. David Roentgen was well connected among those Parisian artisans who were immigrants from the German-speaking lands. Two people in particular were in a position to write him letters of recommendation to help make contacts. One was his former apprentice Adam Weisweiler; the other was his collaborator Januarius Zick, who introduced him to the famous and influential engraver Johann Georg Wille. See the entry on Weisweiler in *Encyclopædia Britannica*; Le Blanc, *Catalogue de l'oeuvre de Jean Georges Wille*; Duplessis, *Mémoires et journal de J.-G. Wille*, 1:577; and, on Zick, *Allgemeine deutsche Biographie*, 43:257–60.

73. An important difference between the Parisian and the German cabinetmaking culture that pertained to Roentgen and his standing in both places was that in Paris, a cabinetmaker could become an official purveyor to the court and hold the status of an esteemed artist. A precious piece of furniture was similar in status to a painting or a sculpture. In Germany, in contrast, cabinetmakers were on the same level as ordinary artisans. Greber, *Abraham und David Roentgen*, 117–18.

74. *Nouvelles des la république des Lettres et des arts* reported this in February 1779 (51) and March 1779 (57–58). The so-called *Vossische Zeitung* (a nickname for the *Königl. privilegirte Berlinische Staats- und gelehrte Zeitung*) reported on 1 July 1779 (issue 78) from Frankfurt on David Roentgen's "recent successful trip to Paris," announcing that he had sold the king a desk for 80,000 livres and that he had received the "patent" *ébéniste méchanicien* for it. See also Greber, *Abraham und David Roentgen*, 121n171 (and a reproduction of the *Vossische Zeitung* on 125). *Affiches, annonces, et avis divers, ou, Journal général de France* of 8 January 1781 announced that David Roentgen had founded a branch of his business in Paris. See also Greber, *Abraham und David Roentgen*, 177 and 192.

75. Friedrich Melchior Baron von Grimm, born in 1723, contributed a great deal to the spread of French culture throughout Europe. He had connections with many influential people, including Denis Diderot and Catherine the Great. Grimm, *Lettres de Grimm*; Grimm, *Correspondance littéraire*; *Allgemeine deutsche Biographie*, 9:676–78.

Maintaining business relations with the French royal family became a key priority for David Roentgen after this first successful introduction, and building the dulcimer player was a part of this plan. Roentgen continued traveling for the latter part of 1779, mostly on the European continent. He organized the building of furniture for the French court after he returned to Neuwied from Paris, planning to offer this next batch to the royal couple on a third visit to Paris sometime in the following few years. Included in this batch of furniture built between 1779 and 1784 and presented to the French court early in 1785 was the dulcimer-playing automaton. Roentgen himself, however, stayed only briefly in Neuwied after returning from Paris in early 1779; in May 1779 he went to Brussels to the court of Prince Karl, then to Berlin to Prince Frederick William of Prussia.[76] He and his work, now becoming well known, were featured, for example, in periodicals in France and in the Holy Roman Empire.[77]

Roentgen could not have been a regular presence in his workshop after the year 1779, and he could not have designed and built "with his own hands" the pieces of furniture, or the dulcimer-playing automaton, that were being delivered to the royal courts in this period of his greatest fame and commercial success. His work during this period must have been mainly directing and maintaining his international firm, which meant handling correspondence, client relations, marketing, accounting, external suppliers, and artistic strategies.[78] An anonymous observer commented explicitly in an essay of 1786 on the issue of David Roentgen's role in his manufacture and his status as his manufacture's "master," "principal," or "inventor."[79] This report claimed that Roentgen did not make "the least" contribution, either artistic or me-

---

76. Greber, *Abraham und David Roentgen*, 124.

77. *Nouvelles des la république des Lettres et des arts*, February 1779, 51; March 1779, 57–58; *Vossische Zeitung*, no. 78, 1 July 1779; "Comments on a Journey through Germany and the Netherlands in the Years 1779, 80, and 81," in Bernoulli, *Sammlung* (1781–87), 14:265–344, esp. 331–32; *Affiches, annonces, et avis divers*, 8 January 1781. See also Greber, *Abraham und David Roentgen*, 192.

78. Roentgen paid much more attention to acquiring and tending clients than traditional workshop artisans did. See Heal, *London Furniture Makers*.

79. The anonymous report was published in a periodical on literature and anthropology (*Litterattur und Völkerkunde*): "Auszug eines Schreibens aus Neuwied." Johann Wilhelm von Archenholz edited this successful periodical (it appeared from 1782 to 1791) and worked with a rather modern understanding of journalistic work, aiming to keep his readers up-to-date about political news and employing neutral, nonpartisan ways of reporting. The report itself was dated 10 November 1785. It was an anonymous letter by someone who had recently visited an exhibition of a large and extravagant batch of David Roentgen's furniture in Neuwied, before it was shipped off to courts in Russia and Sweden. We know of an exhibition in October 1785, on the occasion of the inauguration of a new prayer hall for the Brotherhood. See Fabian, *Kinzing und Roentgen*, 69.

chanical, to the furniture made in his manufacture in the 1780s, referring to him only as the manufacture's "owner and salesman," who had around eighty employees working for him.[80] The report stated that Roentgen was in charge of selling the furniture and acquiring new clients and that, being a cosmopolitan and well-rounded man, he was very talented at these tasks. However, this observer wanted to correct the impression that Roentgen was the main inventor or principal mechanic at his manufacture. Instead, the report claimed, the main inventor and master in Roentgen's manufacture was a man named Johann Christian Kraus.[81]

The report's impressions and numbers agree with other evidence about the operation of David Roentgen's manufacture and his travel activity at that time. In June 1779 Roentgen himself reported to Count Alexander that he had twenty-eight masters, apprentices, and workers at his Neuwied manufacture. Counting six masters at supplier firms, with one to two apprentices each, adds another eighteen, bringing the total number of people working in and for the Roentgen manufacture to forty-five or fifty.[82] Roentgen's manufacture was substantially larger than the workshop of any guild-bound artisan or any ordinary Parisian *ébéniste*.[83]

The fact that we cannot take for granted the presence of David Roentgen in his own manufacture in 1779 and throughout the early 1780s has consequences for assumptions that we make about the origins of, and goals attached to, the dulcimer player in the Roentgen firm, and about the making of android automata in general. Since the dulcimer player was built (and perhaps even invented) at a time when the manufacture's nominal principal was not physically present, there is no obvious way to gauge Roentgen's motivation or intentions for this automaton. Rather than attempting to determine a singular idea or concern behind it, we must instead consider it in as many dimensions (economic, cultural, and political) as the Roentgen business existed in at the time.[84]

---

80. "Auszug eines Schreibens aus Neuwied," 688.

81. The author's primary purpose was to increase recognition of the mechanic Kraus, who, in the author's view, was not sufficiently recognized by published reports in print media or by David Roentgen.

82. The list of his apprentices and cooperating firms (*fabriques*) in Neuwied from the year 1779 is in FWA 103-96-21. David Roentgen submitted the document to Count Alexander for tax reasons. A transcript of this document is in Fabian, *Abraham und David Roentgen*, 351.

83. Fabian, *Kinzing und Roentgen*, 23. An exception was the Parisian *ébéniste* Riesener who allegedly produced about seven hundred pieces of furniture for the Crown alone, obviously not all by himself. See Nemnich, *Tagebuch*, 230.

84. See my parallel observations for the Jaquet-Droz case.

The Clock-Maker and Mennonite Peter Kinzing

One important dimension of the making of the dulcimer-playing automaton was Roentgen's relation with the clock-maker family Kinzing. This family had a substantial history in Neuwied as well, and it intersected in crucial ways with the Roentgens'. The collaboration between David Roentgen and Peter Kinzing enabled them to produce high-end pieces that were ornamented and enhanced by mechanical machinery, unique and lucrative commodities on the luxury market of the international court culture. Most importantly for my concerns, the clockwork mechanism inside the dulcimer-playing automaton was built in Peter Kinzing's clock-making workshop in the early 1780s.

The first known clock-maker named Kintzing in Neuwied was Christian Kintzing, the father of the Peter Kinzing who became David Roentgen's collaborator. Christian Kintzing was born in 1707, son of a manorial miller (father and son Christian and Peter Kinzing used different spellings of their last name). Christian probably learned the milling trade from his father and may well have worked for a while as a miller. He founded his first workshop together with his brother Peter, an organ-builder and piano-maker. Their professions became manifest in their firm's products: clocks, musical instruments, and combinations of the two.[85]

The Kintzing brothers had been admitted into the Neuwied community of Mennonites in 1738, and they applied for permission to build a house and workshop soon afterward. After their application was approved in August 1740, they built a large house on grounds that had been a gift to them from the Count of Neuwied. Christian's brother Peter died in 1743, and Christian continued the business of making clocks, musical clocks, and musical instruments and repairing clocks.[86]

85. There were many clock-maker families in Neuwied at the time. In accord with Christian's and Peter's spellings, I spell Christian's last name with a *t*, and when I refer to both, I use Peter's spelling with a *z* only. Christian was first mentioned in the count's records in August 1740 in a building permit for his and his brother's house. FWA 26-9-3, sheet 103; also transcribed in Fabian, *Abraham und David Roentgen*, 320. The eighteenth-century traveler Joseph Gregor Lang reports that Christian was a miller who taught himself clock-making. Lang, *Reise auf dem Rhein*, 252. See Christian's professional description also in the record of his request for permission to get married, in FWA 95-1-8, Proclamationes et Copulationes, 1743, sheets 39 and 40, where he is called an organ maker ("Orgel-Macher"). The document is transcribed in Fabian, *Abraham und David Roentgen*, 406.

86. The August 1740 building permit for the brothers Kinzing is found in FWA 26-9-3, sheet 103. Various bills and receipts for deliveries and repairs for the count of Neuwied over the fol-

Over the following decades, until the early nineteenth century, several generations of the Kinzing family had clock-making and instrument-making workshops in Neuwied. Along with Christian's, which lasted from 1741 until 1785, brother-in-law Hermann Achenbach, who was initially Christian's apprentice, operated his own workshop next door from 1753 onward. In 1777 Christian's son Peter, David Roentgen's future collaborator, started another workshop, which lasted until 1816. Because he had his own house and workshop, he was able to set up a legal partnership on equal terms with David Roentgen in the 1780s.[87] From the third generation of Kinzings onward, the brothers Christian III and Carl ran a workshop during the last decades of the eighteenth and first decades of the nineteenth century, near David Roentgen's house.[88] That is to say, Kinzing clock-making workshops existed somewhere in Neuwied almost continuously between 1741 and 1838. In the last third of the eighteenth century, they employed about fifty people. The boundaries between the workshops must have been rather porous, especially for members of the family. Many employees in Christian's workshop were related either by blood or by membership in the Mennonite community. Some of the Kinzing employees were also employed at the Roentgen manufacture, as the two firms collaborated over a period of thirty years, mostly for the purpose of building elaborate wooden cases for clocks.[89]

Two religious groups thus shaped the economy of the town of Neuwied and played a significant role in its rise as a center of artisanship: the Moravian Brothers and the Mennonites. In a manner similar to that of the Moravians, the Mennonite community in Neuwied had established itself under princely privileges that had been issued in the late seventeenth and early eighteenth centuries by three generations of counts as attempts to rejuvenate and repopulate their territory. Friedrich III issued his first set of privileges to Mennonites in 1662.

---

lowing twenty years are found in FWA, Ausgaben 1744 no. 624; FWA, Ausgaben 1746, S 189, no. 956; FWA, Ausgaben 1747 no. 1045 (also transcribed in Fabian, *Abraham und David Roentgen*, 321); FWA, Ausgaben 1760, S 224, no. 1272 (transcribed in Fabian, *Abraham und David Roentgen*, 323; FWA, Renteirechnungen 1762, no. 1319 (also transcribed in Fabian, *Abraham und David Roentgen*, 324).

87. Until 1777, when Peter ran his own business, it must have been Christian, Peter's father, who was David's (and Abraham's) business partner. In the 1780s, the signatures on the clockworks and furniture suggest that the Kinzings were working jointly together, and Peter Kinzing was Roentgen's main business partner, even accompanying him on his long journeys.

88. Fabian, *Kinzing und Roentgen*, 47, 51, 62.

89. Ibid., 80.

Religious privileges continued to be a major attraction for religious migrants and sects in later periods.[90]

The Mennonite community in Neuwied was always smaller than the Moravian Brothers' community. Unlike the Moravians, for whom the organization and architecture of settlement was a part of their self-understanding, the Mennonites never had a fixed settlement or neighborhood in Neuwied. And even though Peter Kinzing's relationship to his religious group was, based on what we know, less fraught than David Roentgen's with his, there were important parallels between the Mennonites and the Moravians. Both groups refused to swear an oath to their ruler, both refused military service, and both embraced similar economic principles and ethics with respect to work, labor, prices, quality, profit, and trade.[91] And although in general each religious group was a closely knit independent community, in Neuwied there were economic and family relations between the two.

Peter Kinzing was about two years younger than David Roentgen. He was born in 1745 in Oberbieber, about five miles northwest of Neuwied. He learned the art of clock-making from his father and, from an early age, developed unusual skill in the design and construction of pendulum clocks, astronomical clocks, and flute and harp carillons. In this specialization, he was not unlike the young Pierre Jaquet-Droz. To build his elaborate clocks and musical devices, including pianos, Kinzing worked with mechanics, organ and instrument makers, and painters, and perhaps also with composers. He was a member of a thriving community in Neuwied: there were at least seven additional piano-makers and instrument-makers in Neuwied at the time.[92]

90. On post-Reformation religious persecution, migration, and mercantilism, see the literature cited above and Stürmer, *Luxus*, 20–28; see Lang, *Reise auf dem Rhein*, 244, on how religious groups coexisted peacefully in Neuwied. Bertola de' Giorgi's later travel report from 1796 mentions emerging problems, saying there had been claims of discord between the various religious groups. Bertola de' Giorgi, *Malerische Rhein-Reise*, 172–74. Friedrich III's privileges included general city privileges from 1662 and 1680 (FWA 26-8, transcribed in Fabian, *Abraham und David Roentgen*, 308); privileges concerning building permits, religious freedom, and tax relief for foreign immigrants 1733 (FWA 26-10-6); specific privileges for Protestant sects in 1739 (FWA 65-11-13); a general concession in the year 1747 that provided tax exemption for ten years to everyone who built a house in the city of Neuwied (FWA 80-2-4); a special privilege for the Freemasons in 1752 (FWA 61-6-2); and a privilege for the Augsburg branch of the Moravian Church (Brüder-Unität Augsburgischer Konfession) in 1756 (FWA 65-11-6). A complete list of privileges is also found in Tullius, *Die wechselvolle Geschichte*, 52–54. See also Fabian, *Kinzing und Roentgen*, 18; and Volk, "Peuplierung und religiöse Toleranz."

91. Fabian, *Kinzing und Roentgen*, 36–37; Correll, *Das schweizerische Täufermennonitentum*.

92. Among his collaborators were the clock-maker Anton Roetig, the carpenter and workshop warden of the Roentgen manufacture Johann Christian Kraus, the organ and piano maker Johann Wilhelm Weyl, and that man's brother Johann Christian Weyl, an organ and instrument maker. The painters Januarius Zick and Johann Juncker, who worked with Roentgen in his manufacture,

In its variety, skill, ambition, and size, Peter Kinzing's workshop was, if taken together with the other Kinzing family workshops, comparable to David Roentgen's. Kinzing substantially expanded his workshop in 1782.[93] This made his clock-making workshop the largest in Neuwied, with sixteen to eighteen employees. The extension of Kinzing's manufacture coincided with the period of Roentgen's success in the first half of the 1780s. It must have been around 1784 that the dulcimer player's mechanism was built in Peter Kinzing's expanded workshop. The supposed warden of Roentgen's workshop, Johann Christian Kraus, was also a supervisor for Kinzing. He may well have been the technical coordinator between the clock-making workshop and the furniture manufacture.[94]

Neuwied was an artisans' town characterized by an accumulation and concentration of skill, patronage, and resources similar to those features of La Chaux-de-Fonds. The situation in Neuwied was different in structure and in terms of the specific skills and resources featured, but Neuwied was similarly unusual and productive in its concentration of conditions that enabled extraordinarily productive artisanship.

The Dulcimer-Playing Automaton

We do not know many details about how the dulcimer player was built, but the production of furniture in Roentgen's workshop usually started with a drawing made either by Roentgen himself or his two painters, Januarius Zick and Johannes Juncker. From this draft was produced a technical drawing that indicated the exact proportions and sizes of the piece's components and its inner construction. Measures were taken for the many individual parts, and they were compiled in timber lists.[95] These lists went to the woodcutter, whose work required a great deal of experience and complete familiarity with the design of the finished

may also have worked as draftsmen in Kinzing's workshop. Fabian Dietrich reconstructs these relationships from bills, receipts, and church registers: Fabian, *Kinzing und Roentgen*, 78–80.

93. Peter Kinzing's first payment for his house and workshop (Bunte Straße 51) is in FWA 109-5-28 (Register über Grunzinß-Gelder 1778). Records for the enlargement of this house in October of 1782 are in FWA 80-2-13, Acta Grundzins: Peter Kinzing documented to the count the new parts of his house, and the count's administration recalculated the tax. Both documents are also transcribed in Fabian, *Kinzing und Roentgen*, 403, 415–16. See also Fabian's comments in Fabian, *Kinzing und Roentgen*, 69.

94. Fabian, *Kinzing und Roentgen*, 69, 78. On Kraus, see once again "Auszug eines Schreibens aus Neuwied."

95. Technical drawing was not systematized until the eighteenth century. Feldhaus, *Geschichte des technischen Zeichnens*, 14; Nedoluha, *Kulturgeschichte des technischen Zeichnens*, 64–69.

piece.[96] The wooden pieces were delivered to the actual joiner's work-shop. Here, the cases were built, the interior constructions put together, and the exterior surfaces treated by grinding and then polishing.[97] The final step was fitting all components together: mechanical interiors, locks and handles, and movable parts.[98] All mechanical interiors built into Roentgen furniture—clocks, moon calendars, pianos, flute and harp chimes, and so forth—came from the Kinzing workshop.

Inspiration for the dulcimer player's design may have come from several sources. The image and cultural practice of a piano-playing woman was widespread (as I discuss in chapter 4), and it was (liter-ally) embodied by aristocratic women including Marie-Antoinette her-self, for whom the automaton was made as a gift. There were paint-ings, educational tracts, and musical pieces for the specific audience of women who played keyboard instruments.[99] David Roentgen may also have been inspired directly by the Jaquet-Droz harpsichord player, al-though this connection is highly speculative. When Roentgen was on his second journey to Paris in 1779 (the one on which he received his first audience with the king and queen), the Jaquet-Droz firm's harp-sichord player had already been through the city a few years earlier and may have made a return visit.[100] The inspiration for the dulcimer player could also have occurred during Kinzing's time as an appren-tice, through an ancestor and relative of his. A contemporary source in fact claims that the dulcimer player was a copy of an original that had been constructed many years earlier by a man named Josef Möllinger, a descendant of one of the original three Mennonite families to settle in Neuwied and later a mechanic, piano-maker, and court clock-maker

96. Greber, *Abraham und David Roentgen*, 177–83; and engravings and descriptions in Boit, *Fassliche Beschreibung.*

97. Several apprentices worked together in this process. A separate technical drawing pro-vided the information for the selection, cutting, and coloring of the wooden pieces. Greber, *Abraham und David Roentgen*, 185.

98. Some of the engravings and drawings were taken from contemporary manuals and "text-books" for cabinetmakers, in particular Chippendale's widely circulated *Gentleman and cabinet maker's director* (first edition in 1754, second in 1755, third in 1762). Greber, *Abraham und David Roentgen*, 158.

99. The engraving on the title page of Stefan Zweig's biography of Marie-Antoinette, for example, is an oil painting entitled "Marie-Antoinette at the Clavecin." Depictions of Marie-Antoinette's body also played a key role in the politics of the French Revolution. See Lynn Hunt, "Many Bodies of Marie-Antoinette," 119, and the other essays in the same volume. The dulcimer player that Roentgen and Kinzing created could be considered one of these "many bodies."

100. Pierre and Henri-Louis Jaquet-Droz and their automata probably departed their home-town in the second half of 1774, and we know definitely that they were in Paris in January 1775. See chapter 2; Bachaumont, *Mémoires secrets*, 7:273, 282, from 2 and 9 February 1775.

at the court of Duke Christian IV in Zweibrücken. Peter Kinzing was related to Josef Möllinger through an uncle of Kinzing's, whose first marriage was to a sister of Peter's father. The source suggests that Kinzing copied the model when he was receiving training from Joseph Möllinger in Zweibrücken.[101]

Another look at David Roentgen's activities during the period when the dulcimer player must have been built, starting with his second trip to Paris, will illuminate more about his life as an artisan and entrepreneur and help to explain the position that the automaton occupied in his life.

The sales at the French royal court in 1779 made a significant contribution to Roentgen's business, but they were not enough to run the manufacture to capacity. For this reason, Roentgen went back to his old plan to explore the furniture market in eastern Europe. By this time he had learned how to tend to clients who lived far away and how to deliver furniture over long distances. He no longer needed to move his business to reach the eastern European market, as he had planned ten years earlier. Russia, the most important eastern European state, was the main target in his plans. Tsarina Catherine II had reigned over the Russian Empire since 1762. In the course of her regular correspondence with a wide range of leading intellectuals and politicians in Europe, she had heard of Roentgen and he had been recommended to her.[102]

In Russia, David did not face anything like the competition he had encountered in Paris. The journey was a risk, however, because of the distance, the freight of costly furniture, and the cultural and linguistic barriers. There was a small settlement of the Moravian Brothers in St. Petersburg, and he was hoping to receive support from them for his

101. This source is the same anonymous report on which I relied to understand David Roentgen's role in his manufacture and which challenges his role as his manufacture's master. It also challenges his role as the dulcimer player's inventor. "Auszug eines Schreibens aus Neuwied," 694. Other passages in the report on the dulcimer-playing automaton do not make it seem a reliable source of information, however. The passage describing the dulcimer player is vague, brief, and flawed, referring to it as "the piano that came to Paris and where the doll stepped forward and played" (694). This suggests that the author had likely never seen the automaton. Möllinger left Zweibrücken in 1770 to move to Frankenthal, where he died in 1772, which indicates that the duplication must have happened between 1760 and 1770, that is, before Kinzing was twenty-five years old. *Allgemeine deutsche Biographie*, 4:173–74; Fabian, *Kinzing und Roentgen*, 64, 71, 73–80.

102. Catherine was the epitome of an "enlightened despot." Dixon, *Catherine the Great*, 23–40, 64–87, 113–40; Donnert, *Katharina II*, 92–123; Alexander, *Catherine the Great*, 17–61. She corresponded with Frederick the Great, the emperor Joseph II, Voltaire, and the Baron von Grimm, among others. The latter mentioned Roentgen to her in one report and thus recommended him to her. Greber, *Abraham und David Roentgen*, 195. See also Grimm, *Lettres de Grimm*.

endeavor. This settlement did eventually became a temporary home for Roentgen.[103] He met Catherine for the first time late in 1783. The visit was a great success, since she bought his whole cargo of furniture that he had brought unannounced and uncommissioned. Catherine became his most important client, and he visited her residence again soon afterward, in March 1784, and several more times in subsequent years. Over the years, Catherine bought several hundred pieces of furniture from him.[104]

The production of the dulcimer-playing automaton must have coincided in part with David Roentgen's trips to St. Petersburg in 1783 and 1784. For late in the year 1784, he went on his next major journey to Paris, with a substantial cargo of furniture and the dulcimer player to present to the French court.[105] Marie-Antoinette was reportedly impressed with the automaton's technical and artistic precision. François Lassone, her physician, who presumably attended the presentation, wrote a letter to the Académie des Sciences about the automaton on 4 March 1785, describing its appearance and announcing the queen's plans to donate it to the Académie.[106] I discuss this correspondence at the end of this chapter.

Soon after the journey to Paris, in the fall of 1785, Roentgen embarked on another trip to Russia and remained there until the following spring. The next shipment of furniture from Neuwied, the largest and most valuable to date, arrived in St. Petersburg in March 1786. Roentgen went on a fourth trip to Paris in June 1787, and soon after that he prepared for his fourth trip to St. Petersburg.[107] He stayed there until the next year, delivering more furniture to other dignitaries in Russia.

---

103. The documents pertaining to the Brotherhood's settlement in St. Petersburg are in UA R 12. C. no. 11 a. The pastors of the community reported regularly to the headquarters of the Moravian Brothers in Herrnhut, and these reports are rich sources on Roentgen's whereabouts, his business efforts, and his concerns. See also Greber, *Abraham und David Roentgen*, 196.

104. In a contemporary report about the delivery of furniture to Catherine the Great, David Roentgen is described as a mechanic who was known "even in Versailles and Paris." See Meusel, *Miscellaneen artistischen Inhalts*, 24. See also Greber, *Abraham und David Roentgen*, 196.

105. Other furniture was also delivered to the king's brother and other dignitaries at the time. David's former apprentice Adam Weisweiler had himself been a *maître-ébéniste* since 1778, and David visited him. From Wille's diary we know the exact dates of this visit. See Duplessis, *Mémoires et journal de J.-G. Wille*, 2:109. See also Greber, *Abraham und David Roentgen*, 199; and Fabian, *Kinzing und Roentgen*, 131.

106. J. M. F. Lassone to the Académie, 4 March 1785, Archives of the Académie des Sciences, Paris. See also *Histoire de l'académie royale des sciences*, 1785.

107. On this Paris trip, see Duplessis, *Mémoires et journal de J.-G. Wille*, 2:146. The deliveries to Catherine's court were now specifically commissioned pieces, no longer large batches of "Roentgen's finest." See Huth, *Abraham und David Roentgen*, 57, document no. 60.

Roentgen also continued to deliver to his German princely and royal clients.[108] The deliveries to the Russian court in 1784 and 1786 were by far the largest and most valuable ones: Roentgen's manufacture was at the peak of its production capacity just when the dulcimer player was being built and a little afterward. The production of the dulcimer player had to have taken only a fraction of the production capacities of the Roentgen and Kinzing manufactures at the time.

The dulcimer player was produced in the context of David Roentgen's major court projects. The reason for his success in the court society of Old Reich Europe was that he was in a position to acquire land-owning clients who were becoming wealthier in this period and desirous of ornate furniture. Consumption at this level was not only about the luxury of a rich upper class, but also about the representation of newly obtained power and land. Michael Stürmer's studies of the economic conditions of the Roentgens' manufacture in the Holy Roman Empire demonstrate a significant consequence of the princely privilege: in exempting the Roentgens from guild regulation, it also forbade them to sell their furniture within the borders of the county of Neuwied, making the clients in the international court society their only available market. The demands of this group could hardly be satisfied by ordinary masters organized into guilds. In quite a remarkable way, David Roentgen, as a craftsman and an entrepreneur, existed at the threshold between old artisanship and new factory industry and between the court world of the old Reich and the emerging bourgeois society.[109] The dulcimer player was a product of that positioning, and the Jaquet-Droz family's harpsichord player was, too.

News about the French Revolution reached David Roentgen, and the employees and family accompanying him, in St. Petersburg during a trip in the late 1780s. The revolution had tremendous consequences for his firm. The entire clientele in Paris no longer was in a position to buy luxury items, so the Paris branch was left with no viable economic

108. Those clients included the new Prussian king Frederick William (he became king in August 1786; Roentgen had known him since he was a crown prince and had ordered one of Roentgen's large cabinet-desks in 1779 and more furniture in the 1780s), the duke of Saxe-Coburg, the duchess and duke in Weimar, the princes of Neuwied, and the count of Hesse-Rumpenheim. Greber, *Abraham und David Roentgen*, 208–10.

109. The court society was defined as those who legally, or at the verge of legality, were allowed to purchase "princely goods." That would be the monarch himself or herself, the princely family, the princes' mistresses, ministers, court employees such as secretaries, the aristocracy, and the patrician families in the Reich's Free Cities. Furniture was also an important complementary part of the architecture of palaces. Stürmer, *Luxus*, 8, 29–30.

basis. Business with the tsarina Catherine also came to an end. In December of 1790, and Roentgen and his entourage returned to Neuwied to a profoundly changed business. The Neuwied of the 1780s had seen the culmination of Roentgen's success and productivity, but in 1791 it was taken up by thousands of immigrants from France. The political revolutions in Europe, the loss of his best customers, and the demand from the Brotherhood to give up his "big" business all contributed to the decline of the business in the early 1790s.[110]

Reports on the Dulcimer Player

Whereas the Jaquet-Droz family's harpsichord player was first put on display in their hometown of La Chaux-de-Fonds and then likely taken on tours over the European continent and to London, Roentgen's dulcimer player was built to undergo one single tour and to be presented to one single audience: the French court. Therefore, fewer texts were written about the dulcimer player than about the harpsichord player. There is one letter about the dulcimer player written by Marie-Antoinette's physician François Lassone in March 1785 to the Académie des Sciences, and there is one article on the dulcimer player, which was printed in a German weekly newspaper in July 1785.

François Lassone most likely attended the presentation of the dulcimer player to Louis XVI and Marie Antoinette at the French court in January of 1785. His letter introduces the automaton by saying, "The queen recently acquired a small automaton woman figure that is about 18 to 20 inches high and that plays very beautiful arias on a dulcimer in the form of a small harpsichord."[111] He goes on, "This figure, whose features, proportions, and adjustments are very elegant, hits the different cords of the instrument in rhythm with two small metal hammers which she holds in her hands, and they move with a great deal of accuracy and precision." The letter then explains, "She has, by the way,

---

110. In the late 1780s, Roentgen tried to be readmitted to the Neuwied branch of the Brotherhood; see UA R. 7. G. b. 4., June 1787, Protokolle der Ältesten-Conferenz. The Council of Elders (Unitäts-Ältesten-Conferenz) suggested after Neuwied's Prince Alexander died in August 1791 that Roentgen find a settlement of the Brotherhood elsewhere in Europe or abroad. In October 1788, the protocol of the Council of Elders indicated that David had offered to cut down his business and to cut himself off from his business connections to dignitaries. The protocol is also cited and transcribed in Fabian, *Abraham und David Roentgen*, 368–69.

111. 1 pouce (thumb or inch) = 2.54 centimeters. J. M. F. Lassone to the Académie, 4 March 1785.

while she is playing the tunes, movements of the head and a varied expression in her eyes and gaze, which are very pleasant and a surprising illusion."[112] The final part of this description states, "She is seated on a chair that is mounted on a table made of marvelous wood." Lassone also mentions, after describing the automaton, that the machine was constructed "in Germany specially for the queen by skilled artists, who already made for the king of France a large secretary of amazing perfection, and other works."[113]

Lassone's letter provides to the secretary of the Académie an overall impression of the automaton's size, function, and presentation. It emphasizes not only the precision of the automaton's musical play but also the movement of her head and the varied expressiveness of her eyes and gaze, calling these a "surprising illusion." This observation is similar to the text about the harpsichord player that Pierre and Henri-Louis Jaquet-Droz printed in their brochure. Both texts are the first authoritative written records about their respective automata. Both describe a full display of a sentimental musical performance that is mechanically replicated through a woman automaton. Lassone's letter communicates to the reader how the automaton's musical play and bodily motions worked together to create a pleasant impression, and Pierre and Henri-Louis Jaquet-Droz likewise emphasized in their brochure the harpsichord player's musical play and bodily motions. In chapter 4 I discuss how the automata's mechanisms are designed precisely to yield this effect of musical-sentimental play.

Lassone's letter to the Académie also details a few facts about the dulcimer player after its presentation in the royal household. Lassone relates that the queen requested that the "automaton figure" (*cette figure automate*) be examined by a committee of the Académie and that, if the figure was found to be a worthy candidate for the Académie's machine cabinet, the queen would be inclined to donate it as a gift. Lassone himself, he says, has been authorized by the queen to contact the Académie so that they can conduct their investigation. Since the automaton is to be delivered to his house within the next few days, he invites members of the Académie to dine at his home the following week. He tells the academy, "It would suffice if you arrived at din-

112. I use the pronoun *she*, as the French text uses *elle*, in consistency with the paragraph's subject, *la figure*.
113. This descriptive passage contains one more sentence: "The table also supports the dulcimer, and the whole mechanism is contained and hidden in the table top." J. M. F. Lassone to the Académie, 4 March 1785.

ner time as the investigation of the machine takes only little time."[114] Lassone thus does not assume that members of the Académie will find the automaton so complex or novel that they would spend a great deal of time on it or that they would be unusually captivated by it. In the letter, dinner seems as important a social event as the examination of the automaton.

The dulcimer player received two more short mentions, in the proceedings of the Académie, both in 1785. In the *procès verbaux* of March 5, 1785, just one day after Lassone wrote the letter, the secretary reports that, in response to the letter, members have been nominated for a commission to inspect "an android that the Queen would like to offer to the academy's cabinet."[115] The *plumitif* of March 16, just a few days later, states that the four-man committee reported on the android that they saw at Lassone's house. The report was accepted and sent to Lassone.[116] And the *procès verbaux* of April 20, 1785, mention a letter read by the secretary about reasons that the delivery of the automaton donated by the queen was delayed.[117] Nevertheless, the automaton soon went into the collection of the Académie des Sciences, and this collection of machines and models was eventually given to the Conservatoire national des arts et métiers in 1807, which is where the automaton currently resides.[118]

The first and only printed article on the dulcimer player was published six months after the automaton was presented at the French court. It is one paragraph long and appeared in the *Dessauische Zeitung* in July 1785. The *Dessauische Zeitung* was a weekly magazine, founded and edited by the writer, editor, bookseller, and tutor Rudolf Zacharias Becker in Gotha in 1782. Becker was deeply committed to Enlightenment ideas of education, cosmopolitanism, and the dissemination of useful knowledge to a large population. His periodical and other publications were his instruments in this endeavor, and they were solidly grounded in the tradition of the *Moralische Wochenzeitschriften*

114. "Il suffiroit d'arriver seulement pour l'heure du diné, parce que l'examen de la machine n'exige que peu de temps." Ibid.

115. Procès Verbaux, 5 mars 1785, folio 46. The commission that the Académie put together to investigate the automaton consisted of four people: the natural historian Mathurin Jacques Brisson, the geologist Desmarest, the astronomer abbé Rochon, and the mathematician Adrien-Marie Legendre.

116. "Accepter et écrire à M. de Lassone." Plumitif, 16 mars 1785. The archives do not hold a manuscript of this report. It may have been an oral report.

117. Procès Verbaux, 20 avril 1785, folio 67.

118. The Académie's collection of natural history specimens was given to the CNAM in the same year, and another round of similar dispersion occurred in the year 1824. Saunders, "Archives of the Académie des Sciences," 700.

("moral weeklies"), which I discuss in chapter 4.[119] The articles in the *Dessauische Zeitung* were mostly short, one paragraph or so, and they reported political news and cultural gossip from the German lands and abroad. Common datelines included Paris, Berlin, Vienna, and Brussels.[120] In all regards, Becker's weekly was a common serial of its kind in its time.

The article about the dulcimer player appeared under "Miscellaneous news," a section found at the end of each issue, and was sandwiched between news from Madrid about the Inquisition and a brief report about the invention of a new mill by a miller from Silesia.[121] The article was headlined "Neuwied," and stated:

The famous mechanic, the *Commerzienrath* Roentgen, has, in cooperation with the royal French clock-maker Peter Kinzing, built a work of art, a description of which our readers will appreciate.[122] It is a figure of a woman (*Frauenzimmer*)[123] that is about 18 inches (*Zoll*) high, which is called Mamsel Du Bois by the artist.[124] The figure is seated in front of a piano or grand piano[125] and plays eight melodies with the help of two hammers that it is holding in its hands, one in the right and one in the left hand.[126] The *Mamsel* not only obeys greatest accuracy in musical rhythm; she also replicates (*nachahmen*) all other motions with her head, eyes, and hands in a very natural way. Before she starts playing a tune, she pays the audience a rather polite compliment [bow], during which the motions of the head, hands, and eyes are in best harmony with one another.[127] The motion of the eyes in particular suits

119. *Allgemeine deutsche Biographie*, 2:228; Schulz, *Rudolph Zacharias Becker*.
120. See *Dessauische Zeitung*, 1785, 1:35; 11:81; 30:234.
121. *Dessauische Zeitung*, 30:239–40.
122. *Commerzienrath*: councillor.
123. The slightly pejorative term *Frauenzimmer* for "woman" is a term that Johann Richter ends up using in his satire "Humans are machines of the angels," a text that I analyze in chapter 5. Richter does not use the word regularly in any of his other works at the time, so it is likely that he picked it up from this article.
124. This naming is reported nowhere else and is a unique and odd reference. The author either picked it up from another source (unknown to me) or fabricated this detail to make the article more humorous and more appealing. *Mamsel* is a German diminutive and short version of *mademoiselle*. The grammatical gender of *Mamsel* is feminine, so the corresponding pronoun is "she."
125. This is very inaccurate: with the hammers, the dulcimer player would never be able to play an ordinary piano, since she has no fingers. The author might be mixing up details from the dulcimer player and the harpsichord player.
126. *Frauenzimmer* is a neuter noun in German, and the corresponding pronouns are neuter, too.
127. The dulcimer player's bow is also an incorrect detail: the dulcimer cannot bow, as we will see in its mechanical analysis in chapter 4. The Jaquet-Droz family's harpsichord player, made in Switzerland, *does* bow, though, and it seems likely that details of the two women automata become mixed up in this article, even though the article does not specifically mention either the Jaquet-Droz family or the harpsichord player.

the expression of each melody: slow and alert during an adagio, and joyous and animated during an allegro; in short, the figure seems to be feeling (*empfinden*) the stimulus (*das Reizende*[128]) of her own music herself.[129]

The one-paragraph article closes with one more sentence: it says the value of the piece of art had been estimated by the Royal Academy of Sciences to be 800 *Louis d'or* and that the queen had presented it to the Academy as a gift.

This article is interesting for several reasons, most of all because it adds key details that were not present in Lassone's letter: it mentions the number of melodies that the dulcimer player can play ("eight"), it tells the automaton's supposed name ("Mamsel Du Bois"), it gives more specific details about the automaton's eye movements and their correspondence to the music, it estimates the automaton's monetary value, and it (incorrectly) states that the automaton performs a bow before her play. These details suggest either that there was an additional source of information, perhaps another article or some other conduit of gossip, available to this text's author, or that the author simply invented some of the additional details (such as the name) to make this short article more humorous and more engaging. Other details, such as the mistake about the bow, possibly another effort to make the account more vivid, might stem from a confusion of the dulcimer player with the Jaquet-Droz family's harpsichord player.[130]

The article in the *Dessauische Zeitung* was printed about four months after Lassone sent his letter to the Académie. It would be reasonable to assume that the magazine's editor, Rudolf Zacharias Becker, was the article's author, as was common practice in periodical publishing at the time. At least one can assume that Becker had far-reaching control over the contents of his magazine. In either case, the article does not indicate with certainty whether its author actually witnessed the automaton playing music. The restricted conditions of presentations at the French court and the faulty details in the article suggest that he did not.

---

128. *Das Reizende* could be translated as "the appeal," too, and the text works well with this ambiguity.

129. "Kurz, die Figur scheint das Reizende ihrer eigenen Musik selbst zu empfinden." *Dessauische Zeitung*, 30:240. As I discuss in chapter 4, the eyes actually move merely through inertia (the head moves "around" the eyes and thus creates the impression of moving eyes). The observation that the eyes move differently with different types of tunes is an optical illusion.

130. It would be interesting to locate and identify the individual sources of the erroneous details of the dulcimer player's performance, but I have found no trace of any other account or text.

Even though some of the details that the author includes are flawed or fabricated, the account does provide one crucial piece of evidence for my study's analysis of women automata's role in the Enlightenment's sentimental culture: the article makes the critical connection between the musician's moving eyes and her feelings, a connection that I explore in depth in chapter 4. The key sentence in this regard is the penultimate one (the quote above ends with it): "in short, the figure seems to be feeling the stimulus [or the appeal] of her own music herself." In a matter-of-fact, even casual way, the author states that the eyes' motions tell us how the figure feels the charm of her own music-making. Lassone's letter had merely talked about "movements of the head and a varied expression in her eyes and gaze" that made a "pleasant and a surprising illusion." The article in Becker's *Dessauische Zeitung* thus goes further and assumes that the automaton's moving eyes tell the spectator that the automaton is moved and stimulated by her own musical play and that, crucially, the automaton's feeling is *visible* to the spectator through the mechanical motion of her eyes. The sentence also assumes in a rather offhand way ("in short") a familiarity on the reader's part with this connection between the body and eye motions of a musician, on the one hand, and the musician's inner sentiments, on the other, suggesting that the musical-sentimental scenario that the dulcimer player exhibits was familiar to the general reading public at the time. In this context it is important to note that while Rudolf Zacharias Becker was committed to all the central tenets of the European Enlightenment, he was particularly interested in its sentimental dimension.[131]

I discuss in chapter 5 how the young German poet Johann Richter took up, through this text, the connection between a woman musician's body and eye motions and the sentiments that the woman harbors and explores this connection literarily in a 1785 satire entitled "Humans Are Machines of the Angels." The article in the *Dessauische Zeitung* is the one account that we know with certainty Richter read. Richter himself also most likely never observed either of the two music-making women automata. However, both the automata themselves and the very early texts about them made their moving heads, eyes, and sentiments prominent enough that these details could travel through the textual and cultural worlds of the European continent and get

131. His key commitments included cosmopolitanism, utilitarianism, sentimentalism, and being a libertine. *Neue deutsche Biographie*, 1:721.

picked up by Richter in his own poetic account of automata and sentiments.[132] After the *Dessauische Zeitung* article, there were no further descriptive texts about the dulcimer player in popular media.[133]

Conclusion

The dulcimer-playing automaton was a local product of an international business. As with the harpsichord player, this recognition helps put the reasons the automaton was built in perspective, and it helps make both women automata part of specific universes that had local and international elements. Both Neuwied and La Chaux-de-Fonds were unusual artisans' towns with an unusual accumulation of skill, patronage, and resources. They differed in structure and in the kinds of skills and resources they featured, but they both enabled extraordinarily productive artisanship. The current chapter and the previous one provide a juxtaposition of the Jaquet-Droz and the Roentgen manufactures and bring into focus the parallel and differing historical conditions that rendered automaton-making possible in the eighteenth century. Looking at the two automata through the lenses of rural Switzerland and a small Rhineland principality illuminates crucial details about how and by whom they were conceived, how and by whom they were constructed, and what goals they were supposed to accomplish. Both automata required a tremendous amount of resources in their respective workshops. Furthermore, the two women automata occupied a variety of places on their respective creators' agendas; they were not always at the top. This fact calls into question the assumption that spectacular

132. Richter excerpted in his notebook a passage from the article of the *Dessauische Zeitung*. It reads: "The mechanic Roentgen made in Neuwied a woman (*Frauenzimmer*) that plays eight melodies with two hammers, and pays a compliment to the audience before, and moves the eyes slowly during an *Adagio* and fast during an *Allegro*." See Müller, *Jean Pauls Exzerpte*, 159. This is a greatly reduced excerpt, but it is accurate and in places verbatim (see complete quote above).

133. For the sake of completeness, I want to mention a brief passage from the anonymous report about Roentgen's furniture manufacture on which I have relied several times in this chapter. The description of the dulcimer player in this report is, as I have said, brief and faulty (see note 101), but when discussing the dulcimer player's invention, the report mentions a "safe journal" in which the dulcimer player's invention was ascribed to David Roentgen: "ein Werk, das in einem sichern Journal vor Original des Herrn von Roentgen ist zugedacht worden." "Auszug eines Schreibens aus Neuwied," 694. We do not know whether this "safe journal" was the *Dessauische Zeitung*, or whether there was another printed article about the dulcimer player in the 1780s. It uses the singular, so that there is at most one other reference that this report's author had in mind.

objects such as android automata were the chief devotion of artisans and businessmen in eighteenth-century Europe.

The reasons artificial humans were built in the eighteenth century exceeded the desire to simulate human bodies mechanically or to demonstrate their mechanical constitution. I found equally important cultural and socioeconomic conditions that made possible mechanical replicas of human bodies—or even required them, from a business viewpoint. As chapter 4 will show, the artisans that I consider *did* have an ambition to replicate specific human qualities in their machines, but not the human body's anatomy or physiology. Rather, a comprehensive cultural practice specific to the late eighteenth century—the expression of sentiments by keyboard-playing women through bodily motions—is reproduced in these automata, and it was unusually advantageous economic conditions that enabled the artisans to actualize this idea. The remarkable mechanical deliberation with which the Jaquet-Droz family and Roentgen and Kinzing designed and built their automata to perform precisely those bodily motions that were crucial in the making of eighteenth-century sentimental society will be my concern in chapter 4.

The strategies of the Jaquet-Droz and Roentgen families to furnish their automata with mechanical motions of the upper body in addition to the ability to play music, and thus to create a comprehensive cultural scenario, paid off. This chapter and the previous one have shown how, in crucial texts, these additional mechanical details received significant space and elaboration: the harpsichord player's ability to move her head and torso is noted often and prominently, and the dulcimer player's moving eyes are a central point in Lassone's letter and, more so, in the article in Becker's periodical *Dessauische Zeitung*. These mechanical and cultural details subsequently "traveled" well through the eighteenth-century landscape of correspondence and popular print media. One can follow their leads through newspapers and periodicals, and I even found how details about the two automata were confused with each other because of their similar design. Moving bodies and expressive eyes were a key part of the automata's performances as well as in the sentimental body techniques of eighteenth-century musical culture. The automata were designed and built at a time when a critical mass of literate people were familiar with the culture of sentimental music-making and were able to comprehend the idea that a musician's sentiments were visible in his or her bodily motions. The Jaquet-Droz and Roentgen families thus produced objects that did not perform something new to an audience, but rather something familiar: an already existing cultural and social technique of sentimentality.

The two women automata were mentioned in a variety of texts and used for a variety of purposes. Comments about them were formulated and published in recurrent patterns, and they served specific economic functions in the worlds of textual production in the late eighteenth and early nineteenth centuries. This holds in particular for the Jaquet-Droz family's harpsichord player. I discussed short notes that witnesses of the two automata wrote in diaries and letters, and I showed how these were documents of contemporary practices of watching publicly staged mechanical spectacles and socializing during these events. The accounts on the women automata fit neatly into the general environment of spectacles and textual production, and there is, in this regard, nothing specific about the women automata vis-à-vis other gossip, novelties, anecdotes, spectacles, or natural philosophical observations. Both the automata themselves and texts about them were parts of existing networks of socializing, watching, listening, copying, translating, and paraphrasing. It was only later, in the nineteenth century, that automata increasingly became the subject of general debates on the mechanical constitution of selfhood, the creation of artificial life, or the threat of a scientist or mechanistic worldview. The wonder or concern that we might expect contemporaries to have expressed about automata is not brought up until later, more industrial periods, as I discuss in chapter 6.

Spectacular automata such as my two women were perfect fodder for the eighteenth-century publishing industry, which was in need of a never-dwindling supply of gossip and curiosities. The way texts were composed and used implies that the publishing industry did not need precise or detailed information—and that held for texts on automata as well. There is a curious coincidence of artisanal, preindustrial production of unique and elaborate technical artifacts, on the one hand, and an industrial publishing world whose markets and audience needed mass-produced texts, on the other. Historians of media have explained that the substantial expansion of watching, listening, socializing, reading, writing, and publishing in the second half of the eighteenth century occurred not on the basis of new technical means (unlike, for example, the invention of book-printing or the Internet). Rather, the basis was a shift in consciousness and self-understanding of the people participating in public communication.[134] In the age of the

---

134. Fischer, Haefs, and Mix, "Einleitung," 9.

Enlightenment, there was for the first time an idea of the power and efficacy that media possessed for social and political change. The texts on the women automata were not among the intellectual engines of this change, but the ways in which they were composed and published shows the inextricable link between the production of individual texts and the machinery of textual production at the time.

There were other texts in the economic and literary landscape of the time that engaged the automata in deeper, more detailed, and more deliberate ways. They asked questions about the automata more explicitly, used them as motifs, and made them part of far-reaching intellectual and poetic agendas. They used piano-playing women automata as examples and backdrops to create complex stories about encounters between humans and their mechanical counterparts. Understanding these agendas and understanding what piano-playing women automata became part of, and were used for, is the purpose of my exploration of two literary texts in chapter 5.

FOUR

The Design of the Mechanics; or, Sentiments Replicated in Clockwork

The artisan environments in Switzerland and the Rhineland that produced the two women automata were at the peak of their productivity in the 1770s and 1780s and thus nurtured the creation of the innovative and complex clockwork systems that enabled the automata to perform their nuanced musical and gestural programs. The mechanics inside the two automata move the figures' arms, heads, and torsos and coordinate the individual motions into elaborate choreographies. Both automata's bodily motions go well beyond the hand, arm, and finger movements necessary to produce the music. The automata also move their eyes, heads, and—in the case of the harpsichord player—torso to watch their hands' motions on the instrument, to nod, to breathe, or to bow. Father and son Jaquet-Droz, and Roentgen together with Kinzing, built mechanisms that were significantly more complex than would have been necessary for "mere" music-playing androids.

Musicians' bodily motions had a larger significance at the time, because they were intended in eighteenth-century music-making to communicate affects and sentiments from the musician to the audience during a recital. Such attempts to cultivate and share affects in social relationships were part of further-reaching ambitions in the "sentimental" age in Enlightenment Europe: political and cultural vanguards sought to establish a new type of social order—civil society—to replace the traditional estate society. Sentimental sociability was a key element in the conception and creation of this new social order, and it was practiced in such diverse settings as music-making, the reading and writing of literature, friendship, travel, letter-writing, scientific inquiry, child-rearing, and theorizing about marriage.

Pierre and Henri-Louis Jaquet-Droz and Roentgen and Kinzing deliberately designed their automata to conduct specific, ambitious, and elegant performances. They did everything that was mechanically necessary to accomplish this. If they had wanted merely a harpsichord-playing or dulcimer-playing android, they could have left out many of the mechanics contained in their women automata. Instead, they invested additional resources to design motions that captured spectators' attention, motions that carried significant and widely known social and political meaning during the transformations of Enlightenment Europe. In chapters 2 and 3 I discussed how these efforts paid off: texts in the eighteenth century mentioned the automata's full range of bodily motions. That is, the motions were "visible" in the texts, but also in each automaton's mechanics, as this chapter will demonstrate.

Eighteenth-century artisans tended not to leave behind large numbers of written documents, but their material objects themselves serve as important sources. I investigate the two automata's mechanics in this chapter in two ways. First, I analyze their clockworks to uncover correspondences between individual mechanical modules and the effects they bring about in the moving android. This analysis also gives insight into other parts of the automata's history, such as the exceptional skill that went into designing and building them, key similarities and differences between the two, choices that the artisans made as to where to devote most of their technical skill and effort, conflicts between artistic and mechanical principles that the artisans solved, and social and economic functions that the automata served in their makers' lives. Second, I explore the automata's mechanical performance as a cultural scenario in its own right, with its own history in the eighteenth century. I examine this scenario's connotations and valences in three contexts: I look at music-making women in the eighteenth century, then at the role of sentiments in music, and finally at the general social and political relevance of sentiments. The question of keyboard-playing women who move their bodies was, after all, a question of moral philosophy, aesthetics, and (marriage) politics in the eighteenth century.

The Automata's Mechanics

The Jaquet-Droz family's harpsichord player is an almost life-size figure of an approximately fifteen-year-old girl seated on a stool in front of a small organ (see fig. 1, page 3). The seated figure is about four feet high and is placed on a flat pedestal.[1] The instrument that she originally played was called a *clavessin organisé* or *clavecin organique*, a combination of harpsichord and organ that was primarily developed and used in the 1770s and 1780s in Vienna.[2] Such combined instruments were developed to amplify the sound in musical performances in domestic, semipublic, and public spaces. Father and son Jaquet-Droz may have chosen this instrument for the same reason. There are no eighteenth-century images of the harpsichord player, but nineteenth-century drawings suggest that the automaton was playing either a small organ or an organ-related instrument.[3] The harpsichord player's repertoire consists of five pieces that were most likely composed by Henri-Louis Jaquet-Droz. Their formal structure (two stanzas and da capos) as well as their embellishments follow contemporary French and Italian fashion.[4]

Roentgen's dulcimer player is arranged on a tabletop (see fig. 2, page 4). The figure is about one foot and nine inches high and supposedly depicts Marie-Antoinette, who was thirty years old when the automaton was presented to her. The instrument itself is large in relation to the figure: two feet and four inches long and one foot and two inches wide.[5] It consists of a trapezoid-shaped sound box with twenty-

1. A documentary video on the Jaquet-Droz family and their automata shows the harpsichord player playing her instrument and three couples (human adults) performing contemporary court dances, dressed in contemporary court clothes. The scene takes place in a room that is furnished like an eighteenth-century manorial building. See *Les androides Jaquet-Droz*.

2. The French terms were used in texts from the 1770s and 1780s to describe the android's musical instrument. Maunder, *Keyboard Instruments*, 6–9, 110; Hubbard, *Three Centuries of Harpsichord Making*. It is most likely that the kind of instrument she plays changed over time from a harpsichord to an organ.

3. See, for examples, the images in Beyer, *Faszinierende Welt der Automaten*. Nowadays, when the automaton is presented to visitors in the Musée d'art et d'histoire at Neuchâtel, she plays a small organ that is furnished with a clockwork mechanism and two bellows to make the organ flutes produce the sound. The instrument was especially constructed for the automaton, although it is unknown when it was made and for how long it has been used. See Carrera, Loiseau, and Roux, *Androiden*, 69–71, 83. A woman keyboard player, supposedly made by the Maillardet family in the early nineteenth century, was represented in images as a harpsichord player. Ord-Hume, *Clockwork Music*, 42.

4. Complete analyses of the pieces are found in Carrera, Loiseau, and Roux, *Androiden*, in a separate, unpaginated booklet that accompanies the record.

5. Inventory of the Conservatoire national des arts et métiers. The dulcimer player has the inventory no. 07501-0000.

three pairs of metal strings stretched over it. The automaton strikes the strings with two hammers, which are curved into a hook shape at their end. The instrument's front end is shaped like a harpsichord to facilitate the automaton's playing.[6] The strings are parallel to the automaton's forearms, with the bass notes arranged on the left and the treble on the right, in the same way as a piano keyboard. The automaton's repertoire comprises eight melodies, seven of which are unidentified simple pieces in two voices; one is an aria from Gluck's "Armide." The overall ensemble, consisting of the comparatively small musician, her large instrument, and the table, is about four feet wide, two feet deep, and four feet high. This makes its overall height the same as the harpsichord player's, so that both are at the eye level of a seated spectator.[7]

While the two women automata replicate similar cultural scenarios, there are differences between them in their size and geometry, in the degree of mechanical complexity of their inner machinery, and in the functions they served for their respective makers. The automata were designed around two different types of motion and two different types of musician-instrument interfaces and within two different sets of business relations. The dulcimer player was the inspiration of a cabinet-maker and was meant to promote his economic interest, not that of a clock-maker. This is manifest in the dulcimer player's design as a music-playing piece of *furniture* that happens to take the form of a woman automaton. Father and son Jaquet-Droz, in contrast, built a complex piece of *clockwork* in the form of a woman playing the harpsichord.[8] The dulcimer player, furthermore, was produced to be presented to Marie-Antoinette, similar to the way David Roentgen presented many other pieces of furniture to her. Soon after she received the automaton, she gave it to the Académie des Sciences, and it was probably never shown in public until the twentieth century. The Jaquet-Droz family,

6. Normally, dulcimers can take any shape of a rectilinear or trapezoid type. The range of the automaton's two hammers describes two semicircles, and the location of the hammers' touching the string is not variable. A rectilinear keyboard is geometrically the easiest way to make sure that the player correctly strikes the strings.

7. My account of the harpsichord player's mechanism relies in most details (unless otherwise stated) on the account in Carrera, Loiseau, and Roux, *Androiden*, as well as on conversations with the curator and restaurator at the Musée d'art et d'histoire in Neuchâtel. My account of the dulcimer player's mechanism relies on conversations with Madame Aung Ko, curator at the Conservatoire national des arts et métiers in Paris, and in most details (unless otherwise stated) on the account in *La restauration musicale*, 30–32.

8. At the time, the dulcimer was not as commonly played as the organ or the harpsichord. Roentgen's choice of instrument is peculiar, although it would be plausible to assume that he chose it because the instrument required less complex mechanics than would be needed for an organ or harpsichord player.

in contrast, presented their automaton in public and traveled with it for a while, before exhibiting it for a few years for charity purposes at hospitals in Geneva and La Chaux-de-Fonds and eventually selling it.[9]

Each automaton produces a captivating scenario of music-making. The harpsichord player does so on the basis of a life-size performance and realistic motions of arms, hands, and fingers, while the dulcimer player's primary appeal resides in her miniature effect and the contrast between the instrument's larger size and the small figure that minutely and precisely plays on it. The furniture surrounding the dulcimer player creates this miniature effect and emphasizes the delicacy of her musical play and motions. The music that the automata play is, in each case, a two-voice dance piece standard for the time from the genre of the suite, which had its origins and exemplary form in the period of seventeenth-century baroque.[10] The musical compositions came out of preexisting expertise in the two workshops: Christoph Willibald Gluck had worked with Roentgen before, and Henri-Louis Jaquet-Droz had studied music and design for a few years in Nancy just before returning to his father's house and helping build the harpsichord automaton.

The harpsichord player is technically and artistically more sophisticated than the dulcimer player. She is bigger, has more movable parts, and has a larger gestural and musical repertoire. The hammer-wielding dulcimer player does not require the same minute technical detail, since she needs no finger-moving mechanics. She also does not move her upper body. The most distinct moving parts apart from her arms, as pointed out in the letter from Marie-Antoinette's physician to the Académie des Sciences, are her head and eyes.

Among the most important similarities between the two automata are the modules that conduct their musical play, a studded barrel and a camshaft. In both automata these two modules are combined in typical fashion by cutting the studded barrel into two half cylinders and putting the set of cams in the middle (figs. 4 and 5).[11] The two mod-

9. See chapters 2 and 3.
10. Pearl, "The Suite in Relation to Baroque Style."
11. The combination of studded barrel and levers was an established technique at the time, applied in water mills and hammer works. See Protz, *Mechanische Musikinstrumente*, 16; Ord-Hume, *Clockwork Music*, 17; Bedini, "Role of Automata," 35. Until the first half of the eighteenth century, barrels were pegged in such a way that the barrel could revolve only once for one musical piece. Athanasius Kircher already in the seventeenth century used the technique of moving the barrel by the width of one stud row after a musical piece to allow more than one musical piece to be encoded on the barrel. See his *Musurgia universalis*, 320, and a reproduction of that page in Protz, *Mechanische Musikinstrumente*, 9. For an image of a barrel-maker's workshop, see Simon, *Mechanische Musikinstrumente*, 2. Vaucanson developed (and patented) in 1738 a way to peg

FIGURE 4 The harpsichord player's back and her mechanism: pegged barrel and cam shaft. Photograph by Adelheid Voskuhl.

ules execute two different types of motion. While the cams direct the forearms' horizontal, rotational motion and thus move the hand to the correct position above the keyboard, the studs on the barrel conduct the arms' and hands' vertical motion to produce a tone, in one case making the harpsichord player's finger press a key and in the other making the dulcimer player's hand release the hammer to hit a string and set it into vibration.

The cams and the studs, using different types of mechanical gears, encode different types of information, which are also decoded and transmitted in distinct ways. The cams, on the one hand, have encoded in their profiles information about the pitch, guiding the selection of the correct key or string to be played. As the cams revolve around their axis, levers positioned over the cams' edges (the "cam followers") function as readers and decipher the information on the cams' profiles, which consist of continuous lines of crests and valleys. In the case of both automata, the lever's motion on the cam's crest is transformed into a rotation of the upper arm's vertical axis. The resulting rotation in the elbow causes a sweeping motion of the forearms above the harpsichord's keyboard or the dulcimer's strings. Where the cam's profile is

the barrel in a helical way so that one musical piece could last several revolutions of the barrel. Wolf, *Kaufmanns Trompeterautomat*, 16. See also Riskin, "Defecating Duck," 605.

FIGURE 5 The dulcimer player's pegged barrel and cam shaft. © Musée des arts et métiers-Cnam, Paris/photo P. Faligot.

more inclined, it produces greater motion; where it is flat, the motion stops (fig. 6). The ratio between the motion on the profile and the motion of the forearm is largest for the dulcimer player at 1 to 10, which means that a vertical displacement of 1 mm on the cam is translated into a lateral displacement of the hammer of 1 cm.[12]

The studs on the surface of the barrel, on the other hand, encode information about the music's rhythm, that is, when and for how long the finger hits the key or the hammer the string.[13] The studs direct the

12. Carrera, Loiseau, and Roux, *Androiden*, 52; Chapuis and Droz, *Les automates des Jaquet-Droz*, 23; *La restauration musicale*, 30–32.

13. The "how long" strictly speaking only holds for the organ, where the size of the studs determines the length of time the key is pressed. For the case of the dulcimer, the distance between two studs—rather than the actual studs' size—determines not how "long" the receptive

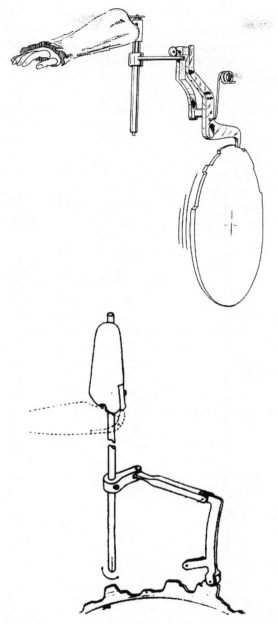

FIGURE 6 Cam followers "read" the harpsichord player's and the dulcimer player's cams (harpsichord player above, dulcimer player below). Carrera, Loiseau, and Roux, *Androiden*, 74; *La restauration musicale*, 17.

downward motion of each of the harpsichord player's fingers to press a key or the dulcimer player's hand to make the hammer strike a string (fig. 7). Unlike the continuous motion of the cams, the studs translate only into two discrete states: motion and rest. Fifty rows of studs, twenty-five on each half cylinder, conduct the harpsichord player's fingers' motion (fig. 4), and five levers on both sides (discernible just above each half cylinder) "read" five individual sequences of studs to control each finger. The cylinder automatically moves sideways by the width of one row of studs after each musical piece and thus makes possible the change of melody and the automaton's repertoire of five musical pieces.[14] The dulcimer player has sixteen rows of studs, eight on each side, one for each melody. This mechanism can play more melodies than the harpsichord player's, with fewer studs and only one lever on each half cylinder, because there is no need to direct individual fingers' motions.

The motions transmitted by the profiles on the cams and the studs on the cylinder combine their effects in the automata's respective elbows, guiding in tandem the forearms' horizontal and vertical motion and making the musical play possible (fig. 8). The cams prepare the correct position above the correct string, and the studs make the finger press the key or the hammer strike the string. It is crucial that the studs be exactly synchronized with the flat part of the cam, so that the finger or hammer moves downward only when the forearm is at rest and in the correct position.

The art of clock-making and mechanical engineering in the early modern period consisted at its core of transforming an engine's original motion (in the case of the automata, a spring) into other types of motion.[15] The two systems used here, spring-cam-follower and spring-cylinder-peg-reader, were two distinct eighteenth-century ways of solving this problem, as they transformed the circular, continuous motion of a spring into two other distinct types of motion: a linear,

---

string is hit but when the next string is hit, so that this, ultimately, determines the length of an individual tone.

14. Two steel rods on top of the cams move the cylinder sideways while it is turning, and thus make possible the automatic change of melodies. See note 11 and Carrera, Loiseau, and Roux, *Androiden*, 74.

15. In the history of clock-making and engineering before the late eighteenth century, the spring drive and the weight-driven clocks were by far the most often used devices. After the development of the spring drive in the seventeenth century (by Robert Hooke and others), weight-driven and pendulum clocks became more specialized gadgets of running mechanical objects (for precision measurements, for example), while the spring became the most versatile tool for running small mechanical machinery such as clocks or music boxes. Protz, *Mechanische Musikinstrumente*, 23–25; Buchner, *Mechanische Musikinstrumente*.

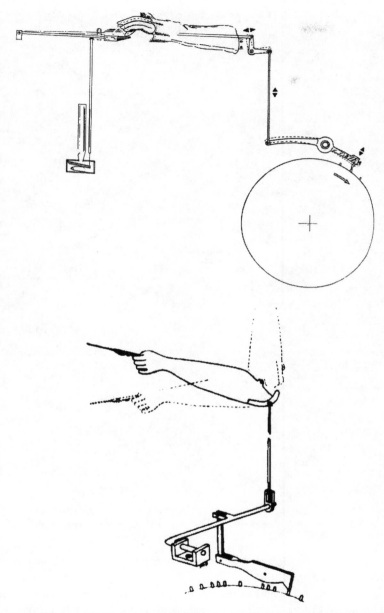

FIGURE 7 The harpsichord player's and the dulcimer player's pegged barrel and relay motion for hands and fingers (harpsichord player above, dulcimer player below). Carrera, Loiseau, and Roux, *Androiden*, 74; *La restauration musicale*, 17.

FIGURE 8 The motions encoded on the cams and the studs combine in the automata's elbows (here shown for the case of the dulcimer player). © Musée des arts et métiers-Cnam, Paris/photo P. Faligot.

discontinuous motion of the automata's forearms above the keyboard and a discrete up-and-down motion of the arm or finger. In the former case, the transmission functions through a system of rods, axles, and hinges and in the latter case through a set of relay motions. The combination of barrel and cams, as well as the art of transmitting and transforming motion via lever-hinge-axle systems, was by the mid-eighteenth century well known to specialist mechanics.

Eighteenth-century literature on clock-making and mechanical engineering reflected and transmitted these principles. As the mechanical trades began to become systematized, the speed and depth of this modernization varied widely over Europe. Master clock-makers and scholars developed standards of training, formalized knowledge and skills, and wrote reference works for establishing and disseminating knowledge. There were treatises dealing with clock-making principles, instructions for designing clocks, and guides for repairing them. Some texts combined instruction in the mechanical arts with histories of the trade. While the centers of clock-making and writing about clock-making were in France and England, key works were translated into other languages and edited with comments.[16]

Some of the complex machinery in the Jaquet-Droz and Roentgen families' automata stemmed from innovations specifically in the trade of clock-making, while other elements came from smaller and more specialized branches of the mechanical arts, such as the making of mechanical musical instruments or the mechanical interiors of luxury furniture. Musical mechanisms included flute-playing clocks (containing not flutes, per se, but rather flute-toned organ pipes), clockwork spinets, carillons of various sizes (from those for cathedrals or town halls to precious miniatures), music-playing automata, tobacco boxes, pendulum clocks, and singing-bird devices. Most musical devices followed the model described above: a stored musical program, an engine, and a sounding device. Between the engine and the program, there was normally a mechanical system to read and transmit the information, and between the program and the sounding device there was a mechanism that received signals and transmitted them to the sounding device. The standard device to provide the program was a barrel with some form of projections (normally studs and blocks) that corresponded to the notes to be sounded, their relationship to each other, and their duration, as in the case of both women automata. The standard mechanism to provide the transmission was some kind of a key frame that was positioned so that the barrel projections would raise a small lever, which would oper-

---

16. The principal example for the general ambition of disseminating artisan knowledge (together with all other knowledge) was Diderot and d'Alembert's *Encyclopédie*. Among the key treatises on clock-making in the first half of the eighteenth century were Sully, *Regle artificielle du tems* (1714); Le Paute, *Traité d'horlogerie* (1755); Thiout, *Traité de l'horlogerie* (1741); Derham, *The artificial clock-maker* (1696); Berthoud, *Essai sur l'horlogerie* (1759); Cumming, *Elements of clock and watch-work* (1766); Le Roy, *Avis contenant les vrais moyens* (1719). Translated and annotated treatises include Leibniz, *Remarques sur le discours de M. H. Sully* (1714); Geißler, *Alexander Cummings's Elemente* (1802); Vogel, *Ferdinand Berthoud's Anweisung* (1791).

ate valves (for organs), or pull-downs (for bells), to produce the musical sounds. A substantial part of the expertise and skill for these systems came from the making and maintaining of cathedral clocks and mechanical theaters, as well as the musical and mechanical spectacles in royal gardens, of which mechanics and scholars such as Caspar Schott and Salomon de Caus were in charge.[17]

The other main mechanism in the Jaquet-Droz family's harpsichord player, aside from the music-making one, is the "life mechanism," which independently drives the automaton's eyes, head, and breathing and is located underneath the barrel. The life mechanism is one of the most complicated in the automaton, not only because of the variety of effects that it is in charge of, but also because it is designed to run for one and a half hours. In a demonstration, the life mechanism can be started first to "prepare" the audience: the musician starts moving her head, eyes, and chest long before she starts playing the harpsichord. These preparatory motions are slow and regular, and they convey the rather captivating impression that the automaton is waking up and preparing her audience for the concert.[18] The slow motion is achieved by a gear reduction in the transmission, which needs, above all, space and robust material for the resulting strong forces.[19] It must have been one of the highest priorities in the Jaquet-Droz workshop to make this mechanism small enough to fit into the automaton: the artisans invested a great deal of effort into making the automaton perform precisely these subtle, slow, and graceful motions in addition to her music-making.

The life mechanism consists of many additional components (fig. 9). The mainspring with a spiral thread and five other revolving devices

17. See Kircher, *Musurgia universalis*; Schott, *Mechanica Hydraulico-pneumatica*; Caus, *Les raisons des forces mouvantes*. Historical analyses include Ord-Hume, *Clockwork Music*, 63; Protz, *Mechanische Musikinstrumente*, 15–22; Chapuis, *Histoire de la boîte à musique*; Feldhaus, *Die Technik der Antike und des Mittelalters*. See also *Wunder und Wissenschaft*.

18. "Waking up" is a term that Carrera uses, an appropriate choice of verb. Carrera, Loiseau, and Roux, *Androiden*, 76. I myself saw a short performance of all three automata at the Musée d'art et d'histoire in Neuchâtel, and the harpsichord player's motions started before any of the other automata's actions started. No one had explained the mechanism to me in advance. Even though it would have been plausible for me to infer that two mechanisms working independently can yield the observed effect, the automaton's "breathing" and "looking around" left a lasting impression.

19. Gear reduction means that the driving wheel is unlike most in having more revolutions than the running wheel. Gear reduction makes it possible for a rather small drive wheel to apply a rather large force onto a running wheel. Many revolutions in the drive wheel yield fewer, more forceful revolutions in the running wheel.

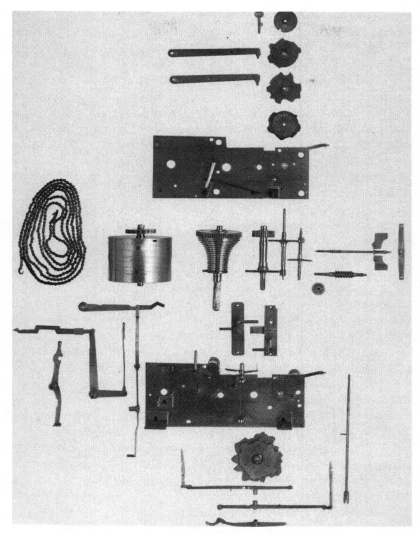

FIGURE 9 Individual parts of the harpsichord player's "life mechanism." Carrera, Loiseau, and Roux, *Androiden*, 79.

can be seen in the middle of the picture.[20] At the bottom is a brass double cam and three levers that produce the nodding and turning of the head. To the left are three levers that make the automaton seem to breathe by raising and lowering her chest at regular intervals—a bodily motion that captured commentators' attention.[21] At the top of the picture are the four cams and two levers that move the eyes: two for moving up and down, two for moving left and right. Only one of the four cams for the eyes' movement has cogs (the one at the bottom of the group). It functions as an interrupter and enables the cams to vary their positions in relation to one another. This feature allows the automaton to never repeat the movements of her eyes over the course of three quarters of an hour.

The dulcimer player also has "added" complexity in the design of her mechanism to make her move her head and eyes: in her case, an extra device is found to the right of the barrel. The figure's vertical head and eye movements are directed through this mechanism, which consists of an extra cam, with a follower reading its information, and a set of rods transmitting the motion upward and to the (left) side, above the cylinder. Here, the motion is translated onto a long vertical rod that reaches all the way into the dulcimer player's head. The synchronicity of motion is accomplished by attaching all motions (musical and nonmusical) to one and the same cylinder. The head's horizontal motion is accomplished similarly through an extra system that translates the cylinder's rotational motion into a rotation around an axis turned by ninety degrees.[22]

In the case of the harpsichord player, one of the most eye-catching parts of her performance is the bow with which she finishes each of the five melodies. That mechanism is set in motion by the main clockwork once a melody has come to an end. It consists of three rotating devices that drive three cams and is located in the lower part of the automaton's body, beneath the barrel (fig. 10).[23] The rods sitting on the cams follow their profiles and transmit the various motions to the upper body and the head. The automaton not only bends her upper body;

20. The spiral spring is in the housing. When, as in this case, the spring housing is furnished with a spiral thread, it serves at the same time the function of the first cogwheel. An alternative construction would be to have an axis attached to the spring that is run by the torque released from the unwinding spring.

21. See chapter 2 and Carrera, Loiseau, and Roux, *Androiden*, 76.

22. *Horloges et automates; Catalogue du musée*, 31–34.

23. This mechanism is similar to the life mechanism in having a spring housing with cogs; thus it does not need an axis to transmit the torque of the spring.

FIGURE 10 The harpsichord player's bow mechanism, located in the lower part of the automaton's body beneath the barrel. Carrera, Loiseau, and Roux, *Androiden*, 75.

she bows her head simultaneously and turns it to the left and the right, while rolling her eyes sideways and up and down.

The mechanism for the harpsichord player's bow is doubly complex: on the one hand, the profiles of the cams of the bowing mechanism are much smaller than the profiles of the other cams; on the other, the rotation speed is very high, which allows the bow—lasting six to eight seconds—to seem rather natural and real but also puts a great deal of strain on the mechanical parts. The gear reduction necessary for this motion therefore entails difficulty over space and robustness similar to that we saw in the case of the life mechanism. The bow mechanism is also subject to further stress and strain. The levers and cams not only have to lift the long rods in the automaton's upper body, enabling the bow, but they also have to release the functions of the other body parts. In addition, it takes a significant amount of force to make half of the automaton's body bend forward slowly. Two springs make the upper body move upward again after the bow.[24]

The automata's nodding and eye movement and the harpsichord player's life mechanism, breathing, and extended bow are all evidence of the automata's elaborate mechanisms that their makers designed to allow elegant, appealing, and entertaining movements during their musical play as well as (in the case of the harpsichord player) before and after playing and between musical selections. For each component of their performance, there is a piece of machinery that brings about a particular effect. These individual elements of the automata's routines have important effects in their overall performance, as they distract and charm the spectator and provide compelling instances of verisimilitude, "natural" motion, and grace. The automata's technical architectures demonstrate that the artisans designed their automata to perform well-planned, precise, and minutely conceived musical-mechanical spectacles. Perhaps the key example of the artisans' success in this endeavor is the fact that the one contemporary report about the dulcimer player stresses that her eye movements are in rhythm with the music, while, in reality, her eyes move exclusively through their own inertia. The only driven motion is the motion of the head.[25]

It is not known for certain whether Roentgen knew about the Jaquet-Droz harpsichord player or whether he even saw it, but both are possible. The Jaquet-Droz automata were taken to Paris in January of 1775, and Roentgen was in Paris around this time. It is plausible that he moved in circles that overlapped with the Jaquet-Droz family', given that they were artisans of similar rank, type, and ambition. There were also reports on the Jaquet-Droz automata published in various popular periodicals, and Roentgen may have read about them. If in fact the

24. Carrera, Loiseau, and Roux, *Androïden*, 72, 74–76.
25. See *Dessauische Zeitung* in chapter 3; Müller, *Jean Pauls Exzerpte*, 159. I was subject to the same optical illusion: when I saw the automaton's play, I interpreted her eyes' motions as in rhythm with the music. See chapter 3, note 129.

Jaquet-Droz harpsichord player was an inspiration for David Roent-gen, it was not an immediate one. Six years passed between the time of Roentgen's first visit to the French court and the delivery of the dulci-mer player to Marie-Antoinette.[26]

The two women automata replicate mechanically a captivating and recognizable cultural scenario of the eighteenth century—music-playing women moving their bodies. The cultural and political rel-evance of the scenario predates the professional lives of the artisan families Jaquet-Droz and Roentgen-Kinzing. Imagery and ideas of music-playing women were prevalent during the second half of the eighteenth century and most likely were accessible to both automaton makers, as was the idea of bodily motion to communicate affects. Both sets of artisans were unusually educated and knew about contempo-rary tastes and fashions in the arts, literature, and music. Furthermore, women's musical play and body technique were part of larger political concerns in eighteenth-century Europe. They intersected with sexual politics, sentimental politics, and the politics of social order.

Women and Music

Women occupied a special place in eighteenth-century musical prac-tice, as they did in the Enlightenment as a whole.[27] They were regularly both subjects and objects of discussions about such key topics as uni-versality, equality, rationality, autonomy, and civilization, and sexual difference thus remained a knotty issue for the Enlightenment's cen-tral ambitions and tenets. And while the social and cultural worlds of the Enlightenment often remained the exclusive realm of men, making and listening to music were to some degree, and in specific ways, ex-ceptions to this.

The polarization of the sexes was, like many other aspects of eighteenth-century societies, more varied and less consistent than is often assumed. There remained a great deal of variability in what it meant to be a woman, and even more in what it meant to be a man, much more so than in the nineteenth century, when polarity between men and women was lived and thought in a more rigid way than at

---

26. On the various possible sources of inspiration for the dulcimer player, including accusa-tions that Roentgen and Kinzing plagiarized it from a model that Kinzing saw as an apprentice, see chapter 3, page 114.

27. See the splendid contributions in Knott and Taylor, *Women, Gender and Enlightenment*.

any time before or after.[28] The emergence of a new bourgeois lifeworld in the eighteenth century, outside of the traditional orders of the ancien régime and the estate society, entailed a transformation of virtually all categories of life, including sex and gender. In a momentous shift, sexual difference was no longer justified on the particular basis of estate but instead on the universal basis of nature. In this new schema, the subordination of women was not a confinement but a demand that stemmed from reason, because reason could recognize sexual difference as being constituted in nature.

The emergence of amateur musicianship in the seventeenth-century French aristocracy was an important backdrop to eighteenth-century women's musicianship. Seventeenth-century noble families often entrusted their children (including their daughters) to composers and tutors, generating a tradition of early musical education. Noblewomen sang and played instruments for individual entertainment, and some were in a position to pay for a small staff of musicians (sometimes even including a composer) for larger and more formal settings.[29] The women members of the French royal family, notably the wives and mistresses of the Bourbon kings, thus established a tradition of music-making, chamber concerts, and other entertainments. In the latter half of the seventeenth century, during the reign of Louis XIV, professional performances were established, and as Julie Anne Sadie explains, such practices were "followed at court" and "widely imitated."[30] Musical activities subsequently migrated from France to courts on the entire European continent.

The spread of musical activity created a small but significant economic and professional infrastructure for women, primarily lower- and mid-level aristocrats, as they began to find appointments as singers and, to a lesser degree, instrumentalists in court musical establishments. Marie-Antoinette, allegedly the model for one of my women automata, was particularly well known for her love of and taste in music.

28. Historians in recent years have examined the shifting nature of sexual difference in various periods of the modern age, aiming to give this process a history of its own. Some historians of music explicitly speak in the plural of "polarities" of the sexes. Grotjahn and Hoffmann, "Einleitung," 1–2. Trepp, "Diskurswandel und soziale Praxis," 8, 14–15; Frevert, *Mann und Weib*, 54. In this context, see also Jessica Riskin's analysis of sensibility vis-à-vis reason in natural philosophy. She concludes that the polarity between cognition and emotion familiar to us nowadays does not have a direct correlate in the eighteenth century. Riskin, *Science in the Age of Sensibility*, 7–8.

29. My account here relies on Julie Sadie, "*Musiciennes*."

30. Ibid., 195. In this context, Sadie points out the traditions established by the women who married into the Bourbon family: Ann of Austria, the mother of Louis XIV; Marie Leszcinska, the wife of Louis XV; and Marie-Antoinette, the wife of Louis XVI.

She was educated in the French language and culture and was taught the harpsichord and the harp in preparation for her marriage to the dauphin. When she moved to Versailles after her wedding, she continued to be an active musician and organizer of concerts, as well as a patron of composers.[31]

The number of women involved in domestic music-making increased significantly during the eighteenth century. Bourgeois circles and the expanding middle classes started adopting musical practices from the aristocracy and evolved their own set of ideas and expectations around music-playing women.[32] Music-playing filled leisure time and drove away ennui. A music-making woman, furthermore, was considered socially accomplished, having a kind of currency that reflected on the gentility of her family. In the case of young women, musical ability ideally led to a good matrimonial match. There were other accomplishments, too, that were conducive to marriage, such as drawing and needlework, but music was a favorite since "it could be shown off best while actually being accomplished."[33] Julie Anne Sadie argues that because women were teaching, performing, writing, composing, and advertising their skills in musical almanacs, "music-making had become a socially less self-conscious pursuit," and she observes a "general broadening of society's attitudes" toward women musicians in the course of the eighteenth century.[34] This greater visibility also became manifest in the two women automata.

In general, women played instruments that were considered sufficiently "feminine," such as the lute, the harp, or the harpsichord, and that could accompany the musician's own voice.[35] The social grace displayed in instrument-playing and singing could secure a diverse set of social assets, such as proving marriageability but also playing before

---

31. Sadie, "*Musiciennes,*" 196. Among those composers was Christoph Willibald Gluck, the composer of one of the melodies in the repertoire of the dulcimer-playing automaton. On the specific uses and abuses of representations of Marie-Antoinette and her body, see Hunt, "Many Bodies of Marie-Antoinette," 119; and Goodman's excellent collection *Marie-Antoinette,* of which Hunt's essay is a part. Note also that the engraving next to the title page of Stefan Zweig's biography of Marie-Antoinette (in the English translation by Eden and Cedar Paul) is an oil painting by François Hubert Drouais entitled "Marie-Antoinette at the Clavecin" (see my chapter 3, note 99).

32. A great deal of literature investigates the intersections, rivalries, and coincidences of "aristocratic" and "bourgeois" behaviors, practices, and ideologies in the eighteenth century. In regard to the present subject matter, see, for example, Frieling, *Ausdruck macht Eindruck;* Gall, *Von der ständischen zur bürgerlichen Gesellschaft.*

33. Loesser, *Men, Women, and Pianos,* 268. See also Neuls-Bates, *Women in Music,* 73. The theory of accomplishments lasted well into the nineteenth century, as Jane Austen recorded in her writings.

34. Sadie, "*Musiciennes,*" 195.

35. Hoffmann, *Instrument und Körper.*

members of the aristocracy. A music-playing woman was a cultural scenario with valences that could sometimes transgress the boundaries of class, estate, and sex. These boundaries were indeed crossed when Roentgen presented his dulcimer player to the king and queen: a (mechanical) music-making woman facilitated an artisan's entry into the court society.[36]

Contributing equally to the cultural meanings of a woman's music-playing was the connection of music-making to sentiment, a notion deeply tied to ideas about morals, class, and the new social order of civil society. The "sentimental age" in the eighteenth century took on a variety of forms in France, England, and the German lands. The shared ground among them was an increased attention to the "inner life" of individuals and to the individuation of feeling.[37] This was in contrast to earlier periods when human affects in general and their moral, political, and epistemic relevance had been a concern to moral and natural philosophers.[38] The changing uses of the term *sentimentality* in the seventeenth and eighteenth centuries reflect this history. The French term *sensibilité*, for example, described originally, and more narrowly, the quality of being sensitive to physical impressions, the ability to sense, or something that touches the senses.[39] In the course of the eighteenth century, it gradually took on intellectual and affective denotations and connotations; it came to refer to larger concepts such as "the faculty of perceiving moral impressions" or "the sentiment of humanity, piety, and tenderness."[40] In this broader sense, the adjective *sensible* came to

36. Another example of such a transgression would be Leopold Mozart presenting his music-playing children at aristocratic courts in the 1760s.

37. Peter Hohendahl suggests this formulation in Hohendahl, *Der europäische Roman*, 1. Hildegard Kruchen also emphasizes the connection between sentimentality and individuality in *Europäische Schlüsselwörter*, 2:144–45.

38. See, for example, the contributions about Spinoza, Descartes, and Malebranche in Stephen Gaukroger, *Soft Underbelly of Reason*. See also Michael Heyd's discussion of moral individualism in the context of Calvinist conceptions of original sin from the early seventeenth to the eighteenth century. Heyd, "Changing Emotions?"

39. "Qualité par laquelle un sujet est sensible aux impressions physiques," quoted in Sauder, *Empfindsamkeit*, 1. "Qui frappe les sens (materiel, physicque, en parlant des choses)," in Cayrou, *Le français classique*, 789. See also Jessica Riskin's account of the pervasiveness of sentimentality in the French Enlightenment and its role in the formation of empiricism. Riskin, *Science in the Age of Sensibility*, 1–10.

40. "Faculté de percevoir les impressions morales" and "sentiment d'humanité, de pitié, de tendresse." These meanings had been known and used since the late seventeenth century. See Wartburg, *Französisches etymologisches Wörterbuch*, 461; Sauder, *Empfindsamkeit*, 1. See also Hildegard Kruchen's argument in *Europäische Schlüsselwörter*, 144. The term, in earlier meanings, could also refer to humans and animals, while the later extensions only referred to humans.

describe someone who was able to feel a moral impression or be easily moved, to have human sentiments.[41]

There developed, thus, a decidedly moral and social dimension to the idea of sentiment in the eighteenth century, as it came to describe and prescribe one's ability to participate incisively in one's own and other people's sentiments. In particular the French term *sensibilité* and the German term *Empfindsamkeit* (as well as the adjective *gerührt*, referring to stirred sentiments) came to denote a social movement.[42] German pedagogues, philosophers, and theologians conceived of *Empfindsamkeit* as involving an ability or inclination to have "tender and pleasant sentiments" (a meaning similar to that which developed for *sensibilité* in France), and they extended this meaning to the ability to *enjoy* the participation in one's own and other people's sentimental life, "to feel pleasure in participating in the soul's motions."[43] The most important moral dimension of *Empfindsamkeit*—its connection to duty—was established in the same way. Karl Daniel Küster, a preacher and theologian in Magdeburg, wrote that a sensitive person was someone who had a disposition of mind and heart that would make him receive "quick and strong insights into his duties and feel an effective drive to do good."[44] Tutors and pedagogues were expected to develop

41. "Qui ressent une impression morale" and "facilement ému" and "qui a des sentiments humains." Bloch and Wartburg, *Dictionnaire étymologique*, 585. The adjective *sensible*: "Qui régit vivement aux impressions physique" (since Montaigne); "facilement accessible à certaines idées" (since the mid-sixteenth century), "qui est aisément ému, touché, attendri" (since the mid-seventeenth century). Wartburg, *Französisches etymologisches Wörterbuch*, 461. *Sentire, empfinden*: "Etre affecté de quelque chose d'extérieur (par exemple d'un affront)" (since the mid-seventeenth century). *Sentir* in the context of music: "avoir l'appréciation délicate de ce qui est beau dans une oeuvre" (470). See also Sauder, *Empfindsamkeit*, 1–2. The article *sensibilité* in Diderot and d'Alembert's *Encyclopédie* indicated this breadth of meaning; it included medical, moral, as well as philosophical meanings. In the French context, discussion of the effects of feelings were more often connected to developments in the fields of philosophy and medicine rather than culture and politics, as they revolved around the problem of materialism and the relationship between body and soul. See, for example, Gärtner, "Remuer l'Ame?"; Vila, *Enlightenment and Pathology*; Vincent-Buffault, *History of Tears*. In the English language, the adjective *sensible* became adapted through the French, and a major eighteenth-century use of the term *sentiment* was in the context of "moral sentiment" among philosophers such as Shaftesbury, David Hume, and Adam Smith. See Smith, *Theory of Moral Sentiments*; Raphael, *Impartial Spectator*; and Kruchen, *Europäische Schlüsselwörter*, 146–49, 169–72.

42. Arthur Wilson calls sensibility in France a "social institution." Wilson, "Sensibility in France," 35.

43. Campe, "Sensation," 902.

44. Küster, "Empfindsam," 47. I use the male pronoun in those contexts where I paraphrase texts from the eighteenth century that follow this convention, in order to convey their manner of speaking. On issues of sexual difference, see my arguments throughout this chapter. The French case was similar, as *sensibilité*'s contribution to social virtue was meant to emerge from people's ability to feel other people's misery. The popular *Dictionnaire de Trevoux*, for example, ex-

this sentimental and virtuous disposition in anyone who participated in the common good: princes and princesses, ministers, lawyers, pastors, teachers and peasants.[45]

In this way, the culture of sentiment helped define the emerging fields of pedagogy, popular philosophy, and politics, all of which played foundational roles in the dissemination and implementation of the moral tenets of the Enlightenment. The main medium for this dissemination in the German-speaking context was the "moral weekly periodical" (*Moralische Wochenzeitschrift*), which provided pedagogical and religious reading material for large audiences.[46] There were hundreds of such periodicals in circulation, including in smaller towns. Their number, along with their readership, increased massively during the second half of the eighteenth century. Just as reading had become a common leisure occupation and a mode of sociability in clubs and associations, it also came to be thought of as a way to develop moral and aesthetic guidelines.[47] The sentimental dimension was once again crucial: popular philosophers defined ethics as "teachings of the art of feeling" (*Lehre von der Gefühlskunst*), conceiving of feeling as a condition for humans to be prepared for meaningful and moral action.[48] Because this view raised questions about the (outward) representation of affects, theories of mimicry and acting became relevant, and those theories, in turn, defined the position of a spectator who deduced from mimicry of others' bodily motions the corresponding states of their souls. Such a scenario, in combination with music-making, is what Johann Richter explores in a satire, discussed in chapter 5.

A similar intersection of new forms of social interaction and moral-

plained that "[the sentimental] disposition of the soul makes people virtuous" and suggests as an example that "the sensibility of the heart at other people's misery is the source of social virtue." Both Sauder, *Empfindsamkeit*, 2; and Wilson, "Sensibility in France," 44, quote the *Dictionnaire de Trevoux*.

45. This is Küster, once again, arguing for this task of tutors and pedagogues (*Erzieher*) in his "Empfindsam," 48.

46. Around 1760, the genre underwent change from a forum for literary texts to popular entertainment. Virtually all influential German writers were at one point in their lives involved with a *Moralische Wochenzeitschrift* as writers or editors, a fact that speaks for the periodicals' economic relevance in the history of German authorship. Wolfgang Martens integrates an investigation of the role of publishers, authors, sellers, and audiences with an analysis of the role of the concept of virtue in the bourgeois self-image. Martens, *Die Botschaft der Tugend*.

47. Hull, *Sexuality, State, and Civil Society*, 207–10. Hull says: "In these societies members literally practiced the basic principles of civil society and polished the virtues necessary to create and uphold it" (209). An account of these practices and the "experience" that came with them is found in Dülmen, *Kultur und Alltag in der Frühen Neuzeit*, 226–39; and on philosophers' and pedagogues' views on how behavior, action, or conduct (*Handlung*) could be structured not only by reason but also by aesthetics, see Bachmann-Medick, *Die ästhetische Ordnung des Handelns*.

48. Bachmann-Medick, *Die ästhetische Ordnung des Handelns*, 35–36.

ity was forming around music, whose aesthetic and sentimental powers were perceived to contribute to virtue and the common good. Music, in fact, was considered as suitable a medium as others—such as literature—to make the sentimental motions of the soul available to people, and thus to educate young people and make them empathetic persons. For example, Philipp Julius Lieberkühn, a teacher, wrote a prize-winning essay about ways to provoke the love of humanity (*Menschenliebe*) in the hearts of young people who were destined to hold high office or become land-owning gentry.[49] He considered music to be capable of opening "even the most closed soul" by means of "its tender harmonies" and thus making that soul amenable to moral feelings.[50]

The intersection of musical culture and marriage in women's lives is a particularly interesting focal point for investigating the increasing moral, aesthetic, and political relevance of music-playing women and music-playing women automata. Ideas about music and marriage were explicated in contemporary periodicals that were the equivalent for women of the "moral weeklies." One of them was the monthly *Amaliens Erholungsstunden*, which Marianne Ehrmann (née Brentano) edited in the late eighteenth century. The journal was initially distributed by a small publishing house that she ran with her husband, but after a year it became so successful that they negotiated a transfer to the eminent publisher Friedrich Cotta in Stuttgart.[51] Among other topics, *Amaliens Erholungsstunden* featured articles on "feminine occupations." Ehrmann's concern was to share her life experience and to teach adolescent girls about marriage. She wrote that in order to achieve a durable marriage, several things had to be in place: the right husband, an orderly house, and also reading and friendships. Alongside the manifold duties and virtuous activities of middle-class women, Ehrmann listed music-making as an appropriate activity, since musical education and practice contributed, like reading, to the formation of taste and feeling. She emphasized the importance of feeling by pointing out how easily, in her age, one could be accused of being "without feeling" (*fühllos*).[52] Like many contemporary pedagogues, Ehrmann was concerned with

49. *Menschenliebe* and its relevance to government was already taught in the Early Enlightenment's Leibniz-Wolff school (see theorists below). *Allgemeine deutsche Biographie*, 18:577 (1883).

50. Quoted in Sauder, "Die empfindsamen Tendenzen," 52.

51. In 1792, however, there were so many difficulties emerging with the publisher that the journal was terminated, and both Ehrmann and Cotta continued with different journals. Cotta founded the journal *Flora* and kept many subscribers to the previous journal, and Ehrmann founded the journal *Die Einsiedlerin aus den Alpen* with the publisher Orell, Gessner, Füßli & Co.

52. Ehrmann, "'Über weibliche Beschäftigungen'"; *Amaliens Erholungsstunden* 3, no. 3 (1792): 266–80. See also Frevert, *Women in German History*.

the right balance of feeling, reason, duty, and moderation, and she looked at these issues specifically through the lens of music-making women and marriage.[53] Reading and music-making enhanced a woman's marriageability, because they made her a better companion for her husband, a better mother, and a better instructor for her servants. Being well-read and musically educated could even offset poverty or lack of beauty. Cultural historians describe the flip side of this coin: as soon as the educated woman went too far and became too well-read, she was no longer deemed desirable and became a subject of malicious caricatures. It was a fine line that women had to walk; and the balances that the culture of sentiments required, and the limits required within that culture, were often imagined in terms of women.[54]

The increasing interest in music-playing and musical education also created a market for music teachers and drove sales of printed music, instruction manuals, and poetry or paintings about music playing, including those that utilized the motif of the keyboard-playing woman. There was a growing demand in particular for musical compositions—such as solo pieces for piano, violin, or flute and songs for piano accompaniment—that catered to the new market, often in the form of collections of chamber music dedicated by composers to their aristocratic pupils or individual pieces named after them. Increasingly, women became the addressees of collections advertised as "for the sentimental piano player" or "for feeling souls" or "for heart and feeling." Carl Philipp Emanuel Bach himself wrote a collection titled "Six sonatas for the clavecin to be used by ladies."[55] Johann Martin Miller, a Göttingen-based poet, wrote poems on the same motif, entitled "At Daphne's piano" (1773), "Praise of a girl. At the piano" (1773), and "As Mariane was singing at the Piano: At midnight" (1775).[56]

Contemporary art similarly reacted to this interest in the culture of women's music-playing. Engravers produced prints of young ladies playing musical instruments, and the works depicting keyboard instruments in general were particularly likely to feature women. The majority of Dutch paintings from the fifteenth to the eighteenth century

---

53. *Amaliens Erholungsstunden* 3, no. 4 (1792): 5–7.

54. Martens, *Die Botschaft der Tugend*, 533, 540; Frevert, "Bürgerliche Familie und Geschlechterrollen," 93. Pointed caricatures of women playing keyboard instruments are found in Rieger, *Frau, Musik und Männerherrschaft*, 37, 50; and Hoffmann, *Instrument und Körper*.

55. Bach, *Six sonates pour le Clavecin*; Sauder, "Die empfindsamen Tendenzen," 52; Sadie, "Musiciennes," 191–95.

56. Sauder, "Die empfindsamen Tendenzen," 57.

showing keyboard instruments, in fact, display female players.[57] The harpsichord and the organ remained popular instruments for women well into the eighteenth century, before the piano emerged as an alternative. Such representations were probably how the artisan families Jaquet-Droz and Roentgen-Kinzing became familiar with the motif.

Music and Sentiment

Already in the early eighteenth century, questions revolving around sentiments had been raised in music theory and practice. As the discussion progressed through the century, sentiments were also discussed in relation to the musician's moving body. In the 1720s, as Renaissance and early modern ideas about aesthetics, representation, and metaphysics were being recast, musicians and composers began to devise general theories of music.[58] In the early 1700s it was still accepted that the individual arts (literature, theater, sculpture, and painting) were fundamentally related to each other through the common purpose of mimesis of nature, allowing those formulating theoretical foundations for music to rely on established aesthetic theories.[59] One commonly held tie between literature, music, and the other arts was their role in generating affects. Composers and music theorists explicitly made the

57. An informal survey by Dutch art historians of 266 such paintings concludes that "about 90% of the keyboard players pictured are women and girls." Dijck and Koopman, *Het klavecimbel*, numbers 78 to 266 (no page numbers). Busch-Salmen's discussion of paintings and engravings featuring keyboard instruments also emphasizes the abundance of inspirations for artists at the time to engage this intersection of women and music. Women and keyboards represented a social and cultural reality, but artists also used the motif to make references to figures such as Saint Cecilia (patron saint of music) and motifs of Greek mythology. Busch-Salmen, "Die Frau am Tasteninstrument," 41–45. For paintings, see in particular *Lady Playing the Virginals*, 1653, by Wenceslaus Holler (a Czech etcher [1607–1677]), and Johannes Vermeer's *Woman with a Lute*, ca. 1664. Both images are found in Neuls-Bates, *Women in Music*, 57. Another such painting, by Johann Friedrich August Tischbein, *Lautenspielerin*, is in Krüger, *Das Zeitalter der Empfindsamkeit*, 17. See also Loesser, *Men, Women, and Pianos*; Jackson, "Musical Women," 99–100, 116–18. Richard Leppert explores in a social history of music two further musical instruments relevant in this context, the musettes and the hurdy-gurdies. Leppert, *Arcadia at Versailles*.

58. My account here relies strongly on Neubauer's analysis of the ways in which musical theorists in the eighteenth century integrated affects into their nascent theories of musical expression. Neubauer goes back to the Pythagorean tradition of mathematics, language, art, and affects to explain this. Neubauer, *Emancipation*.

59. This made it easier to justify music as aesthetically equal to the other arts (music was sometimes considered inferior to poetry and painting, especially in regard to its representational powers). See, for example, the widely read Sulzer, *Allgemeine Theorie*, vol. 1, part 3, 225. Sulzer states in particular that music can express and stimulate affects and yet has weaker powers to represent objects (*Gegenstände*).

representation and generation of affects a shared goal of composers and performers. In doing so, they relied on, and developed further, theories of affect from the sixteenth and seventeenth centuries, which had relied on categories from natural philosophy and also had borrowed from hermeticism and ancient cosmological speculation.[60] These early modern theories attempted homologies between the elements, the humors, and the temperaments, while also taking up newly emerging Cartesian modes of thought.[61]

As eighteenth-century theorists took on these traditions, the relationship between affects and musical theory and practice became more complex.[62] In the later eighteenth century, theorists of affect in music came to include composers, critics, journalists, and publishers.[63] Many among them agreed that music should represent and generate passions, but there was less consensus about concrete theoretical and practical implications of this principle.[64] These theorists developed taxonomies in which musical forms (such as intervals, keys, styles, meters, and rhythms) correlated with certain affective values, but it was more difficult to formulate fundamental definitions and categorizations.[65]

Johann Mattheson's 1739 encyclopedic treatise *Der vollkommene Capellmeister* (The perfect musical director) had an influence that was second perhaps only to Rousseau's works on music. Mattheson assembled nearly everything that was available on musical theory and practice at the time, demonstrating an unusual breadth of musical ex-

60. Cardanus, Ficino, Agrippa con Nettesheim, Paracesus, and Giordano Bruno had revived these in the Renaissance. Neubauer, *Emancipation*, 45–46.

61. Ibid., 46. These ideas were also recast in the framework of new mechanistic models for the circulation of blood and other physiological processes. Johann Mattheson, for example, the most prolific and well-known musical theorist of his time, interpreted sadness as the "contraction of subtle parts of our bodies." Mattheson, *Kapellmeister*, 15–16.

62. Neubauer, *Emancipation*, 51. An example here is, once again, Johann Georg Sulzer, who mused on the question of what the artist could do to generate or calm affects, how to increase someone's sensibility, or how to decrease it if someone was already too sensitive. Sulzer, *Allgemeine Theorie*, vol. 1, part 3, 224–25.

63. They wrote handbooks for performers, conductors, and composers and wrote theoretical treatises for philosophers. Among the most prominent were Johann Mattheson (1681–1764), Johann Adolf Scheibe (1708–1776), Friedrich Wilhelm Marpurg (1718–1795), Christian Gottfried Krause (1719–1777), and Johann Georg Sulzer (1720–1779).

64. Neubauer, *Emancipation*, 51.

65. Neubauer even says that these efforts were often half-hearted and vague. He quotes Mattheson once again, who said that key alone could not determine a happy or sad mood. Neubauer, *Emancipation*, 52. In another passage Mattheson states that hope and despondency could be represented with appropriate sounds and meters but that it would be too long-winded to list all motions of the soul. He would therefore list merely the important ones, elaborating on pride, humility, stubbornness, zeal, anger, revenge, jealousy, and hope. Mattheson, *Capellmeister*, 15–18.

perience.[66] He developed in the *Capellmeister* a general theory of music, which he hoped would guide composers in their everyday work. One specific ambition of his, articulated very much in the fashion of the Early Enlightenment, was to constitute the relationship between musical expression and morality, by showing how music could provide pleasure and promote virtue. He formulated the true goal of a melody as being the stirring of the passions of the soul, and he appealed to composers to write only music that moved the listener (this goal is listed immediately next to another fundamental one, expressing the glory of God).[67] Mattheson required that all music express the passions and have a moderating, and thus moral, effect; he claimed that a composer who was moved by passions himself would know how "to express all the movements of the heart . . . as if it were a real speech."[68] For Mattheson, the inclinations of the soul were the true matter of virtue, and virtue was nothing but a well-adjusted and wisely moderated sentimental motion: "Where there is no passion, no affect, there is no virtue."[69] He thus considered music to be part of ethics (*Sittenlehre*) and its true characteristic to be a theory of morals and order (*Zucht-Lehre*) above others.[70] Under his theory, a perfect musician (*Tonmeister*) had to master this art if he wanted to represent virtues and vices with his sounds and instill in the listener's mind love for virtue and loathing for vices.

Musical theorists and popular philosophers who engaged Mattheson's work highlighted different aspects of it. For pedagogues, it was a guide that confirmed the intersection of affect and pedagogy, emphasizing the importance of tempering the minds and souls of children, women (as future wives), future civil servants, and future monarchs. These pedagogues integrated music into their handbooks of popular philosophy.[71] Further musical-theoretical works followed in the mid-

---

66. Petersen-Mikkelsen, *Die Melodielehre*, vii. Bellamy Hosler, *Changing Aesthetic Views*, 72.

67. Mattheson, *Capellmeister*, 5. Hosler also reads these key passages closely, and some of my readings rely on his. Hosler, *Changing Aesthetic Views*, 69–84. Hosler finds another mention of music's triple purpose of affect, virtue, and glory of God in Mattheson, *Capellmeister*, 129.

68. Mattheson, *Capellmeister*, 145, 207–8. The quote on "speech" is also used by Hosler, *Changing Aesthetic Views*, 84.

69. This is a famous quote, which was also echoed by his followers. Mattheson, *Capellmeister*, 15. Mattheson explains that many of the affects that nature gave to us are not the best ones: they need to be curbed (15–16).

70. Ibid., 15. He uses examples of love, hope, sadness, and joy (where joy corresponds to wide intervals and sadness to small intervals) and even connects that to physiology, in the chapter titled "The natural philosophy of sound" ("Von der Naturlehre des Klanges").

71. Examples of such handbooks are, once again, the moral weeklies; see note 46.

eighteenth century. They represented an intensifying interest in and engagement with affect theory in music.[72] Even works in the later eighteenth century, such as Heinrich Christoph Koch's *Versuch einer Anleitung zur Composition*, of 1782, were based on the idea that the subject (*Gegenstand*) of music was the generation and expression of the sentiments.[73]

The connection between art, affect, and virtue was also constituted in general theories of aesthetics. Johann Georg Sulzer wrote an encyclopedia of the arts in the 1770s that became the most widely read work of its kind in the last part of the eighteenth century.[74] He claimed a privileged status for the arts, music in particular, in regard to the cultivation of sentiments and passions.[75] His article on passions (*Leidenschaften*) stated, in a manner by then familiar, that an immediate purpose of an artist was to provoke or calm passions. He talked about the affects, their true nature, the ways they expressed themselves, and their good and bad effects. In order for artists to have precise control over the affects, Sulzer demanded that they have acute knowledge of these affects' nature and origins, echoing Mattheson's and others' call to educate musicians in matters of sentiment.[76] He singled out music among all the arts, arguing that it might be the most powerful because it affected the nervous system most directly and most forcefully.[77] Similarly, Sulzer's article on sentiments (*Empfindung*) resonated strongly with the moral reflections of theorists of musical-sentimental culture. He conceived of sentiments as driving behavior and claimed that the arts, through their sentimental effects, were capable of implanting proper inclinations (love of one's country, virtuousness, honesty, gratitude, true honor, freedom, and humanity) in individuals. For him, art's purpose needed to be an "orderly sentimentality of the heart." Sulzer warned, however, that the measure and moderation of sentiment could

72. Examples are Scheibe's *Der critische Musicus* (1738–40) and Mizler's *Neu eröffnete Musikalische Bibliothek* (1739–54). Scheibe and Mizler were both students of Gottsched, who was an influential theorist of language. Neubauer, *Emancipation*, 51.

73. Koch, *Versuch einer Anleitung zur Composition*, 15. Koch also formulated the controversial argument that music itself already spoke the language of sentiment (*Empfindung*) and was therefore not in need of textual language (poetry, for example) to generate sentiments (31–32). The relationship between language and music was the subject of heated debates in eighteenth-century Germany. Hosler, *Changing Aesthetic Views*; Neubauer, *Emancipation*, chap. 2; Bonds, *Wordless Rhetoric*, chaps. 2 and 4.

74. Sulzer, *Allgemeine Theorie*. See also Hosler, *Changing Aesthetic Views*, 3.

75. Sulzer, *Allgemeine Theorie*, part 2, 53 ("Empfindung"); and part 3, 223–37 ("Leidenschaften"). See also Krause, *Von der musikalischen Poesie*, chap. 4.

76. Sulzer, *Allgemeine Theorie*, part 3, 223–24.

77. Ibid., 226.

not be transgressed, and he mentioned unmanliness as one potential hazard. His target was poets, in particular, since they seemed to think there could never be enough stimulation for the soul. Sulzer's thoughts culminated in his claims about the relevance of art and affect to the common good and social order. He argued that if the sentiments (of honor, honesty, lawfulness, etc.) that sustained social cohesion and prosperity were too weak, then a society would be doomed.[78]

Alongside handbooks on composition and orchestra management, and theoretical and encyclopedic works on the arts, manuals for playing specific instruments also became popular. These books also dealt with the process of generating affects, typically focusing on recital performances. By focusing on practice and catering to a wider audience, instrument manuals contributed to the establishment and dissemination of the culture of affects in music. The expanding bourgeois practice of music-making, bolstered by bourgeois notions about education and cultivation, created a market for such works. Though indebted to rhetorical and aesthetic theory, these manuals were not as academic in their instruction and terminology as the more specialized works just discussed.[79]

In my context of the two music-playing women automata, it is particularly relevant what musical theorists and pedagogues said about performance practice. Among the most influential authors on performance practice (remaining so to this day) were Johann Joachim Quantz and Carl Philipp Emanuel Bach. Both received appointments at the court of Frederick II (later called "the Great") when he became king in 1740. Quantz, a composer and Frederick's flute teacher, was the author of the *Essay of an Instruction to Play the Transverse Flute*, published in 1752. His *Essay* was a foundational work, quickly going through several editions, with multiple translations following from the 1750s to the 1790s. Quantz had received his musical training around 1720 as a young man at the court of Frederick August I (called "the Strong") in Dresden, and he remained faithful for the rest of his life to the theories and practices of music-making that he learned there. Under his control, from 1741 until his death in 1773, the musical life at Frederick II's court remained rather conservative.[80] Quantz was a committed pedagogue, however, and was deeply influenced by the German Early

78. Ibid., part 2, 55–57.
79. Neubauer makes this distinction. Neubauer, *Emancipation*, 37.
80. This is in comparison to the musical activities in Mannheim, Vienna, and Paris, where innovation in composition and theory of forms was under way in the 1760s and 1770s. Kuijken, "Einführung," xi.

Enlightenment. He emphasized not merely mimetic musical play but also intellectual challenge, as well as the development of good taste and judgment, in a musician.[81] The attempt in his "Essay" to systematize musical performance practice was a groundbreaking project at the time, both in regard to musical performance itself and in regard to musical performance's larger culture and economics.[82]

Carl Philipp Emanuel Bach (1714–1788) was one of Johann Sebastian's sons and a—maybe *the*—leading musician and musical theoretician of his time. The two volumes of his treatise "Essay on the true manner of playing the piano" were published in 1753 and 1762, respectively. Like Quantz, he was employed at the court of Frederick II when he wrote his treatise. Bach was perhaps the best academically trained musician of his time, and his treatise on piano-playing became one of three foundational piano tutorials.[83] After François Couperin's work *L'art de toucher le clavecin*, of 1717, Bach's was only the second actual textbook for piano-playing that had ever been written. It had a profound influence on other piano-playing textbooks written in the 1760s, 1770s, and 1780s.[84] It resembles Quantz's work in form and organization, as well as in some details of the content and the phrasing of the chapter headings.[85]

Quantz's essay on flute-playing is divided into eighteen chapters. In the introduction, he outlines the talents and personality traits needed by those who want to devote themselves to music and, by extension, to the common good. The individual chapters deal with topics such as the history of the transverse flute, posture, fingering, scales, embouchure, the musical sign system, the use of the tongue and lips when playing, breathing, embellishments, cadenzas, accompaniment, performance in general, and assessment of performance.

Throughout the chapters that deal with practical matters of playing and performing, Quantz makes explicit the correlation between musical material and affects. For example, in the ninth chapter, on appoggiaturas, in addition to describing their function and proper execution, he explains that certain ornaments, depending on the character of the piece, serve the purpose of inducing cheer or happiness, whereas oth-

---

81. See Quantz, *Versuch einer Anweisung*, chap. 13.
82. Kuijken, "Einführung," xii–xv.
83. The other ones were Marpurg, *Anleitung zum Clavierspielen*; and Türk, *Klavierschule*.
84. Examples would be Löhlein's *Clavierschule* (1765); Wiedeburg's *Der sich selbst informierende Klavierspieler* (1765–1775); and Wolf's *Unterricht im Klavierspiel* (1783).
85. Niemann, "Vorwort," iii.

ers soothe or sadden.[86] In chapter 12, "How to play the allegro" ("Von der Art das Allegro zu spielen"), he explains that "what is funny" is expressed through short notes that move in small and large intervals; that "the glorious" is expressed through long notes that are accompanied with fast movements in the other voices, as well as through dotted notes; and that "flattering" is expressed both through dragged and drawn-out notes that go up and down in small steps and through syncopation.[87] In chapter 15, on cadences, Quantz explains that the purpose of a beautiful cadence is to surprise listeners toward the end of a movement in order to leave a special impression in their hearts, a new and touching (*rührend*) sense of astonishment, thus driving the desired provocation of affects to an extreme.[88]

Quantz's chapter 11 deals specifically with performance practice and is entitled "On good performances in singing and playing in general" ("Vom guten Vortrage im Singen und Spielen überhaupt"). Like other musical theorists before him, Quantz borrows from the theory of rhetorical forms. The chapter's often-quoted opening paragraph formulates the close relation between rhetoric and music: "The musical recital can be compared with the performance of a speaker. A speaker and a musician have one and the same intention . . . , namely: to win the hearts, to excite and calm the passions, and to set the listeners now into this, then into that, affect."[89] Quantz then lists characteristics of a successful performance, including correct and clear intonation, an easy and flowing presentation of the notes, a varied and multifarious performance, and an effort that "light and shadow be continuously engaged." Quantz's final criterion concerns the expression of passions. He says that "a good recital, finally, has to be *expressive and in accordance with each passion that comes up.*" In an allegro, for example, it would be liveliness; in an adagio, it would be tenderness. Quantz explains that in order to generate those affects in the listeners, "the performer must aim to put himself into those . . . passions that he is meant to express," thus emphasizing that the musician must generate the affects first in himself, according to the score's instructions, and then communicate them to his audience.[90]

---

86. Quantz, *Versuch einer Anweisung*, 81.
87. Ibid., 116.
88. Ibid., 153–57.
89. Ibid., 100. See also Neubauer, *Emancipation*, 22.
90. Quantz, *Versuch einer Anweisung*, 106–7 (Quantz's emphasis). The pronoun for the musician in Quantz's text is consistently in the masculine form, and there is no explicit mention of a gender differentiation among musicians. However, playing the piano, as well as other performa-

Quantz then illustrates the progression of these processes over the course of a performance, asking the musician to precisely control the sequencing of affects generated in himself and in the audience. Based on the musician's ability to exercise this control, Quantz eventually derives a criterion for a successful performance: "One would have to set oneself, so to speak, into a different affect with each bar, in order to be now sad, then merry, then grave, and so on . . . He who is able to master this art appropriately will not easily miss out on the listener's approval, and his performance will thus always be *affectively moving* ("rührend")."[91] A successful recital, then, is one that stimulates affective involvement—*Rührung*—in the audience, by means of the musician's own *Rührung* communicated in a precise and well-timed sequence, according to the pattern put forward by the musical piece. This key concept of *Rührung* was a part of the broader German-speaking terminology revolving around the eighteenth-century notion of *Empfindsamkeit*, and it connoted in particular the social, the relational, and the causal relationships that made someone feel something because of her or his interaction with another person or entity. *Rührung* appears at a central point in Quantz's work, in his concluding passage on the relationship between the musician's and the audience's set of feelings. The poet Richter uses the term *Rührung* in exactly the same context in his satire that I read in chapter 5.

Corresponding passages from C. P. E. Bach's work on piano-playing express similar ideas, and they also make explicit the connection between the musician's communication of affects to the audience and his or her bodily motions. Bach's work breaks down into two main parts. The first is a general section on fingering, manners (embellishments), and principles of the musical recital; the second is specifically about accompaniment—one of the key functions of a keyboard instrument. Bach states early in his work that three elements—fingering, manner, and public performance—constitute the "veritable method of playing the piano."[92] Bach's textbook aimed to subject the entire practical art of piano-playing to a theoretical analysis, particularly in his detailed chapters on manners, affects, improvisation, harmony, and basso continuo. This ambition toward a "complete" theory was very much in the spirit of the Enlightenment and contributed to the work's influence.

---

tive cultural techniques, was theorized as having different meaning and directives for men and women, as discussed in this chapter.

91. Quantz, *Versuch einer Anweisung*, 109 (Quantz's emphasis).

92. Bach, *Versuch über die wahre Art*, 1.

The third section of the book's first part is the most relevant to my case. Its title, "On recital" ("Vom Vortrage"), is similar to the title of Quantz's analogous section. Like Quantz, Bach begins by expounding criteria for good (and bad) recitals. According to him, a good recital consists of "nothing else but the ability to make musical thoughts transparent and sensitive in their true content and affect to the ear, either in singing or in playing." His definition of a bad recital is also worth noting: "The subjects of a performance are the strengths and weaknesses of the tunes, their pressure, flicking, pulling, pushing, thrust, breaking, holding, dragging, persisting, and moving forward. He who uses these things either not at all or at the wrong time has a bad recital."[93] This rather odd-sounding enumeration of the tunes' "activities"—it comes across even more powerfully in the original German—vividly illustrates the dynamics of shaping and sequencing musical tunes in a musical performance.

The communication of the musician's affects to the audience is as important in Bach's textbook as it is in Quantz's. Just as Quantz says that the musician has to "set [himself] . . . into a different affect with each bar," Bach equally emphasizes how the musician has to generate affects in himself first in order to communicate them to the audience: "And because a musician cannot move others [affectively] unless he is moved himself, he thus has to put himself into all the affects that he intends to arouse in his audience; he conveys his sentiments to them, and this is the best way to make them feel the same, along with him." Bach demands further that a musician, accordingly, should be dull and sad during dull and sad passages and merry and fierce during merry and fierce passages. According to Bach, this results in a dynamic process in which "once the musician has calmed one affect, he arouses another one. He thus constantly alternates between the different passions."[94]

Ultimately, Bach links the musician's ability to excite and calm the correct affects at the correct times—or, as he puts it, the musician's ability to "appropriate his listeners' hearts"—to the musician's bodily motions. He argues: "That all this could happen without any gestures and motions would only be denied by someone who is forced through his lack of sensitivity to sit before the instrument like a piece of carved wood."[95]

93. Ibid., 82.
94. Ibid., 85.
95. Ibid., 85. Bach uses the verb "denied" when he means to say "claimed." The confusion is in the original text: Bach does not keep track of his double and triple negatives.

Taken together, such passages in Quantz's and Bach's works emphasize that in the course of a musical performance, a good eighteenth-
century musician was supposed to arouse the correct sequence of affects
in the audience according to the program put forward by the musical
piece. The musician was meant to do this by first generating these affects in himself and then communicating them, not least through his
own bodily motions, precisely and effectively to the listeners.

It is significant that, when they explain these dynamics, Bach and
Quantz make the process sound as if the musician could, like a machine, "switch on and off" the affects in himself and in the audience,
as if he could use his moving body as the shifter for such on-and-off
switching of sentiments. The rhetoric of control, and the pedagogical
and paternalistic tone in which the textbooks are written, add a pe-
culiarly "mechanical" flavor to Bach's and Quantz's elucidations and
instructions on how to play musical pieces in public. Despite Bach's use
of the metaphor of a "piece of carved wood" as an example of what *not*
to be, musical performance is portrayed as a rather mechanical business in crucial passages of Quantz's and Bach's pedagogical works.

This ambiguity between expressive musical play and mechanical
motion is clearly exhibited in the automata, while they also challenge
the boundary between sentimental human music-making and machine
behavior. The instability of the music-machine boundary here, akin to
the human-machine boundary, is exploited at great length in the work
of the poet Johann Paul Friedrich, whose satires I analyze in chapter 5.
In the rest of this chapter, I explore the significance of "sentiment" in
the broader society and then return, in the conclusion, to the relevance
of the music-machine ambiguity in the automata and in the pedagogical literature of Bach and Quantz.

Sentimentality in Eighteenth-Century Society

The phenomenon of sentimental music-playing, as outlined, not least,
in textbooks such as Quantz's and C. P. E. Bach's, provides a telling
microcosm of broader efforts to found a new social order, efforts in
which sentimentality played a large role. The sentimental project took
a variety of forms in Europe and was manifest in virtually all realms of
society, culture, and politics: literary production and reception, friendship and letter writing, travel, moral and natural philosophy, music-
making, and conceptions of marriage and child-rearing. Such widespread attention to sentimental activity was an outgrowth of the fact

that, at a theoretical and conceptual level, affects and their theoretical foundation were in one way or another relevant to the entire program of the Enlightenment.[96]

Because the idea of sensibility (French *sensibilité* and German *Empfindsamkeit*) described an attitude involving the ability and inclination to feel and engage with one's own and other people's sentimental motions, there was a decidedly social, moral, and political dimension to it. Sensibility was, furthermore, not only about being easily excited or moved, but also about an interest in one's own "inner self," introspection, and the constitution of one's selfhood. The criterion for one's inclusion in sentimental culture was not the intensity of one's affect as such, but rather one's general disposition toward being easily moved by someone or something else. Peter Hohendahl even calls the sentimental disposition a "dialectical" relationship in which the subject finds and experiences itself. The subject's sentiments, according to this idea, are not "about" the external stimulus that brings them into being, but rather about the soul's reflection on its own motion. Hohendahl suggests that this reflexivity generates a "peculiar refraction" in which the feeling subject encounters itself. The culture of affect, in his view, builds on this type of deliberate self-experience in which it is not feelings as such that are sentimental, but rather the individual who is conscious of them.[97]

Hohendahl's view of sensibility helps us understand more clearly the mechanical reproduction of the subject-formation process. As the two women automata perform motions that are meant to generate affects in their audience, Hohendahl's proposition raises the question of whether the automata also reproduce mechanically their own reflexive relationship to their experience of sentimental motion, and therefore whether such mechanical feeling and corresponding self-awareness would be visible to a spectator. In one direction, this question leads directly to the more modern question of machine consciousness and the various thought experiments with which, since the mid-twentieth century, philosophers and scientists have debated about an observer's ability to tell whether a machine is conscious.[98] I want to pursue

96. Hohendahl, *Der europäische Roman*; Frevert, *Women in German History*; Krüger, *Das Zeitalter der Empfindsamkeit*; Riskin, *Science in the Age of Sensibility*; Wegmann, *Diskurse der Empfindsamkeit*.

97. Hohendahl, *Der europäische Roman*, 2.

98. A significant part of the question of machine consciousness has revolved around the question of how we come to believe that certain machines, when they exhibit certain behaviors, are conscious. This epistemic twist of an originally metaphysical question (whether machines can be

Hohendahl's insight in another direction. My focus is on sentimental culture in eighteenth-century Europe and on the artisanal and musical production of sentimental selfhood. I discussed artisanal production in the first part of this chapter and dealt subsequently with musical literature and production. My contention is that the two types of cultural production together bring into being the mechanical replication of subject-formation, of self-experience, and of spectatorship and thus raise the question of how "real" and reliable this process is in regard to its social and mechanical dissemination. This potential instability of sentimental-reflexive culture, along with the potential impossibility of its engendering a new social order, is also, once again, the subject of my reading of Richter's text in chapter 5.

In the German lands, more than in other European territories, sentimental culture had a political dimension. *Empfindsamkeit* both referred to a pattern of bourgeois communication and self-understanding and expressed a political ambition. Other large European states had developed centralized units governing large territories, but there was no comparable national unit for the German-speaking territories, no overarching organizational structure that would allow political participation for the emerging, and increasingly self-confident, middle classes. The traditional estate society and elements of the court society were still flourishing in most areas, and political rulers continued to govern on the basis of absolutist notions. In eighteenth-century Germany, there was no nation-state and citizens; instead, there were subjects and rulers.[99] Court culture entailed specific models of social relations that reached widely into the rest of society. Conduct at court was structured through restrictions, codes, and etiquette and rarely allowed relationships of continuity and mutual trust outside of these rules. The pressure to participate in the ceremony and hierarchy that represented power at court (and also the fierce competition for the ruler's attention) meant that interaction was ruled by rigid tactics. Having friendships was good and useful, although not as a way of increased sociability, but rather as a strategic move and advantage.[100]

---

conscious or not), as well as rapid innovation in computer technology, made possible in the middle of the twentieth century a surge of creativity in devising thought experiments involving conscious machines. See, for example, Shieber, *Turing Test*; Collins, *Artificial Experts*. For a more comprehensive view on relations between humans and machines and renderings of human-machine interfaces in the twentieth century, see Mindell, *Between Human and Machine*; Mindell, *Digital Apollo*.

99. Dülmen, *Poesie des Lebens*, 19–44; Niethammer, "Bürgerliche Gesellschaft als Projekt"; Gall, *Von der ständischen zur bürgerlichen Gesellschaft.*

100. I owe a great deal to Nikolaus Wegmann's analysis here. Wegmann argues, quoting Norbert Elias, that the precision with which the court society conformed to rules in conduct and ges-

However, in the second half of the eighteenth and the first half of the nineteenth centuries, a "cultural" nation was emerging in the German-speaking lands.[101] It included a specific understanding of the German language and a shared legacy of literature and culture. The practical innovations of the German Enlightenment—in legislation, jurisdiction, agriculture, artisanship, technical progress, hygiene, and pedagogy—went hand in hand with the explosively increasing membership in reading societies and in the readerships of periodicals and newspapers.[102] Inculcated by these and other means, major concepts and principles of the Enlightenment—among them self-responsibility, dissemination of knowledge, improvement of morals, and sanctioned social interaction—created a bourgeois identity that stood in profound tension with the legal and social structures of the estate and court society and the small territories in the Holy Roman Empire that were run as absolutist states. Sentimental and virtuous practices in reading and music were key features of this new bourgeois identity. Indeed, in the eighteenth century feeling, publicness, and intimacy were closely linked in a way that does not correspond to customs in the twentieth and twenty-first centuries.

In this political context, feelings served as a guide to interaction and as the basis of a utopian theory of society. From letter-writing, reading societies, and music-making, sentimental speech and interaction extended to the political sphere, functioning as an alternative to the courtly-feudal social rationality. The sentimental movement did not formulate a developed critique of or challenge to the court society. But its ethical system did slowly and continuously undermine the social rationality specific to life at court. Sentimentality provided a moralizing theory of society that was directed toward equality (at least in theory) and sociability, and it was for a long time not attached to a particular political macrostructure. It was thus a form of interaction that, in a very Enlightenment manner, was considered appropriate to humans in their very nature.[103]

---

ture correlated with the necessity to demarcate court society to lower classes. He shows examples from contemporary novels, where key protagonists are noble people acting as schemers who are uninterested in substantive interaction. Wegmann, *Diskurse der Empfindsamkeit*, 56–70.

101. Some contemporaries and historians have characterized it that way. Contemporary cultural elites, among them Klopstock, Schiller, and Goethe, understood this nation to be a cosmopolitan entity.

102. See chapters 2 and 3.

103. In its idealism, humanism, and desire to find a harmony of man, nature, and society, this was a typical eighteenth-century way of thinking. Members of the emerging bourgeois class used this new ethical system of interaction to discredit the court system as a site of moral

Yet, while the new sentimental social behavior thought itself unconventional and novel, it often also proved schematic, rule-governed, and indeed "mechanical." I identified an undercurrent of this tendency in my readings of Quantz's and C. P. E. Bach's textbooks.[104] Critics of sentimental culture used, among other tools, metaphors of mechanical reproduction to illustrate the sentimental movement's flaws and instabilities. The young poet Richter specifically targets the mechanical nature of sentimental music-making in a satire and uses music-playing women automata as a motif (see chapter 5).

In the Enlightenment, there was not only deliberate thinking but also deliberate feeling.[105] But "feeling" was not a singular activity, since there were considerable variety and complexity in views of sentimentality. For example, feelings were considered neither primarily irrational nor primarily feminine. Men were not always and necessarily "public" and "professional," and women not always and necessarily domestic and sentimental. And feelings played a role not only in friendship, family, and social and political institutions, but also in the social and economic makeup of the actual spaces of private houses, as those became sites for communicative "publicness" in the emerging middle classes (such as in musical recitals). Everyday noninstitutional sociabilities, organized by men and sometimes by women, became crucial sites of bourgeois culture.[106] The private-public boundary, and its relation to gender and the culture of feeling, continued to shift until well into the nineteenth century. The experience of both men and women changed profoundly in the second half of the eighteenth century, in public as well as in semipublic and domestic spaces. And throughout this period, music-making and listening were part of this experience.

---

decline—although not, as Wegmann points out, as a site of political injustice. Wegmann, *Diskurse der Empfindsamkeit*, 63. Isabel Hull coined the term "practitioners" of the Enlightenment to emphasize the deliberate and reflexive manner in which people in the German lands set out to forge and fabricate civil society. Hull, *Sexuality, State, and Civil Society*, 200.

104. On this aspect, see also Wegmann, *Diskurse der Empfindsamkeit*, 56–70. The culture of *Empfindsamkeit*, of internality and feeling, has also been interpreted as one of the rising bourgeois classes' greatest weaknesses, which supposedly became a trait of the entire politically impotent German modernity, resulting in the two world wars and the holocaust. Jürgen Kocka, "Bürgertum und Bürgerlichkeit," 48–54; Krüger, *Das Zeitalter der Empfindsamkeit*, 10.

105. This is Anne-Charlott Trepp's very nice formulation. Trepp, "Diskurswandel und soziale Praxis," 15.

106. Ibid., 15–16; as well as Frevert, *Mann und Weib*; Frevert, *Women in German History*. See also the essays in Düll and Pass, *Frau und Musik*. Jessica Riskin also challenges received notions of the gendering of the relationship between sentiment and reason. Riskin, *Science in the Age of Sensibility*, 8–9.

The eighteenth-century salon, as it emerged in Paris, Vienna, and Berlin, provided a space for gatherings of members of the aristocracy with the bourgeois classes, as well as of intellectuals and music lovers. Furthermore, although the academies founded in the period were male institutions, many salons were organized under female patronage.[107] The class status of musicians within the German-speaking estate society varied a great deal, however. Leopold Mozart, for example, counted himself among bourgeois educated people, such as professors, lawyers, civil servants, and professionals in the service of princes, the church, or local authorities. In this hierarchy, whereas itinerant musicians, private tutors, or military musicians were not desirable company, cantors and court conductors were.[108] For centuries, musicians had been servants in court service or church service, and the roots of self-confident bourgeois music-playing were in the musical societies. These institutions stood for the *Verbürgerlichung* of music, which also entailed the dissemination of music-making into medium and small towns and the emergence of more amateur musicians, processes that are not least manifest in the production of music-playing women automata in rural Switzerland and Rhineland.[109] Such processes changed the economic position of the musician in society, in a way comparable to the position shifts of literary writers, booksellers, and editors at the time. When musical societies were founded, it was assumed that music was for all estates. The difference (or similarity) in status between men and women, however, remained a contested issue.[110] At the core of the problem of musicians, sentiments, class, and gender was also an issue of agency: in the mid-eighteenth century a fundamental change was under way, from mimetic aesthetics to expressive aesthetics. In other words, music was

107. Beci, *Musikalische Salons*. Beci investigates musical salons in relation to reading and literary salons in the seventeenth, eighteenth, and nineteenth centuries (in Berlin, Paris, and Vienna), differentiating bourgeois classes in relation to the nobility and musical salons as women's culture. On salons and sociability, see also Terrall, *Man Who Flattened the Earth*, 3–6.

108. Rieger, *Nannerl Mozart*, 306.

109. The societies for music were called *collegium musicum*. For an analysis of the "devaluation" of the estate of the nobility through the Enlightenment principle of humanity and the resulting universalization of the bourgeois, see Balet and Gerhard, *Die Verbürgerlichung*. Concrete examples for their (rather contested) thesis are all performing members of the Mozart family in their own way: Leopold, Wolfgang, and Nannerl. Rieger, *Nannerl Mozart*, 309ff. Monika Mittendorfer shows the parallels between the lives of Goethe's sister and Mozart's sister; they were only six months apart in age. Cornelia and Nannerl were quintessential women of this age, and perhaps the type that we see represented in the two women automata. Mittendorfer, "Unterdrückte Kreativität."

110. On women instrumentalists, see Rieger, *Nannerl Mozart*, 315–20; as well as Hoffmann, *Instrument und Körper*.

no longer expected to express something; instead, the musician was to express herself or himself in music.[111]

The Ambiguity of the Music-Machine Boundary

The scenarios replicated mechanically by the two women automata bring into focus the problem of self-expression and its reliable communication and reproduction. The textbooks by Quantz and C. P. E. Bach demonstrate that there are in fact two sites of ambiguity for the music-machine boundary in the present case: in the pedagogical literature and in automata. Bach's and Quantz's elaborations blur the boundary between human and mechanical music-making bodies, as do the two music-playing women automata themselves, in a manner likely intended by their makers (and likely desired by their audiences), a manner characteristic for the eighteenth-century automaton as a spectacle.

Producing an accomplished, marriageable woman was one purpose of the cultural technique of music-playing and of making other people "feel." Creating a bourgeois musical culture and, in fact, a bourgeois society, was another. The dissemination of the scenario of the music-playing woman, the organized way in which it was accomplished, and the promise of moral, social, and sexual improvement attached to it point to a peculiar coincidence in the eighteenth century: the coming together of preindustrial, artisan production of technological artifacts with immense cultural dissemination of individuality, or mass-produced subjectivity. It is a great irony, and an interesting contradiction, that, in my findings, it is the technologies that are unique, individual, and artisan-produced; and it is the cultivated and sentimental human selves that are being mass-produced in large numbers. This circumstance preceded, and is distinct from, historical relationships that are more familiar to us, namely the relationships between the industrial revolutions in the nineteenth and twentieth centuries, the emergence of mass culture, and the supposed disappearance of the eighteenth-century preindustrial, unique, individual subject.

The questions and paradoxes emerging from the characteristically modern phenomenon of mass-producing subjects, and our inability to distinguish these subjects from machines, manifest in my study as a blurring of the boundary between humans and distinctly *pre*industrial

---

111. Ruth Heckmann, "Mann und Weib," 19.

machines. Only much later, after the First and the Second World War, did the blurring of the human-machine boundary in the industrial age and industrial warfare emerge as an explanation for the failures of the promises of the entire modern world.[112] In the context of the eighteenth century, however, the confusion of humans and machines, on one hand, and the mass production of this ambiguity, on the other, were instruments that helped develop the promises of mass-producing selfhood in the first place. The poets' texts discussed in chapter 5 explore the shifts and changes that occurred from the eighteenth into the early nineteenth century, when encounters between music-playing android automata and their human counterparts were no longer productive or funny but rather became fatal.

112. See chapter 6.

FIVE

Poetic Engagement with Piano-Playing Women Automata

Many writers in the years between 1750 and 1820 used android automata or other artificial humans to explore the boundaries between humans and machines. Some relied on actually existing contemporary automata, while others created a wide range of fictitious biologically or mechanically constituted entities, often in critical social commentary or in complex narratives of gloomy encounters between humans and their artificial counterparts. These texts proved very popular, and many of them served as inspiration for writers in the later nineteenth and the twentieth centuries. Two writers used specifically piano-playing women automata as motifs to develop broader intellectual and poetic agendas about the boundaries between humans and machines: Johann Paul Friedrich Richter (who later in life went by Jean Paul) and E. T. A. Hoffmann.[1] Both used these automata to destabilize the boundary between humans and machines, challenging readers' notions about what distinguished the two. In this chapter, I discuss Richter and Hoffmann in relation to contempora-

---

1. I have found no other literary texts from the period between 1750 and 1850, or after that, that deal with piano-playing women automata. "Women-machines" more generally are widespread, though, as well as human women protagonists who interact with machines. Examples range from the automaton Olimpia in Hoffmann's famous "Sandman" to Fritz Lang's *Metropolis*. See chapter 6.

neous literary work about automata by closely reading their texts (one by each author) that use piano-playing women automata as a motif.

Richter's text, "Humans Are Machines of the Angels," is from 1785; Hoffmann's "The Automata" is from 1812. The instability of the human-machine boundary is ostensibly comical in Richter's texts: he uses androids as vehicles for satirical social critique in the 1780s and makes fun of contemporary society and culture through machines. In contrast to that, the ambiguity between men and machines is fatal in Hoffmann's texts: in a series of dark stories that he wrote in the years following 1810, the hero is typically a young male student or artist whose encounters with his mechanical counterparts result in catastrophe or in death. Hoffmann's works are to this day among the most influential romantic interpretations of the interrelations between mechanics, art, and selfhood.

Along with keyboard-playing women automata, Richter's and Hoffmann's texts feature practically the entire set of mechanical androids known at the time. This makes them part of a considerable and wide-ranging body of German, French, and English texts in the late Enlightenment and early romantic period in which android figures appeared. Richter's and Hoffmann's work illustrates key elements of Enlightenment and romantic literary programs—which have had such a profound impact on our own twenty-first-century understanding of the human-machine boundary—and the role of piano-playing women automata in them. Just as I tracked in chapters 2 and 3 the representations of the two women automata in short periodical texts, in this chapter I trace how representations of the women automata unfolded in later literary culture and probe the poetic functions they served there.

The short periodical texts discussed in chapters 2 and 3 were also the conduits through which information about the women automata traveled to literary writers. We know specifically that Johann Richter relied on those texts and never personally saw a woman automaton. He was an industrious reader of many kinds of contemporary newspapers and periodicals, where short articles on automata were printed. We can tell from his notebook that he read (and copied passages from) the one article about the dulcimer player that was printed in Rudolf Zacharias Becker's *Dessauische Zeitung* in 1785 (which I discussed in chapter 3).[2] Furthermore, he read enough about the Jaquet-Droz family and their automata to refer to them by name: there are no notes

---

2. *Dessauische Zeitung*, 1785, 30:240; Müller, *Jean Pauls Exzerpte*, 159.

on them in his notebooks, but their names appear in his literary texts. Hoffmann, too, read texts about automata and was socially involved with the automaton-making family Kaufmann from Dresden, whom I discussed briefly in chapter 1. Hoffmann's works of the second decade in the nineteenth century suggest that he knew all the other automata of the eighteenth century and their makers by name. Although there is no source directly confirming that he read about either the Jaquet-Droz or the Roentgen family's woman automaton, his work makes it clear that he was familiar with piano-playing women automata, and no other automata of this kind are known, only those two. And since neither the dulcimer player nor the harpsichord player, as far as we know, was exhibited publicly between 1800 and 1880, public showings were thus not likely to be the source through which Hoffmann learned about them.

Mechanical Arts in the Literature of the European Enlightenment and Romanticism

Literary writers increasingly used androids in their works from the mid-eighteenth century onward, and fictitious androids were as prominent as the ones that were actually known to exist. Fictional depictions included mechanical automata, marionettes, dolls, living creatures such as Frankenstein's monster, mythical figures such as the golem, and other artificially conceived beings.[3] The themes and narratives constructed around these figures typically highlight confusions between "real" and "artificial" human protagonists, as well as paradoxes about mechanical motion and free will in human agency.[4] Marionettes, dolls, and automata, furthermore, were widely used metaphors during the Enlightenment and romantic periods for the variety of dependencies that subjects and individuals experienced vis-à-vis institutions, governments, aristocratic court culture, and one another.[5]

Alongside literary writers, philosophers from the seventeenth century onward also paid attention to the mechanical arts and were interested in machines, clocks, and mechanisms, as I outline in chap-

---

3. Swoboda, *Der künstliche Mensch*, 12; Sauer, *Marionetten*, 19; Wittig, *Maschinenmenschen*, 12.

4. Sauer cites and criticizes literary historians and critics who assume that the automaton as a literary metaphor took over in the eighteenth century the negativity that the marionette metaphor had carried earlier. Sauer, *Marionetten*, 123.

5. Ibid., 268.

ter 1. Descartes, for example, relied on metaphors and models of the clock and the automaton as he developed his rationalist epistemology and dualistic conception of matter. His ideas were imported into the German Early Enlightenment through the work of Leibniz and Wolff, whose belief in the powers of human reason, and in a complementary relationship between knowledge and divine revelation, relied on clock metaphors for support.[6] In political theories too, penned by Thomas Hobbes and Frederick the Great, among others, ideas having to do with clocks, machines, and automata were prevalent.[7] Julien Offray de La Mettrie was perhaps the most prominent among those philosophers who used the human-machine trope to articulate materialist and atheist ideas in eighteenth-century France, as I mention in chapter 1. In his well-known 1748 work with the suggestive title *Man a Machine* (*L'homme machine*), he relied on contemporary mechanical arts such as clock-making and automaton-making as well as on natural philosophy.[8] He has recently been rediscovered as a scholar who is more subtle and ironic than his blunt title might suggest. Indeed, he was much more than a radical atheistic, mechanistic materialist; he used sophisticated rhetorical strategies possibly belied by his reliance on the genre of the pamphlet.[9]

Literary writers in the eighteenth and early nineteenth centuries found the artisan cultures surrounding mechanisms, clocks, and automata just as interesting as the philosophers did, and also helpful for elaborating social, aesthetic, and poetic concerns in their texts. They used a range of artifacts, entities, and creatures, putting them in close and often destructive and fatal dialogues with human protagonists. Marionettes and marionette theaters, for example, were a topic of interest throughout the late eighteenth and early nineteenth centuries. Marionette theaters were part of ordinary people's experience at fairs and on the streets and also entertainment at princely courts. Along

6. On Descartes, see, for example, Descartes, *Meditations on First Philosophy*, 58; and interpretations by Mayr, *Authority*, 82–88; and Sutter, *Göttliche Maschinen*, 41–79. On Wolff, see Wolff, *Der vernünfftigen Gedancken*, chap. 4, sec. 556 (on the "parable" that the world is a clock). On Leibniz's "two clocks example," see Scott, "Leibniz and the Two Clocks."

7. Stollberg-Rilinger, *Der Staat als Maschine*, 49, 65, and 68; Mayr, *Authority*, 130.

8. He mentions all three Vaucanson automata together with an astrolab by Huygens and a watch by Julien de Roi. La Mettrie, *L'homme machine* (1865), 140.

9. Jessica Riskin suggests reading philosophers' texts about "machines" or automata from the time not exclusively as conceptual treatises, but taking into account their pamphlet-type, essayistic-ironic character. Riskin, "Mr. Machine and the Imperial Me"; Riskin, "The Adventures of Mr. Machine." See similar arguments in Jauch, *Jenseits der Maschine*, 171-172; Jauch, "Maschinentraum und Traummaschine."

with providing distraction, marionette plays often featured political satire, and as a result they were subject to censorship and persecution by the police.[10] Such plays typically dealt with contemporary cultures of sentimentality and romanticism, problems of genius and subjectivity, love stories between male humans and female dolls, and problems of marriage and the family. The marionette theater became a site where conflicts between natural and affected behavior, and between people and marionettes, were staged.[11] Perhaps the most famous and influential short story on marionette theaters was Heinrich von Kleist's *Über das Marionettentheater* (On the marionette theater). It gained considerable influence in the second half of the twentieth century, not least because of Paul de Man's famous interpretation.[12] Standard interpretations take Kleist's marionette theater to be a critique of the Enlightenment's model of "instrumental reason," a term popularized in the twentieth century by the Frankfurt School.[13]

Many aspects of the two women automata that I discussed in the previous chapter deal with the creation of bourgeois selves in the second half of the eighteenth century. Texts from the early nineteenth century onward often use marionettes and androids in a contrasting manner, to discuss alienation and exclusion from the emerging and stabilizing bourgeois society. These texts typically feature male protagonists who, for a number of reasons, end up alienated from established bourgeois society, as disowned sons or orphaned foundlings or for other reasons. Such stories center on tragic encounters of artists with the bourgeois world, but also on the crumbling of the bourgeois world itself and its rationalist-pragmatic value system. In these narratives, the bourgeois man is portrayed as limited in his outlook, paying attention exclusively to his social and economic status, while the artist seeks to transgress boundaries and move between multiple economic, aesthetic, and epistemic domains.[14] Artist-heroes in these stories encounter creatures that, as far as they know, may be either humans or machines (or, in some cases, marionettes and wax figures). Ambiguous machine-men

10. See Gunhild Berg's afterword in Iffland, *Die Marionetten*; as well as Kotte, *Theaterwissenschaften*, 244; Kotte, "Iffland, Kleist und das Marionettentheater."

11. Goethe's play *Triumph der Empfindsamkeiten* dates from 1777, Johann Friedrich Schink's *Marionettentheater* from 1778. A comedy in one act entitled *The Marionettes*, by the actor, playwright, and director August Wilhelm Iffland dates from 1808.

12. De Man, *Rhetoric of Romanticism*, 265–80; Jackson, *Harmonious Triads*, 75–82.

13. Vogel, *Against Nature*, 51–69, 101–44; Whitebook, "The Problem of Nature in Habermas"; Alford, *Science and the Revenge of Nature*, 49–69, 119–38.

14. This is normally discussed exclusively in relation to male figures.

specifically serve to represent people who work in the bureaucratic service of a state—a preferred target of criticism in romantic literature.[15] A story by Clemens Brentano about a clock-maker named Bogs explicitly connects an individual's alienation from bourgeois society to mechanical arts and artisanship.[16] In his ordinary life, the clock-maker Bogs makes clocks, instruments to rationally measure time and efficiency. But he also has inclinations toward the fantastic, the dreamy, and the playful. He seeks admission to a quintessentially bourgeois association (Schützengesellschaft), but his attempts fail.[17] The constitution of subjecthood (in aesthetic, moral, and rational ways) was discussed intensely in both Enlightenment and romantic literature. And clocks, automata, and mechanisms, as well as their makers, provided props to hash out the fault lines and contradictions in these discussions.

Some poets of the time, on their often extended journeys, witnessed with their own eyes the remains of eighteenth-century automata, in particular the remains of Vaucanson's automata. Gottfried Christoph Beireis (1730–1809) was a widely known collector, physician, physicist, chemist, and lawyer. He lived in Helmstedt, taught at the local university, and owned a substantial collection of physical and chemical apparatuses, mechanical curiosities, and paintings, among other objects.[18] Beireis was known to be a generous host, although he was often not taken seriously. His collection included the remains of Jacques de Vaucanson's three famous automata from the 1730s, which Johann Wolfgang Goethe saw during an 1805 visit. In his diaries, the *Tag- und Jahreshefte*, Goethe mentions the flute player and the duck (but not the galoubet player). The Vaucanson automata were in a state of dis-

---

15. Ludwig Tieck's *William Lovell*, in ten volumes and first published in 1796, is such an epistolary novel. It is concerned with the demise of the young Englishman William Lovell, who tries to find a living and a life after being disowned. Drux, *Marionette Mensch*. Another example is the short, novel-type text *Nachtwachen*, published in 1804 under the pseudonym Bonaventura. The text deals with sixteen vigils of the night watchman Kreuzgang (meaning "cloisters" in translation, referring to the place where he was found as an orphan infant), who is an outsider, outside bourgeois society, and lives in the dark of the night, where he encounters beings that are on the boundary between humans and machines. Böning, *Widersprüche*; Katritzky, *Guide to Bonaventura's Nightwatches*.

16. Brentano, *Entweder wunderbare Geschichte von Bogs dem Uhrmacher*. Brentano (1778–1842) worked closely with Achim von Arnim, and Arnim ended up marrying Brentano's sister. Brentano and Arnim are considered the two main representatives of the so-called Heidelberg romanticism, the middle period of German romanticism. Brentano also wrote a fairy tale called *Gockel, Hinkel, Gackeleia* (1838), which deals with a mechanical doll. See also Fetzer, *Romantic Orpheus*, 283.

17. See also Schmitz-Emans, *Seetiefen und Seelentiefen*, 106–9; Sauer, *Marionetten*, 151–64.

18. He accumulated them for teaching purposes: *Neue deutsche Biographie*, 2:20; *Allgemeine deutsche Biographie*, 2:293–94.

repair, and Goethe writes that he found them to be "paralyzed."[19] Twentieth-century historians' accounts of Goethe's reaction to Vaucanson's automata are problematic in a manner similar to twentieth-century accounts of eighteenth-century reactions to the Jaquet-Droz automata and Roentgen's dulcimer player. In chapters 2 and 3 I demonstrated that there is neither evidence of a general shared sense of excitement about automata in eighteenth-century Europe nor any recorded reactions treating the android automata as threatening or foretelling how human selfhood was becoming mechanized. Similarly, there is no evidence to support the contention of some literary historians that Goethe, when he visited Beireis's collection, saw "himself mirrored" in the remains of Vaucanson's automata. In fact, Goethe presents a sober, mildly scoffing account, making fun of Beireis's own excitement, and offering no reflections about human self-recognition in an android machine.[20] Achim von Arnim, among the most important figures in German romanticism, visited Beireis a year later, in 1806. He saw the remains of Vaucanson's duck there and described the encounter in his novels *Gräfin Dolores* and *Isabella von Ägypten*.[21]

Mythical and biological androids such as the golem and Victor Frankenstein's monster emanated from their own distinct literary traditions, yet even though these figures originated separately in earlier periods, they became particularly important for twentieth-century interpretations of the android motif and, in that period, merged to some degree with it. The legend about the golem—an artificial human made of clay and brought to life through magic—appeared in the twelfth century and circulated frequently in the sixteenth century and again in the eighteenth century, in part because of the increasing popularity and fame of android automata. The golem attracted renewed interest in the twentieth century. Some literary renditions of the golem became best sellers; cheap editions were popular among soldiers in the trenches of the First World War.[22] Mary Shelley's *Frankenstein* deals similarly with

19. Goethe, "Tag- und Jahres-Hefte," 141.
20. Walter Hettche makes that argument, too: Hettche, "Goethes Sommerreise," 66. He responds to Matthes in "Goethes Reise," 152.
21. Arnim, *Armuth, Reichthum, Schuld und Busse*; Arnim, *Isabella von Aegypten*. He also wrote a shadow play in two acts, *Das Loch*, and a puppet game, *Die Appelmänner.*
22. The golem is a mystical figure to which many legends are attached. The main legend tells of a rabbi who creates an obedient servant out of clay. On the golem's forehead, the word *emeth* (truth) is written. When the rabbi takes away the word's first letter, the remaining word is *meth* (death), and life is taken away from the golem. The most important rabbi connected with the golem legend is Rabbi Löw of Prague in the late sixteenth century. Jacob Grimm mentions the legend in 1808, as does Achim von Arnim in 1812 in *Isabella von Aegypten* (a female golem) and E. T. A. Hoffmann, often in the context of other androids. The creature appears, then, in Gustav

a living being created from no-longer-living matter and brought to life through quasi-scientific (electrical) forces. Among the many themes that Shelley's novel addresses, the destructive relationship between creator and creature is particularly pronounced. Shelley's *Frankenstein*, like the golem, gained tremendous influence through literary and film adaptations in the twentieth century.[23]

Within this field of literary, philosophical, and critical work spanning two hundred years, keyboard-playing women automata are recognizable figures that can be traced in distinct ways, thus making more specific the sometimes amorphous discussion about the role of android automata in modern society and in our understanding of the relationship between humans and machines. Rather than assuming that eighteenth-century automata represent a coherent and focused field of meanings, or assuming that their Enlightenment and romantic commentators' texts form a coherent response, I trace the internal boundaries, delineations, and disruptions in this horizon of artifacts and texts.

Theories of Literature in the Enlightenment and Romanticism

The two texts that I focus on in this chapter were written almost thirty years apart: Richter's "Humans Are Machines of the Angels" dates from 1785 and Hoffmann's "The Automata" from 1812. Those thirty years saw profound changes in the theory and practice of literature, changes that affected notions of genres, readership and authorship, and representation. Innovations in literary production and criticism also overlapped with epochal changes in political and philosophical notions such as those relating to knowledge, rationality, selfhood, and legitimation.[24] New ideas about individuals and society; new sites, functions, and economies of literary production; and new literary agents provoked and promoted new theories and practices of literature. The dissolution of the courtly social order on the European continent, furthermore, created new functions for literature: the bourgeois individual needed a

---

Meyrink's *Golem* (1915) and Stanislav Lem's *Thus Spoke Lem* (1981), in which "Golem XIV" is a machine intelligence of the twenty-seventh century. Meyrink became a best-selling author during the First World War. Twenty-nine authors in the German-speaking areas alone used the legend to write book-length stories in the first half of the twentieth century. On the relevance of android literature during the two world wars, see also chapter 6.

23. See Joseph Pearce's introduction in Shelley, *Frankenstein*, xii; and Hitchcock, *Frankenstein*; see also chapter 6.

24. Cassirer, *Philosophy of the Enlightenment*; Beiser, *Enlightenment, Revolution, and Romanticism*.

stage, as it were, to experience itself as an individual and a place to participate in shaping its identity. In searching for this identity, literature provided forms and media for self-exploration and self-representation, just as the sentimental musical practices replicated in the two women automata did.[25]

The literary theories of the Early Enlightenment had claimed that the aim of literature should be to disseminate Enlightenment ideas about reason, utility, and humanity in pleasing and intelligible ways.[26] Toward the middle of the eighteenth century, resistance to those ideas emerged. Opponents found that model limiting and conceived of literature as the product of an individual creative genius, instead of a rule-bound Enlightened pedagogue. Key themes of this new literary movement were spontaneity, genius, individuality, feeling, and originality. Combining the two strands, the literary theories of the late Enlightenment complemented the strict rationalism of the Early Enlightenment with theories of feelings and individuality as well as with an irrational element that placed it in some tension with the age's mainstream. These theories also radicalized the Enlightenment in a social and economic sense: they made the poet an advocate for the rights of the middle classes, a mouthpiece of the people, and an agent in the struggle for political emancipation.[27] This emancipation, again, has a correlate in the practices of sentimental playing and listening to music, practices that the two women automata perform.

Alongside genres that represented and in turn buttressed bourgeois consciousness, there emerged satirical forms that criticized the sentimental practices and changing forms of such consciousness.[28] Satire was a major instrument of social commentary, one whose use by Richter in the second half of the eighteenth century I discuss below. Satire provided a means for expressing and putting pressure on the contradictions that arose in discussions of subjectivity, social conditions, and alienation. The genre offered an early critique of the Enlightenment, although it should be noted that the Enlightenment had self-critical

25. Ueding, *Klassik und Romantik*; Rolf Grimminger, *Deutsche Aufklärung.*
26. The most important representative was Johann Christoph Gottsched (1700–1766). Möller, *Rhetorische Überlieferung.*
27. This was mainly formulated in the movement of Storm and Stress (*Sturm und Drang*), with Gotthold Ephraim Lessing as one of its spearheads. The movement also marked the beginnings of the counter-Enlightenment. Hill, *Literature of the Sturm und Drang.*
28. In the German-speaking lands, this happened in the wake of Swift's *Gulliver's Travels* (1726) and Voltaire's *Candide* (1759).

elements from its very beginning.[29] The most important genres of the eighteenth century in this regard were the play and the novel, because they functioned in the service of the bourgeois process of introspection and emancipation and were also conducive to the more old-fashioned directives of educating and moralizing. The novel evolved from a moralistic and universalistic genre into a subjective illustration of the individual, however, and subjective and autobiographical elements gained influence in the later decades of the eighteenth century. This shift, together with satirical intervention, added crucial layers of reflection to the bourgeois processes of self-exploration and self-definition.

The romantic era took the challenge to the utilitarian view of literature from the Enlightenment further. Romantic aesthetics aimed to separate literature from the other arts and to emancipate literary texts as systems of representation in their own right, independent of philosophy, pedagogy, commercial interests, and the natural sciences. The romantics formulated, in contrast to the prevalent eighteenth-century view, the principle of the autonomy of poetry.[30] The focus thus shifted from the social and political contexts of a literary work's writer and reader to their subjective relationship to the work. This meant in particular that contemporary political and social tensions were no longer expected to be solved in the medium of art. The resulting contradictions between the self-professed artist and the social, economic, and cultural reality of the bourgeois world are recurring motifs of romantic literature, staged in the form of marionettes or mechanical dolls.[31]

E. T. A. Hoffmann set himself firmly in the romantic tradition of the deliberate use and development of writing about selfhood. His way of using piano-playing women automata (and other automata) for this goal sets him apart from Richter and Richter's use of the motif. Hoffmann and other romantic authors took the late-Enlightenment interest in the irrational further by directing attention to the "dark" side of rational discourse: terror, fear, magic, and madness.[32] Rather than being in opposition to the Enlightenment in this undertaking, however, the romantics understood themselves to be complementary to it. The Enlightenment had drawn up a model in which subjectivity and identity were clearly distinguished from (internal and external) "brute"

---

29. *Deutsche Literaturgeschichte*, 144–53.
30. Kremer, *Prosa der Romantik*, 4.
31. *Deutsche Literaturgeschichte*, 157.
32. Hoffmann was deeply involved in the contemporary natural science, engaging works such as Schubert, *Ansichten von der Nachtseite der Naturwissenschaft*.

nature. The romantics reintroduced archaic structures into the human psyche, desires and wishes such as for the exotic, the adventurous, and the sensual, all of which had been "policed" in the Enlightenment.[33]

Both Richter and Hoffmann were fascinated by the contemporary automaton culture. In the following sections, I sketch elements of Richter's work as a young man, then read closely a selected text of his that deals with automata; and, finally, I read one of Hoffmann's texts on automata as a counterpoint to Richter's and a transition to later periods of literary renderings of android automata. While Richter is a key representative of the print and musical culture that I investigated in relation to the automata, Hoffmann provides us with a way to understand the beginnings of the journey that eighteenth-century mechanical androids took from the romantic period to the end of the Cold War.

Richter's Texts on Automata

All of Johann Richter's early literary works, written between 1780 and 1790, were satires.[34] As a young man, he carefully studied satirical literature from the English eighteenth-century tradition, notably Alexander Pope and Jonathan Swift, and adopted this genre as his own. He processed their inspirations in what he called his "satirical vinegar factory," producing an output of hundreds of satires during the 1780s.[35]

Richter's satires cover a broad spectrum of themes. They contain impressions of the major cultural and intellectual habits of his time, ranging from religious practices, court culture, musical concerts, and educational journeys, to smaller-scale observations about spectacles, works of art, and curiosities. Recurring topics include theological sermons, natural philosophy, alchemy, and medicine, as well as human foolishness, genius, and love, including idiosyncratic engagements with quotidian practices such as eating, beard-grooming, and going to church on Sundays.[36] Especially noteworthy is Richter's fondness for

---

33. Ultimately, romantic thinking also entailed a rejection of the bourgeois world and modern civilization and the embracing of lyrical introspection and previous forms of life and times such as the Middle Ages. *Deutsche Literaturgeschichte*, 175.

34. In 1780 he was seventeen years old. He published his first novel in 1793.

35. Richter collected a substantial number of these satires in two volumes and published them in the 1780s. The titles were *Die Grönländischen Prozesse* (1783) and *Auswahl aus des Teufels Papieren* (1789) (usually referred to as *Teufelspapiere*). Both volumes were published under the pseudonym J. P. F. Hasus. Lindner, "Jean Paul als J. P. F. Hasus," 411–12; Schlaffer, "Jean Paul," 394.

36. His literary renderings typically engage "snug domestic scenes" and "the humble and the homely." This spot-on phrasing is in Casey, *Jean Paul*, 3–4.

self-reflexive themes: he likes to write about the process of writing, relationships between writers and critics, and relationships between writers and readers.

Richter's satires are fragmentary and often enigmatic collages, usually with no clear narrative structure underlying them. Common challenges for his reader include his "quirky vocabulary, wayward syntax and conceits, [and] endless metaphors."[37] They also have an "open-ended fragmentariness" and a quickly shifting narrative mode that are in remarkable contrast to classical Enlightened literature, which was typically "objective, ordered, disciplined, contained."[38] Music-making women automata, and android automata in general, were among Richter's snapshots of cultural and domestic activity. Indeed, android automata and humans usually live together in the same social and cultural spaces in Richter's literary texts, thereby blurring the boundaries between humans and machines.

Five of Richter's satires deal with machine-men and artificial humans and have the long-winded titles that are typical of him: "Humans Are Machines of the Angels," "The Machine-Man and His Properties," "Simple but Well-Disposed Biography of a New, Pleasant Woman Made of Pure Wood Whom I Invented and Have Married Long Since," "Most Humble Introduction of Us, of All Players and Talking Women in Europe, against the Establishment of von Kempelen's Playing and Speaking Machines," and "Personal Data on the Servant and Machine-Man."[39] These texts amount to complex assemblages of metaphors and references, often swiftly juxtaposed and with little detailed elaboration. "Humans Are Machines of the Angels" offers the reader a little more continuity and deeper reflection on android automata in general and piano-playing women in particular. I will read this text most closely, after introducing the others.

Richter was familiar with the automata that were built in the course

---

37. Ibid., 6.

38. Ibid., 5. In the history of eighteenth- and nineteenth-century literature, Richter is normally grouped in a separate section together with others who were not part of a mainstream working circle or literary movement. It is usually a group of three, he together with Hölderlin and Kleist. See Sørensen, *Geschichte der deutschen Literatur*, 330–42; *Deutsche Literaturgeschichte*, 186–94; Borries and Borries, *Deutsche Literaturgeschichte*; Hoffmann and Rösch, *Grundlagen, Stile, Gestalten*, 213–27.

39. The earliest text is the one on humans, machines, and angels. It is part of the satire collection written between *Grönländische Prozesse* and *Teufelspapiere* (in the mid-1780s); the "Most Humble Introduction," the "Woman Made of Pure Wood," and the "Machine-Man" are in *Teufelspapiere* and were thus written in the second half of the 1780s; and "Personal Data on the Servant and Machine-Man" is a revised version of the "Machine-Man," published as part of Richter's *Palingenesien*, a set of stories published in the late 1790s.

of the eighteenth century, as well as with other contemporary spectacles. His texts make references to such existing contemporary automata and their makers (often by name), but also to others whose names or characteristics are fictitious.[40] Depending on the contexts in which they appear in his peculiar literary worlds, his automata write, speak, pray, serve their masters, make meteorological observations, conduct musical orchestras, or preach sermons in church—just as his humans do.

One of the five satires, "The Machine-Man and His Properties," explicitly mentions the name Jaquet-Droz in the context of music-making automata.[41] The text is a collection of impressions and comments typical of Richter's style, and the references to automata are mostly allusive and passing. The text is not a full-fledged treatise and does not sustain a consistent philosophical argument. I emphasize this because other scholars have read Richter's satires more as systematic, propositional texts, similar to eighteenth-century philosophical texts.[42] For my purposes, I do not follow this strategy for reading Richter. I find his texts fragmentary and open-ended in their expression, both when they talk about philosophy and when they talk about concrete automata. Hence, I regard them as snapshots of and commentary on specific historical situations in the eighteenth century, situations in which humans and android automata meet. Furthermore, I treat his texts as

---

40. The list of automata mentioned in Richter's texts matches the list of automata that existed in the eighteenth-century: we can assume that he knew all of them.

41. "Machine-Man" is part of *Teufelspapiere* and was written in the second half of the 1780s.

42. Monika Schmitz-Emans, for example, holds that Richter's satires indicate that the machine is for Richter the epitome of "dead, soulless immanence" and that the "process of mechanization is," in Richter's view, "a cipher for the displacement of freedom and spontaneity through law and order, of the individual through the norm." The general tenor of Richter's satires, she says, is that "machine" is man-as-docile-being and product of norms, conventions, and unreflected rituals. Schmitz-Emans, "Georg Christoph Lichtenberg," 75. Frank Wittig, in a historical study of "machine-men" as a motif, explicitly talks about an "aversion" that authors such as Richter and E. T. A. Hoffmann harbor against the idea of an artificial human, and he claims that in the early nineteenth century an artificial piano player was "ostensibly still a threat." Wittig, *Maschinenmenschen*, 58. He states in a different context that "behind an automaton emulating human behavior, there is always the lurking danger of the human becoming an automaton" (80). Lieselotte Sauer interprets Richter's satires on machines as saying that the mechanistic philosophy does "nothing but . . . turn the entire animate world into machines," and she interprets his use of the automaton motif as his way of "debunking the threatening individual and social machine-ness and thus debunking hypocrisy and banality." Sauer, *Marionetten*, 73–74. Peter Gendolla argues similarly that, through Richter's satirical exaggeration of automata's human qualities and capacities, a characteristic tendency built into automata becomes "ever so clear," namely the tendency that the individual is robbed of his or her autonomy in favor of the "large machine into which the social (*das Soziale*) transforms itself." Gendolla, *Anatomien der Puppe*, 98–99. On the status of this type of eighteenth-century text vis-à-vis full-fledged systematic treatises, see, once again, Jessica Riskin's "Mr. Machine and the Imperial Me" and "The Adventures of Mr. Machine."

compositions and reconfigurations of themes, figures, and questions—rearrangements that bring out certain themes in novel ways and generate novel philosophical and cultural agendas. These agendas, not least, tie back into the themes I discussed in chapter 4: selfhood, music-making, and sentiments.

In "The Machine-Man and His Properties," the narrator's tone is playful, casual, and ironic—rarely serious or focused. Early on the narrator explains his relationship to the "machine-man": he says he greets the machine-man in the morning and the evening but otherwise "can't stand him, because of his damn follies: he does everything through machines."[43] No further explanation attends this declaration, and the subsequent narration is never resentful or critical in a sustained way. There is thus no consistently strong support for the assertion that Richter's narrator, much less Richter himself, as some critics claim, is resentful of the machine-man, or of machine-men as such.[44] The narrator describes the many machines that the machine-man brought into his household, running through a substantial list of eighteenth-century automata: a replica of the "the writing-machine of the emperor" that had been shown to the machine-man in Vienna, the calculating machine of the "Reverend Hahn," the speaking machine by von Kempelen, praying machines, and machines for eating and chewing.[45] Other machines mentioned in the text include an alarm clock that makes fire and opens curtains in the morning, a maid, and even a machine wife.

"The Machine-Man and His Properties" is a good example of how Richter's texts, even if they are not deliberate or focused philosophical treatises, are, taken as a whole, a veritable compendium of the big issues, ideas, and spectacles of his time, a kind of microcosm of the last two decades of eighteenth-century Europe. For example, there is an important, but rather unexpected, reference to materialism in the text on the machine-man. The narrator states that, in this world of machines and machine-men, machine-men would not even keep their "I" (*Ich*)

43. Jean Paul, *Werke*, vol. 2, part 2, 447.

44. See, once again, the works by Schmitz-Emans, Wittig, Sauer, and Gendolla quoted in note 42.

45. Jean Paul, *Werke*, vol. 2, part 2, 447. The praying machine and the eating and chewing machines are probably fictitious: no automata of these kinds are known or are referenced as really existing, as far as I know. The writing machine was by Friedrich von Knaus. See the folder "Friedrich von Knaus, Alesschreibende Wundermaschine," Archives of the Technisches Museum, Vienna (see also chapter 1, note 54). Richter's text says that the machine makes "double and manifold copies," which is not an accurate representation of this machine. Rather, Richter's text takes the idea of a writing machine and adds further ideas to it. The calculating engine is a reference to a device by Philipp Matthäus Hahn. See chapter 1, page 36, and Munz, *Philipp Matthäus Hahn*, 39–40.

but would "have a materialist carve one" for them.[46] The text makes fun of materialism, but no more than of other philosophies. There is no sustained examination of materialism, no open or obvious hostility toward it, and thus little basis for reading Richter's text, or Richter himself, as condemning it.

Richter's specific reference to the Jaquet-Droz family comes about in the context of a wintertime concert organized by the machine-man, a machine-concert in which "neither the composer nor the music copyist nor the conductor nor the musicians were alive."[47] The narrator states that the musicians, who "did wonders on the flute, the piano and on an organ with card-board pipes," were "carpentered partly by Vaukanson and partly by Jaquet Drotz and son."[48] At the end of the concert, the machine-man states, "I can flatter myself that nowhere else is there a music band, a concert hall, an orchestra in which nothing but machines *play*," and the narrator responds: "But I have sat in those concerts where nothing but machines were *listening*, and where a sentimental drum roll *moved* [*rührte*] the human hearts that were present."[49] This is a brief reference, in theme and vocabulary, to those tenets of sentimental music-making that I discussed in chapter 4. A longer reference to the sentimental age appears in the text on humans, machines, and angels that I discuss later in this chapter.

The satire entitled "Simple but Well-Disposed Biography of a New, Pleasant Woman Made of Pure Wood Whom I Invented and Have Married Long Since" is, like the machine-man piece, in the satire collection *Teufelspapiere* and was written in the second half of the 1780s. At twenty-nine pages, it is a comparatively long text, and it is, like the others, a collage with no issue or theme pursued in depth. The text contains references to a rich assortment of androids and artificially made humans, alluding in passing to the speechlessness of women and to von Kempelen's speaking machine in this context.[50] The text

46. Jean Paul, *Werke*, vol. 2, part 2, 452.
47. Ibid., 449. Some did not even look like humans, the text says. The composer was a set of dice, with the help of which the machine-man threw together music according to fashion and rules of composing. The copyist was an "improvising machine" (*Extemporisirmachine*), on which he played the diced-out melodies, and it wrote them down.
48. Ibid., 449. The text mentions the name "Vaukanson" (in this spelling) once again, on page 452, in the context of artificial "doves, eagles, flies, and ducks" (it mentions Vaucanson together with Archytas and Regiomontanus).
49. Ibid., 447 (Richter's emphases).
50. Ibid., 403, 417. The first-person narrator dresses his wife in Parisian clothes and shows her in the window, so that the whole "female town" would imitate her style. Hoffmann's story "The Sandman" bears some resemblance to this scene, as Olimpia, too, gets displayed in a window.

entitled "Most Humble Introduction of Us, of All Players and Talking Women in Europe, against the Establishment of von Kempelen's Playing and Speaking Machines" is also in *Teufelspapiere* and was thus written in the second half of the 1780s.[51] Richter deals in this text with issues surrounding labor and industrialization, especially the substitution of machines for human laborers. He starts with common examples such as spinning and weaving machines and then extends the theme by imagining gaming machines, civil-servant-administrator machines in high government offices, churchgoer machines, thinking machines, and book-writing machines (that put writers out of business and salary), as well as legal courts where humans and speaking machines sit on the judge's bench together and arrive at unprejudiced verdicts. The narrator trusts the machines more than the humans to exercise good judgment.[52] Descartes also receives a brief mention—in the context of pets: the narrator purports to be relying on Cartesian philosophy when explaining that animals are "well-designed machines that, like all machines, are much better at certain human actions, such as hearing, smelling, seeing, loving, and hating."[53]

Richter's Texts and His Writing Techniques

Richter's work habits directly influenced his literary production. In particular, the fragmentary form of his texts grew out of reading and writing habits that were typical of the printing and literary culture of the late Enlightenment, such as the practices of copying and pasting in calendars and periodicals, discussed in chapter 2. Richter was an obsessive writer and reader, from a young age filling numerous notebooks with thousands of informative, fanciful, and sometimes random ex-

---

51. The text has substantial length (eighteen pages), making this one and the one about the wooden woman the two longest ones. The beginning is rather grand and claims that "it is well-known that recently two strange machines made a big tour through Europe; one of them played, the other spoke." Again, a text by Hoffmann shows some resemblance to Richter's: the beginning of Hoffmann's story *The Automata*, discussed later in this chapter, resembles this opening. Richter's text gives no description of the two machines introduced at the beginning, and it remains unclear what kind of machine the "playing" machine is, whether existing or fictitious. The speaking machine is a clear reference to the one made in the eighteenth century by the court mechanic and court secretary Kempelen (who was employed by Maria Theresa). See chapter 1, page 35.

52. Jean Paul, *Werke*, vol. 2, part 2, 168–73. He mentions also a factory where female speaking machines are built.

53. Ibid., 170.

cerpts from his reading, along with his own commentary on them and on his own experiences.[54] He started the first notebook in 1778, when he was still a schoolboy, and the last one dates from 1825, the year of his death. During the intervening decades, he created sixty such notebooks. A local pastor had originally suggested that he excerpt books in preparation for his university education in theology, not least because he was too poor to afford a personal library, a condition that persisted into his adulthood.[55]

These notebooks provided him the topical allusions, metaphors, and aphorisms for his satires. Music-making women automata were among the many "cultural quotations" from this repository.[56] Richter's fragmentary, seemingly arbitrary style was grounded in his reliance on a catalog of presorted references that were at hand. The excerpting, one critic argues, first *dis*sociated texts, while the catalog and Richter's writing subsequently *a*ssociated texts.[57] Richter's compositional practices went hand in hand with a self-ironic mode of narration in which he plays up "his own presence and [parades] his artistry and, indeed, artificiality of the work." He constantly interrupts his characters' stories. At any moment in a Richter text, as Timothy Casey says, one might meet him "wandering through the book."[58]

Richter's eccentric writing style was not a veneer, literary critics have argued, or mere self-display and indulgence. Rather, it was an expression of fundamentally eccentric life experiences, a real sense of dislocation and disorientation.[59] Richter was born in 1763 into the family of a pastor and organist in a poor and provincial area, and he remained isolated throughout his life, living far from the centers of European intellectual activity.[60] He had difficulty supporting himself and repeatedly changed towns to avoid creditors. Indeed, scarcity in almost all re-

---

54. In just a few years, this enabled him to develop a profound understanding of the contemporary intellectual landscape. See Müller, *Jean Pauls Exzerpte*; Blair, "Humanist Methods in Natural Philosophy"; Klappert, *Die Perspektiven von Link und Lücke*.

55. Casey, *Jean Paul*, 6; Schweikert, *Jean Paul*, 1. Richter's manuscript collection is held at the *Staatsbibliothek* in Berlin and is being cataloged. See Goebel, *Der handschriftliche Nachlaß Jean Pauls*. Berend calls Richter's collected manuscripts an insight into the poet's "workshop." Berend, "Prolegomena zur historisch-kritischen Gesamtausgabe," 3. Among his contemporaries, Richter was known for the excerpts. See, for example, Hegel, *Ästhetik*, 1:289.

56. Schweikert, *Jean Paul*, 1. Many allusions were not comprehensible even to his contemporary readers. Berend, "Prolegomena zur historisch-kritischen Gesamtausgabe," 3, 21.

57. Müller, *Jean Paul im Kontext*, 78–79.

58. This presence of authorship is normally considered to be typical for later periods, in particular romanticism. Casey, *Jean Paul*, 6.

59. I am following Casey here again. Ibid., 5.

60. For more than fifty of his sixty-two years of life, he lived in villages and small towns. Hoffmann and Rösch, *Grundlagen, Stile, Gestalten*, 215.

spects shaped his literary expression: scarcity in economic resources, in intellectual exchange, and in social companionship. Richter's obsessive reading and writing from his early youth onward, the massive, almost obtrusive, erudition in his work, and the creation of his own inner, literary universe were in many ways responses to these deprivations.[61]

Richter kept outside of the literary mainstreams of his time, in particular the Weimar classicism that was run by Goethe and Schiller in Weimar and the various romantic schools in Berlin and Heidelberg. Nevertheless, even during his lifetime he reached a status and place equal to those of his romantic and classicist contemporaries.[62] He was in contact with his contemporaries—eminent writers, publishers, and critics—but repeatedly turned down invitations to join any of the larger literary circles.[63] This self-imposed isolation did not harm his success. Even in his youth, Richter was one of the main social and political commentators and satirists of his age, and his novels (written in the 1790s and after) made him even more famous. There was a time when he was the most popular novelist in Germany, his work considered more influential than even Goethe's.[64] He also became known and appreciated in England and North America through the writings of Thomas De Quincey, Henry Wadsworth Longfellow, Samuel Taylor Coleridge, George Eliot, and Thomas Carlyle, and in France through Madame de Staël.[65] His novels continued to receive the praise of literary writers in the later nineteenth and twentieth centuries.[66]

Since Richter's work is such a unique and comprehensive microcosm—an encyclopedic index—of the intellectual and cultural landscape of late-eighteenth-century Europe (including the calendars and

61. His life was the complete opposite of, for example, Goethe's model of a scholarly identity: no cosmopolitanism, very little travel, and little intellectual and cultural interaction. Schlaffer quotes Goethe from a letter of 1795 to Schiller, in which Goethe indicates how isolated Richter was. Schlaffer, "Jean Paul," 389–90.

62. *Deutsche Literaturgeschichte*, 186.

63. Goethe invited him on several occasions to join him and Schiller in Weimar. He also knew, and received support from, influential publishers such as Friedrich Nicolai and Karl Philipp Moritz. Hoffmann and Rösch, *Grundlagen, Stile, Gestalten*, 214–18.

64. His first successful novel, published in 1793, received a great deal of praise from Karl Philipp Moritz, whose support helped to get it published. This work was the first of his to be published under the name Jean Paul. This pseudonym was a sign of his admiration for Jean-Jacques Rousseau. His next novel appeared in 1795 and made him famous overnight: it was considered one of the biggest literary successes since Goethe's *Werther*. Frenzel, *Daten deutscher Dichtung*, 1:263, 267.

65. Richter was regarded among non-German-speaking readers as a representative of the supposedly German characteristic of inwardness. Casey, *Jean Paul*, 4.

66. Writers such as Arno Schmidt and Heinrich Heine considered him their role model. Ueding, *Jean Paul*.

periodicals discussed in the preceding chapters), his satires about machine-men and machine-women provide insight into the place of android automata in this landscape. His writing style and the conditions under which it developed intervened in significant ways in the convoluted debates in German-speaking literature in the late Enlightenment, debates that Wilhelm Schmidt-Biggemann once called a "syndrome" of motifs, metaphors, and forms of argumentation. Not only that, but Schmidt-Biggemann takes Richter's satires to be "proto-typical" for this syndrome.[67] Richter's excerpts became more elaborate after 1782. In their entirety, they are a depot of ideas that is unparalleled in German literary history in terms of its range, content, and originality. His work characteristically blurred boundaries between note-taking, collecting excerpts, and creative writing. Given his prolific writing, his large horizon, and his sharp eye, it is valuable for me to analyze how he "processed" in his satire factory snippets about piano-playing women automata together with other social and philosophical concerns of his time.

The Satire "Humans Are Machines of the Angels"

Richter's 1785 text "Humans Are Machines of the Angels" brings together all the eighteenth-century phenomena I discussed in chapter 4: keyboard-playing women automata, musicians' moving bodies, and the communication of sentiments during musical performance. His text integrates, furthermore, the cultural scenario that is the subject of my study—women playing the piano and communicating sentiments—with the unstable boundary between humans and android automata.

"Humans Are Machines of the Angels" is a short text, about four pages long, and it is among the earliest of Richter's writings about automata.[68] The text opens with the narrator's claim that we had to be enlightened for a long time to come to realize that the world does not exist because of us and that it will take even more Enlightenment to make us realize that we actually live here because of higher beings, which we call angels. They are the true inhabitants of the earth; we are just furniture. The narrator attempts to offend the reader further

---

67. Schmidt-Biggemann, *Maschine und Teufel*, 15.
68. The original title reads: "Menschen sind Maschinen der Engel." Jean Paul, *Werke*, vol. 2, part 1, 1028–31. The editor Eduard Berend most likely gave the story this title in his edition of Richter's complete works in 1927.

by stating that since so few of our activities here on earth seem to contribute to our well-being, we should have had doubts a long time ago about whether they actually serve our own purposes. Indeed, to him it seems "obvious [that] our industriousness, which works against our happiness, is conducive to other beings' happiness, whose hands conduct ours as their tools." The narrator concludes that it is therefore not a poetic saying, but the "bleak, naked truth" that we humans are "mere *machines*" that serve higher beings who were first chosen to inhabit the earth.[69]

What kind of machines are humans, then, according to the logic of this text? And how did they come into being? The narrator explains that when the angels first entered the earth, they "did not yet have the numerous human-machines on which they can now congratulate themselves." Rather, it was only piecemeal that they invented "machines, or, as we call them, humans." Humans are thus machines that were built by the angels for various purposes; now, says the narrator, there are enough machines on the earth to serve "all needs" of the angels.[70]

In the following passage, the narrator introduces an example. He talks about an angel who built chess-playing machines, for the sake of "curiosity and pleasure," rather than for "utility."[71] Richter's choice of chess-playing machines here is not accidental, since it has an important historical correlative. The Hungarian nobleman Wolfgang von Kempelen had in the late 1760s built a chess automaton, which quickly became famous and remained a topic of discussion in subsequent decades. It depicted an almost life-size Turkish man, dressed in a cloak and a turban, sitting behind a wooden cabinet and playing chess. The cabinet contained a clockwork mechanism that supposedly enabled the automaton to play chess against human players. The Kempelen chess automaton inspired a large number of legends, including the claims that it had played chess (and won) against Frederick the Great, Napoleon, Benjamin Franklin, and Charles Babbage. But the automaton's performance was based on a hoax: a human chess player, hidden in the body of the cabinet, conducted the moves of the game.[72]

Even though the example of chess-playing machines is not de-

---

69. Jean Paul, *Werke*, vol. 2, part 1, 1028 (Richter's emphasis).
70. Ibid., 1029.
71. Ibid., 1029.
72. The history of the chess automaton is well researched from a variety of perspectives. See, for example, Heckmann, *Die andere Schöpfung*, 219–30, 258–62; Schaffer, "Enlightened Automata"; Standage, *The Turk*. See also chapter 1, note 6.

scribed in great detail, it has important consequences for the scenario that Richter's text is developing. The chess automata disclose the crucial fact that the machines of the angels (that is, us humans) in their turn also build machines that "look like" or "pass as" humans and perform human tasks. The passage on chess-playing automata imparts details about the ways in which humans and machines relate to each other in the text's logic, and yet it is in form and content a source of profound confusion for the reader. In the following sentences, the narrator artfully exploits the double-machine-building by talking alternately, in his typically cryptic and casual-ironic way, about angel-made chess machines (that is, chess-playing humans) and human-made chess machines (such as the Turk made by von Kempelen) without consistently distinguishing between the two: "All my readers must have seen creatures of this kind, which play chess without any help of an angel, just through a mechanism in their head; they move their right arm by themselves, and they even shake their head . . . upon a wrong move of their opponent; and once the king is in checkmate, they won't make another move under any circumstances." Not only is the reader invited here to remember occasions on which she or he has watched other humans play chess; but Richter follows this passage with a suggestion that the reader will readily recognize how similar these chess machines are to the "well-known chess machine invented by Mr. v. Kempelen."[73] In doing so, the text comes full circle in its deliberate confusion of chess-playing humans and chess-playing machines. The juxtaposition of angels' and humans' machine-building, the similarity between their products, and the consistently unclear references in the text leave the reader continuously confused about the status of the story's actors, as well as the reader's own status, as machine or as human.[74]

The buildup of these confusions eventually leads the text to the fundamental question of the difference between humans and machines. In the closing sentence of the passage about the chess-playing machines, the narrator makes the first brief reference to this theme. After describing von Kempelen's chess machine as a "copy" of human chess players, he says: "Notwithstanding all this, there will still always remain a tremendous difference between the two kinds of machines, and the work of the angels will always stand out by far against the

73. Jean Paul, *Werke*, vol. 2, part 1, 1029.

74. Hans-Walter Schmidt-Hannisa has investigated the relationship between author and reader in Richter's work. Addressing the reader and engaging in a conversation with him or her is a key feature of Richter's writing in general. Hardly any of these addresses to the reader are without irony, as Schmidt-Hannisa points out. Schmidt-Hannisa, "Lesarten."

work of a human."[75] This is an unusually clear statement on the part of the narrator. While in the earlier parts of the text, the narrator described humans as being machines with little further qualification, in this passage he is willing to admit the superiority of humans over their own machines. The claim, however, still appears in the context of a profoundly ambiguous passage, and the ostensibly serious tone here is conspicuous given the directly mocking tone the narrator has adopted elsewhere in the text.[76] After this last sentence of the chess-playing passage, the narrator does not comment any further on the difference between humans and machines until the very end.

In the final sentence of the text, the narrator says this about the difference between humans and machines: "The machines of the earth must almost always step back behind the machines of the angels, and one does not really offend the former in claiming that they are . . . mere replicas and weak copies of those machines which the angels contrived: this woman, for example, who plays the piano is at most a fortunate copy of those female machines who play the piano and who accompany the music with bodily motions, which obviously seem to betray affective involvement."[77] In this concluding sentence, the narrator thus recasts the question of the difference between humans and machines in terms of "female machines," bodily motions, and the "betrayal" of affective involvement. The German word that I translate with "affective involvement" is *Rührung*, and it and the verb *rühren* connote both a person's state of being moved and the act of being moved by something or someone else. A musician's performance, his or her inner emotions, his or her bodily motions, and his or her ability to communicate these affects to an audience are drawn together in this term and its connotations.[78]

75. Jean Paul, *Werke*, vol. 2, part 1, 1030.

76. Schmidt-Hannisa explains more specifically that ironic ambiguities coupled with offenses against the reader are among the foundational aspects in Richter's way of construing the author-reader relationship. Schmidt-Hannisa, "Lesarten," 51.

77. This passage reads as follows in the original: "Die Maschinen der Erde müssen fast alzeit den Maschinen der Engel Vorrang lassen und man thut jenen nicht zu viel, wenn man behauptet, daß sie . . . blosse Nachahmungen und schwache Kopien der Machinen sind, die die Engel erdacht: ienes Frauenzimmer z. B., das Klavier [spielt], ist höchstens eine glükliche Kopie der weiblichen Maschinen, die das Klavier schlagen und die Töne mit Bewegungen begleiten, die offenbar Rührung zu verrathen scheinen." Jean Paul, *Werke*, vol. 2, part 1, 1031.

78. An important parallel between the French and the German cultures of sentiment is that in both cases, sensibility and sentiment are not only moral and epistemological categories; they also describe an affect in response to a sensation. See Riskin, *Science in the Age of Sensibility*. On the gendered aspect of "the fainting woman and the mechanical man," see Vincent-Buffault, *History of Tears*, 42; and my chapter 4.

The bewilderment generated by previous passages is encountered again in this final sentence. The scenario described in the second half of the sentence—female machines and motions that "seem to betray affective involvement"—is ostensibly a supportive example for the claim in the first half of the sentence, that machines of the earth have to "step back" behind the machines of the angels. However, there is more obfuscation than illustration going on in this passage. The illustration of the distinction between piano-playing women and piano-playing automata is so grammatically convoluted that it is impossible for the reader to determine correct referents in this supposed clarification of the difference between humans and machines. As a result, the sentence itself becomes an illustration of the confusion of humans for machines and vice-versa.

I discussed the motif of piano-playing women in its mechanical, artisanal, cultural, and economic manifestations in the preceding chapters, noting how common it was at the time and how it embodied numerous preoccupations as a motif, an aesthetic ideal, and a cliché. The two women automata perform mechanically, by means of their clockwork, a reflexive practice of bourgeois selfhood. For a *spectator* of the women automata, this mechanical reproduction raises the question of whether the reflexive dimensions of this subject-formation are visible in the machines, that is, whether a spectator can "see" the sentimental motion and the subject's self-experience in the automata's musical play. The automata thus raise questions about whether processes of subject-formation are "real," "mechanical," or "fake." Richter's text creates a scenario in which the reader becomes a spectator and is confronted with piano-playing women and piano-playing women automata; and both of them "betray" to him or her their sentiments while playing music, by moving their bodies.[79]

Richter's text integrates the uncertainty over the human-machine boundary with the uncertainty over the "sentimental project" in the European Enlightenment. Just as the text (and the women's performance) leave no conclusive clue to make the distinction between humans and machines, the text asks about the sentimental practice of music-making itself, whether it can be reproduced mechanically and

79. The verb "verrathen" has two meanings, parallel to the English "betrayal" and "to betray" that I use to translate it. Both terms carry the meaning of disclosing something that is hidden from view, in this case a musician's sentiments. The terms also carry a connotation of deliberate deception and conveying "fake" information. Richter's text—with its uncertainty about the status of feelings in a piano-playing woman android—uses the term's ambiguity to destabilize further the human-machine boundary.

whether it can be faked. The threat that humans can be like machines, and machines can be like humans, is in this context not about a mechanization of the body and soul, but more specifically about the contemporary culture of feelings and sentiments in music-making and in literature: about the self-consciousness, idealism, and deliberation on the part of the practitioners of civil society in Germany. Richter's satire brings focus to the fact that the cultural practice of generating affects could be replicated in a mechanical artifact, raising the question of whether the product of that sentimental cultivation—the bourgeois self—was unfailingly real, recognizable, and reliable. This uncertainty coincides with the generic uncertainty over the human-machine boundary, implying two things: that the human-machine boundary is historically specific, and that the human-machine boundary is distinct from its alleged sole source, the mechanistic worldview.

There is a broader social criticism in Richter's account of real and mechanical subject-formation, namely a concern over whether the idea of creating an equal and just social order on the basis of shared sentiments was desirable and feasible. Many critics, writers, and philosophers in the last three decades of the eighteenth century asked precisely this question: was the idea of basing a new, modern social order on a culture of feeling a promising idea, and would it yield the desired stability and reliability in a society and its citizens?[80] Hoffmann's rendering of piano-playing woman automata picks up from this question and integrates the motif into different encounters between humans and androids. To ponder further the relationship between sentimental music-playing women and the unstable human-machine boundary in the modern era, I read as a corollary Hoffmann's text and its engagement with a piano-playing woman automaton.

Hoffmann's Novella *The Automata*

E. T. A. Hoffmann, eminent literary writer and critic of German romanticism, used the automaton motif in a number of his stories in the first and second decades of the nineteenth century. He was familiar with contemporary automata through reading as well as through personal friendships with musical instrument makers and musicians. He was, in

---

80. Numerous other prominent cultural commentators at the time complained about excesses in the sentimental movement of the German Enlightenment. The culture of affect was, from its very beginning, a celebrated as well as a contested practice.

particular, friends with the Kaufmann family, who built the trumpet-playing automaton between 1810 and 1812, as described in chapter 1 (pages 5–6). We can also tell from Hoffmann's papers that he was taken with Kleist's essay on the marionette theater (first published in 1810), and he noted in his diary and in a letter that he would like to build an automaton himself.[81]

In the 1812 novella *The Automata*, he uses a piano-playing woman automaton as a background prop to discuss not a later era's concern with artificial intelligence but his own era's concern with the effect of music and biographies on bourgeois and artistic selves.[82] While the main automaton in Hoffmann's story resembles the chess-playing Turk by von Kempelen, a considerable part of his story is about the general automaton culture and musical automata from the eighteenth century. Hoffmann stated, as he submitted the essay to the *Allgemeine Musikalische Zeitung*, that it was an "attempt to engage everything that is called automaton, in particular musical objects of this kind."[83]

The Automata is a multilayered story, about thirty pages long. The text starts out by stating that there was an automaton in town, a speaking Turk, which caused a great deal of commotion, and everybody—young, old, rich, and poor—flocked to see it and hear oracles that were whispered by the automaton to anyone who would ask a question.[84] The whole construction and presentation of this automaton was of a kind, the narrator says, that distinguished it from other spectacles, such as those seen at fairs. Everyone "had to" be able to see this difference and be attracted to it.[85] During a showing, spectators could whisper a question into the figure's right ear. It then turned its eyes and head to the person who had asked the question. When it answered, it

---

81. Hoffman to Hitzig, 1 July 1812, in Hoffmann, *Briefwechsel*; diary entry, 2 October 1803, in Hoffmann, *Tagebücher*. More automata are mentioned in Hoffmann's diaries, as he saw some in 1801 in Danzig and in 1813 in Dresden. He also presumably read an article written by Carl Maria von Weber about the Kaufmann trumpeter automaton in the *Allgemeine Musikalische Zeitung* in 1812; and Hoffmann was familiar with Johann Christian Wiegbleb's twenty-volume work *Unterricht in der natürlichen Magie*, whose second volume (column 231) lists contemporary automata. See the Kaufmann papers in the Archive of the Deutsches Museum, Munich; and see chapter 1.

82. Hoffmann's text *The Automata* has been interpreted as one of the intellectual origins of modern research in modern artificial intelligence. Emily Dolan challenges this. See Dolan, "E. T. A. Hoffmann."

83. See a letter of 16 January 1814, in Hoffmann, *Briefwechsel*, 1:436. See also Dolan, "E. T. A. Hoffmann," 7. The Turk by von Kempelen was well enough known to stand for any kind of automaton.

84. Hoffmann, *Die Automate*, 397.

85. Ibid., 396. The Turk was a large, nicely designed figure and was seated on a stool next to a small table.

felt as if the answer came right from the figure's interior.[86] After a few answers had been given, the operator would apply a key to the figure's left side and wind up the clockwork with a great deal of noise. The artist would also open up the automaton, upon request, and one could see in the figure's interior a mechanism with many cogs: the wheelwork took so much space, the narrator explains, that there was no way a human being could fit somewhere inside the figure.[87] People came up with an array of speculations about the source of the automaton's messages, examining nearby walls and furniture and adjacent rooms, but all in vain.

The narrator states that the public's interest would have faded away soon had it not been possible for the artist to continue to attract spectators in a different way. This attraction was, as the narrator states, entirely due to the answers that the Turk gave. Each answer was given with a deep insight into the personality of the person who was asking. They were often witty, dry, or humorous and always to the point, sometimes a mystical gaze into the future, sometimes in a different language, but always in one that the questioner knew.

Every day there were new stories about the automaton, and it was also the subject of discussion of an evening society in which two students, Ludwig and Ferdinand, were members.[88] The two had not seen the automaton yet.[89] Ferdinand explained why he found the Turk interesting and deemed it different from other mechanical automata: having heard the stories about the automaton, he concluded that the figure's good looks and his moving eyes and head were only meant to distract the audience, that the key to its mystery was somewhere else. Much more startling for Ferdinand was the mental power of the unknown human whom he assumed must be behind the automaton, giving answers and penetrating into the depths of the questioner's soul.

The two students decided to go and see it. At first, everyone in that evening show received uncharacteristically unsatisfying, uninspired,

---

86. Ibid., 397. One could even feel the automaton's breath.

87. Ibid., 397. The narrator says that the cogs and mechanics did not seem to have any influence on the automaton's speaking ability, but he speculates that it might be the motions of the head and the arms that the clockwork controlled.

88. Ibid., 399. Evening and academic societies came into being in Germany during the Enlightenment and were among the new types of sociability in the emerging civil society that I discuss in chapter 4.

89. Ibid., 399. In fact, Ludwig was not too excited about the idea of an automaton. He explains to his friend at length how even as a young boy he did not like wax figures and how he thought this automaton was going to haunt him in sleepless nights.

and dull responses from the automaton. But just before leaving, Ferdinand said that he wanted to ask one more question. He whispered it into the automaton's ear, but the automaton did not want to answer. Ferdinand insisted, and finally received an answer, which caused him to turn pale and become visibly disturbed. He told the audience that he did not want to disclose the answer, and the party soon dissolved "in a gloomy mood."[90] After the show Ferdinand told Ludwig that he was convinced that the mysterious being that communicated to him through the Turk was capable of knowing his most secret thoughts. Ludwig suggested that Ferdinand himself may have asked something very suggestive and was now overinterpreting the ambiguous response that he had received. This interpretation is typical for the figure of Ludwig in the story: he regularly assumes a strong "projective potential" on the part of individual humans, a strong tendency of humans to use the world or other humans as screens or mirrors for their own interior.[91]

But Ferdinand disagreed with his friend Ludwig and told a story from his past to explain why he found the automaton so terrifying. It turned out that in order to give the answer the automaton had given, it (or the speaker communicating through it) would have had to know crucial parts of Ferdinand's life story. A few years earlier, Ferdinand explained, he had been traveling and encountered a mysterious woman with whom he fell deeply in love, but he had no opportunity to speak with her. He painted a miniature portrait of her when he returned home and framed it in a golden locket that he wore around his neck. He said no one had ever seen it and he had told no one about it. On the evening of the automaton show, Ferdinand asked the Turk automaton whether there was ever going to be another time for him like the one during which he had met this woman and been happiest. This was the moment when the Turk was unwilling to respond, but when Ferdinand insisted, the automaton said: "I am looking into your chest, but the glitter of the gold, which is toward me, distracts me. Turn the picture around." The picture of the woman was indeed on his chest. Then the figure said, in a sorrowful tone: "Unfortunate man! At the very moment when next you see her, you will lose her forever."[92] It was terrifying for Ferdinand, he told Ludwig, to find the secret of his romantic love un-

---

90. Ibid., 403.

91. Hoffmann is known to be a pioneer of making this "interior" a visible and productive element of literary texts.

92. Hoffmann, *Die Automate*, 408.

covered by a fearful and unknown power and then be threatened with sorrow and death.

About a week later, the two students were again at a meeting of the evening society, and people were still raving about the oracle automaton and its inventor. An elderly gentleman stood up and said that the automaton was not invented by the person who was exhibiting it. The originator was, rather, a local professor, namely Professor X. The automaton had been shown in public for a couple of days a few weeks earlier, the elderly gentleman said, but nobody had then taken particular notice of it. Except that Professor X had seen it early on. When he had heard one or two of the Turk's answers, he supposedly took the exhibitor to the side and talked to him. The exhibitor then closed his exhibition, and nothing more was heard of the Turk for a fortnight. Then, new bills and announcements were posted throughout town, and the Turk was staged with a new head and all the other known arrangements. It was only since then that his answers had become so clever and interesting. The Professor himself was also said to possess many most extraordinary automata at his home, most of them musical automata (the professor's other pursuits were chemistry and natural philosophy).[93]

Ferdinand and Ludwig were impressed by the old man's revelations and wanted to find out more about Ferdinand's mystery. They visited Professor X, hoping he could throw light on their many questions. He was an old man, the narrator tells us, with small, unpleasant gray eyes and a sarcastic, not very attractive, smile. His voice was also unpleasant: a "high-pitched, screaming" tenor.[94] In this context, a piano-playing woman appears: Professor X showed the two students his automata, which were on display in an elegantly furnished hall. There was a piano in the middle of the room on a raised platform; beside it, on the right, was a life-sized figure of a man with a flute in his hand; on the left was a female figure, seated at an instrument somewhat resembling a piano; and behind her were two boys with a drum and a triangle. In the background there was an orchestrion, and musical clocks hung on all the walls. Hoffmann here displays for the reader almost the complete cast of eighteenth-century automata: Professor X had them in his living room, and they serve as a backdrop to the story of the young student

---

93. Note Mary Shelley's story about Victor Frankenstein here. He was taught at university by two professors, one in the scholastic tradition, the other in the scientific tradition.
94. Hoffmann, *Die Automate*, 417.

who finds himself in such terror. Both Vaucanson's flute player and the Jaquet-Droz family' harpsichord player receive explicit mentions.

In the story, the professor then proceeded to play the piano, "an andante in the style of a march . . . once . . . by himself; . . . the second time the flute player put his instrument to his lips, and took up the melody; . . . one of the boys drummed softly . . . the other . . . touched his triangle. . . . Then the lady came in with full chords sounding something like those of a harmonica, which she produced by pressing down the keys of her instrument . . . ; the musical clocks came in one by one, the boy drummed louder . . . and lastly the orchestrion set to work."[95] The Professor concluded the performance with one final chord, and all the machines finished also, "with the utmost precision."[96] Ludwig and Ferdinand left hastily before the Professor could offer any more performances, angry that there were no answers to their terrifying mystery. At the end of the story, Ferdinand is called home by his ailing father, leaving the reader with an open ending.

In Hoffmann's text, the music-playing woman automaton is part of a threatening theater made up of music-playing machines. The boundary between human and machine is drawn in a different place in this scenario than in Richter's story. It is not located "in" the automaton, but in a mediating agent. As Ferdinand struggles with an agency that is hidden from his view and grasp, and as he converses with his friend Ludwig about the matter, he wants to find out what kind of agencies are at work when android automata threaten human selfhood.

In greater detail than Richter's text, Hoffmann's text differentiates between types of human-machine boundaries and types of uncertainties arising from their instability, while also introducing agencies and forces outside of humans and machines. Mechanical reproduction of subject-formation also figures in Hoffmann's text: the hidden agency of Professor X, manifesting itself both in the speaking Turk and in his house concert, raises in typical romantic manner questions of forces, attractions, and connections between subjects, citizens, and individuals whose constitution was made possible during the Enlightenment in the first place. In regard to the question of whether the human-machine boundary took on increasingly threatening and fatalistic connotations

---

95. Ibid., 417. On the orchestrion, see also Dolan, "E. T. A. Hoffmann"; and Jackson, *Harmonious Triads*, 79–83. See also chapters 2 and 4 on the fact that the Jaquet-Droz harpsichord player plays the instrument "by pressing down the keys herself" with her fingers.

96. Hoffmann, *Die Automate*, 417.

in later, more industrial and more "modern" periods, Hoffmann's text suggests that our ambiguous understandings of human agency in relation to machine agency manifest themselves both inside and outside of actual human-machine encounters.

Conclusion

Texts on piano-playing women automata in the eighteenth century and the early nineteenth century have complicated histories of their own, much as the actual objects have. Piano-playing women automata get integrated into different economic and intellectual agendas, and these processes go on over long time periods and in different places. In Richter's text on humans, machines, and angels, piano-playing women automata serve the role of linking the ambiguous human-machine boundary to questions about the reality and reliability of a new, evolving sentimental social order. Piano-playing women automata serve in Hoffmann's text as the backdrop for elaboration on the figure of a professor who collects automata and seems to possess supernatural powers, with the help of which he manufactures clairvoyant androids. These androids, in their turn, make a young student encounter questions of past happiness, lost love, and invisible and dark forces. In the shorter, more descriptive texts discussed in chapters 2 and 3, in contrast, harpsichord-playing automata served to provide material and fodder for an ever-expanding media industry in the eighteenth century.

To summarize: In chapters 2 and 3 I discussed the origins of my two keyboard-playing women automata in protoindustrial artisan cultures and presented contemporary short texts about them. Chapters 4 and 5 then have dealt with the cultural, political, and aesthetic aspects of these remarkable technical artifacts. I explicated in chapter 4 how bodily motions that communicated sentiments were a technique in the eighteenth century that was meant to generate virtuous and emancipated selfhood and social order and how the activity performed by the two women automata was therefore a microcosm in which the formation of modern politics and selfhood was mechanically reproduced. I emphasized in chapter 4, furthermore, the self-reflexivity of such sentimental techniques, building on an argument by Peter Hohendahl. Eighteenth-century sentimentality was a process in which the subject found and experienced itself, as much as it was a process of communication with others and a process of experimenting with a new social order. The emphasis on the subject's *self*-experience in the process of music-making raised the question of whether the automata reproduced mechanically this self-reflexive part of musical sentimentality, too— that is, whether the women automata encountered, as it were, their own sentimentality.

For a spectator who sees the automata perform their mechanical business, or, for that matter, for a reader of Richter's and Hoffmann's texts, this question takes on yet another form. The spectator is not only uncertain whether these entities are human or machine. She or he is also uncertain about whether the automata replicate mechanically the entire spectrum of sentimental subject- and society-formation. When we observe keyboard-playing women automata playing music, we not only are unsure whether they can communicate affects to the audience (that is, us) as a human musician would. We are also not sure whether the automata are in a self-reflexive relationship to their own sentimentality. Richter's and Hoffmann's texts turn us readers into spectators and make us wonder whether mechanical reproduction of self-reflexive music-making challenges humans' clear demarcation from machines, and whether the texts also challenge our attempts at being self-reflexive, modern subjects.

In chapter 6 I discuss what "happened" to the eighteenth-century automata and the Enlightenment android during the two hundred years of industrialization in Europe and North America. Between 1780 and 1980, artificial humans undertook a remarkable journey in which they became metaphors or motifs for practically all promises and perils of the modern machine age. To this day, they function as symbols of industrial and postindustrial human-machine encounters. Both the meanings attached to Enlightenment automata and the meanings' transformations over time are vast in their size and depth. My study in the previous chapters, in comparison with the journey in the following chapter, demonstrates the enormous capacity of mechanical humans to absorb, and subsequently communicate successfully, any imaginable aspect of the long history of establishing modern industrial selves and society.

The "Enlightenment Automaton" in the Modern Industrial Age

In this final chapter, I follow the path taken by the figure of the "Enlightenment automaton" in the nineteenth and twentieth centuries. The journey demonstrates how versatile the figure was and traces how the automaton became a symbol of human-machine relations in the industrial age. I outline broad developments, such as the formation of historical and literary disciplines and engineering education in the Second Industrial Revolution, and I also look in detail at individual texts, such as passages from Marx's *Capital*, Thomas Carlyle's "Signs of the Times," and Norbert Wiener's *Cybernetics*, to bring out nuances in the struggle over the cultural significance of the automaton. This history and these texts, taken together, help explain the origin of current conceptions of Enlightenment automata (such as the flawed ideas discussed in chapter 1) and why they have proved desirable and useful despite the considerable historical and conceptual contradictions between these ideas and the findings of my study. This chapter's journey also amounts to a short history of the industrial age itself, viewed through the lens of the Enlightenment automaton.[1]

1. This chapter is a variation on numerous other survey histories of the "Enlightenment automaton" in the nineteenth and twentieth centuries. Minsoo Kang, for example, has recently presented a history of the automaton from the

The android automata made between the 1730s and the 1780s— the two that I have focused on and the others that I reviewed in chapter 1—attracted the attention of many influential commentators in later years. Already beginning in the 1790s, important texts used details of the androids for purposes of entertainment and education. Among these texts were technical encyclopedias, engineering textbooks, histories of technology and invention, and manuals for clockmaking and other mechanical arts.[2] Such texts were produced in increasing numbers at the time as a result of school reforms, increasing literacy, and changing cultural and economic functions of reading. Interest in technological education came with the Enlightenment's general commitment to education in the "useful" sciences, as well as with a growing middle class that could afford to buy and read books and was interested in institutional learning.

The production of educational technological literature continued during the nineteenth century's Industrial Revolution in Britain and on the European continent. Scholars and state administrators added technology to university curricula, and there was increasing demand for reading material for the educated classes and for newly founded polytechnic educational institutions.[3] The texts produced to fill these needs proved influential not least because their authors wrote and published widely and prolifically and were often interested in larger social and political issues. The references in these texts to eighteenth-century automata as illustrations and examples kept their images and names alive throughout the nineteenth century.

Close connections in this literature between technology, pedagogy, economics, and natural sciences were typical, and they allowed the automaton motif to travel between genres and audiences. A prominent example is the works of Johann Heinrich Moritz Poppe, an engineer, mathematician, educator, and historian of technology. The son and

---

ancient period to the twentieth century, and he devotes a special excursus after his eighteenth-century chapter to the "post-Enlightenment legacy of the automaton-man." Kang, *Sublime Dreams*, 166–74. I rely in this chapter on Dotzler et al.'s superb bibliography *MaschinenMenschen*.

2. An example of such an encyclopedia is Gabriel Christoph Benjamin Busch's eight–volume *Versuch eines Handbuchs der Erfindungen* (1790–98). It devoted separate entries to specific automata such as the harpsichord player, Vaucanson's duck, and the hoax chess player. Another popular genre of educational books was works on "natural magic," which contained instructions for experiments in electricity, magnetism, optics, chemistry, and mechanics. Passages on "mechanical tricks" described Vaucanson's flute player and the chess automaton, such as Martius, *Unterricht in der natürlichen Magie* (1779–1805), 1:283, 2:231. On the relationship specifically between encyclopedias and technological artifacts and invention, see Fröhner, *Technologie und Enzyklopädismus*.

3. Ashby, "Education for an Age of Technology."

apprentice of a university mechanic of the University of Göttingen, Poppe studied mathematics, physics, and state sciences (cameralism) in the 1790s under eminent teachers such as Georg Christoph Lichtenberg and Johann Beckmann. He wrote in a large variety of genres from the 1790s on, and virtually all of his works contained extensive references to automata.[4] Johann Beckmann himself was a well-known professor of cameralism with an interest in technology and invention, and he published just as widely.[5] Another teacher, scientist, and cameralist was Johann Konrad Gütle, who wrote textbooks in the 1790s on "magical mechanics," on mechanical experiments, and on industry and the state—all of which contained references to automata.[6] Such authors—as well as people like Gabriel Christoph Busch, a pastor in a leading administrative position in Thüringen, and Charles Hutton, a mathematician and school teacher—wrote and edited dictionaries between 1790 and 1830, relying on and copying from one another and thus building up a critical mass of literature that continuously imparted histories of eighteenth-century automata.[7] Authors from the later nineteenth century used and expanded these encyclopedia entries for their own accounts of automaton history.[8]

4. In 1797 and 1801, Poppe published histories of clock-making, between 1803 and 1810 an encyclopedia of "all" mechanical engineering and machinery, between 1807 and 1811 a history of technology in several volumes, and in the 1830s a history of "all" inventions and discoveries in science, technology, and the arts. All of these works contained references to eighteenth-century automata, and one book in particular kept their names circulating into the nineteenth century: a work of 1824 on mechanical "wonders," which listed the names of the main automaton makers from the eighteenth century by name in its title (*Wunder der Mechanik: oder, Beschreibung und Erklärung der berühmten Tendlerschen Figuren, der Vaucansonschen, Kempeleschen, Drozschen, Maillardetschen und anderer merkwürdiger Automaten und ähnlicher bewunderungswürdiger mechanischer Kunstwerke*).

5. Beckmann, *Beyträge zur Oekonomie* (12 vols.); Beckmann, *Beyträge zur Geschichte der Erfindungen* (5 vols.). On the relevance of Poppe and Beckmann as early historians of technology and automata, see Bayerl, "Die Anfänge der Technikgeschichte."

6. Gütle, *Beschreibung* (1790); Gütle, *Zaubermechanik* (1794); Gütle, *Nüzliche Versuche* (1796).

7. Busch wrote a handbook on inventions that had eight volumes in its first edition (1790–98) and twelve volumes in its fourth edition (1805–22). See also the entry "automaton" in Hutton, *Mathematical and philosophical dictionary* (1795–96), 1:176. On the ways in which all these dictionaries rely on one another, see the textual similarities of the entries "automaton" in Gehler, *Physikalisches Wörterbuch* (1798); Beckmann, *Beyträge zur Geschichte der Erfindungen*; and Brewster, *Letters* (1832), 317–45.

8. The entry "Automaten" in a new and expanded edition of *Johann Samuel Traugott Gehler's Physikalisches Wörterbuch* (1825), 1:649–60, is among the most complete and well-researched entries. It is chronologically structured, like the others of this kind, and it cites extensively other encyclopedias on which it relies. It describes father and son Jaquet-Droz, saying that their automata "exceeded" those by Vaucanson "by far" (652). There is no mention of Roentgen's dulcimer player, but regarding the Jaquet-Droz family's harpsichord player the entry says that, besides moving her head and body, she "also plays the piano" (653).

The Industrial Revolution

The automaton motif took on novel and multifaceted connotations in the course of the nineteenth century. It continued to feature in debates about materialism and the "man-machine," for example, which, since the seventeenth and eighteenth centuries, had been concerned with elemental topics in theology, ethics, and metaphysics, as well as in physiology and anatomy, as I discussed in chapter 1. The nineteenth-century materialism debates touched as much on questions about matter, consciousness, and God as on newly emerging issues about the evolution of species. Furthering these debates, new editions of eighteenth-century works on materialism were printed in the nineteenth century, histories of materialism were composed, lectures on materialism were delivered, and biographies of eighteenth-century materialists were written.[9] This sustained interest kept the eighteenth-century android automata visible, since they were habitually quoted in tandem with references to eighteenth-century materialism, not least because they supposedly had been the inspiration for La Mettrie's *Man a Machine*, whose materialist theory was the eighteenth-century's best known. As I discussed in chapter 1, La Mettrie did rely on Vaucanson's automata, calling him a "new Prometheus."[10]

The automaton motif was also employed in reflections on accelerating industrialization, changing working conditions in factories, and general changes in the pace of life in industrializing areas—a discussion as prominent as the one about materialism. The scientific and philosophical literature of Victorian England, the first society to face the rapid and profound social changes of the Industrial Revolution, aimed in particular to understand these changes and promote the scientific and technological culture of its time. Enlightenment automata

9. La Mettrie's *Man a Machine* was still of interest to historians and philosophers in the nineteenth century. Examples include Poritzky's *Julien Offray de Lamettrie*, of 1900; Emil Du Bois-Reymond, "La Mettrie," a lecture that he gave in 1875 at the Akademie der Wissenschaften; and Lange, *Geschichte des Materialismus* (1866–73), which went through many editions and translations. New editions of La Mettrie's *Man a Machine* with introductions, translations, and additional historical materials were written and printed in the later nineteenth and early twentieth century: in France in 1865 and in the United States in 1912, 1927, and 1961.

10. La Mettrie, *L'homme machine* (1865), 140. Vaucanson's work was perhaps promoted most effectively in the eighteenth century through the article "L'automat" in Diderot and d'Alembert's *Encyclopédie*. Jakob Elias Poritzky's 1900 study of La Mettrie is an important example of an interpretation that groups all eighteenth-century androids into one set of supposedly homogeneous and uniform artifacts. Poritzky, *Julien Offray de Lamettrie*, 174–75.

provided symbolism and vocabulary that helped these Victorian writers make sense of their predicament.[11] The two debates about materialism and industrialism converged when authors lamented, for example, the loss of religious faith or the loss of humanity in the modern factory world. The sustained impact of such Victorian scholarship and the attendant sensibilities contributed crucially to the automaton's visibility in the twentieth century.

In later Victorian decades, Britain's industrial preeminence was threatened as other nations gained strength, and there was a shift in political balance in Britain caused by the decline of agriculture and the ascendancy of workingmen's votes, at the expense of landowning gentry and nobility. Efforts increased once again at this time to grapple with the Industrial Revolution and its consequences, including efforts to conceive of its history as a whole and bring into focus the evolving ideas and practices attached to it. The term *Industrial Revolution* was coined at this time, too, to encompass the profound social changes coming with it and, more specifically, the sense that humans were changing into machines.[12]

Thomas Carlyle's influential 1829 essay "Signs of the Times," in which he deployed Vaucanson's mechanical duck, was written in the thick of the British Industrial Revolution. Carlyle paints a foreboding picture of many key aspects of industrial modernity and characterizes his age as the "Mechanical Age" and the "Age of Machinery."[13] He outlines the increasing influence of mechanization on public political and economic life, integrating this concern with the metaphysical debate around materialism and the nature of the human mind. He uses Vaucanson's duck when discussing materialist ideas that thought and imagination have material origins in the human body.[14] David Brewster also moved in prominent Victorian scientific circles and was a leading figure in the founding, in 1831, of the British Association

11. Altick describes the Victorian age as a "crucial moment in modern history," in which people "found themselves living in a world whose novel demands they were wholly unprepared to meet." Altick, *Victorian People and Ideas*, 73.

12. It was mainly Arnold Toynbee's lectures of 1884 that helped popularize the term *Industrial Revolution* in the Anglophone world. Toynbee, *Lectures on the Industrial Revolution*. On Toynbee's influence, see Mokyr, *British Industrial Revolution*, 7, 130, 162; Berg, *The Age of Manufactures*, 70, 321. In Germany and elsewhere, the term had been brought into circulation earlier by Friedrich Engels, under the impression of the industrial changes in Britain. Engels, *Die Lage der arbeitenden Klasse* (1845), 14.

13. Carlyle, *Essays*, 2:317. The essay has been described as marking the beginning of the Victorian age.

14. Ibid., 2:319.

for the Advancement of Science (BAAS). In his 1834 *Letters on Natural Magic*, Brewster devotes one entire chapter to the history of automata.[15] He explains in this book how apparently supernatural phenomena rely on optical illusion or mechanical and chemical tricks.[16] In the chapter on automata, Brewster rehearses a familiar history, written in many reference works in the eighteenth century, which stretches from the Greeks via Regiomontanus to Louis XIV.[17] Like Carlyle, he integrates the history of automata with the debate around materialism, vitalism, and mechanicism, adding to the standard encyclopedia automaton histories the claim of a profound change in automaton history from the middle of the seventeenth century onward: he states that at that time mechanics and philosophers aspired to mimicking "functions of vitality" in their mechanical systems, mentioning Vaucanson as the important example.[18] This claim has survived to the present day as a characteristic that supposedly applies to all eighteenth-century automata; as I explained in chapter 1, it is one of the general assumptions about eighteenth-century automata whose universality my study challenges. At the end of his history, Brewster makes the crucial connection that reflects the preoccupations of his time: he connects automaton history to the history of the steam engine and industrialization. He states that the eighteenth-century automata were made by their makers "to astonish and amuse the public" but that this was not their only result: the "ingenious" mechanical devices, he says, "contributed in future years to purposes of higher import," for they "re-appeared in the stupendous mechanism of our spinning-machines and our steam-engines."[19]

Another powerful Victorian figure, Thomas H. Huxley, joined the

15. It is the book's eleventh letter, out of thirteen. Brewster, *Letters*, 317–45. Automata figure elsewhere in Brewster's book, too: in the letter on sounds (letter 8), for example, he mentions "acoustic automata," among them an early one by P. Jaquet-Droz from the 1750s, one by Maillardet, Vaucanson's flute player, and von Kempelen's talking machine. Brewster also mentions "Henri-Louis Jaquet-Droz's draughtsman" (on page 282) and states that he received this information from Thomas Collinson's article "automaton" in Hutton, *Mathematical and Philosophical Dictionary*, 1:176.

16. The tradition of natural magic that Brewster relied on came out of the seventeenth and eighteenth centuries and blended conjecture and magic with scientific experimentation. It was promoted and sustained by textbooks and itinerant lecturers.

17. He relies on, and cites, Beckmann's *Beyträge zur Geschichte der Erfindungen*; and Gehler's *Physikalisches Wörterbuch* (see my note 7).

18. No other authors of automaton histories at the time claim the existence of this supposed shift in the mid-seventeenth century. Brewster, *Letters*, 268.

19. Brewster, *Letters*, 286. There's an article almost identical to Brewster's letter that shows how these standard accounts were circulated and copied in the nineteenth century. It is entitled "Symptoms of Ingenuity Misapplied" and appeared in *The practical mechanic and engineer's magazine*, vol. 1, part 5, Glasgow, 12 February 1842.

discussions about animals, automata, mechanicism, and materialism in the 1870s. His essay "On the Hypothesis That Animals Are Automata and Its History" was first delivered as a lecture at the BAAS meeting at Belfast in 1874. Huxley engages Descartes's work in depth and emphasizes its importance for nineteenth-century physiology and anatomy and for questions about animal consciousness and the constitution of the mind and body. He uses Vaucanson's automata to raise the question of whether a human's bodily motions are accompanied by consciousness, and he actually suggests greater similarity between humans and animals than Descartes does, proposing that both humans and animals are automata.[20] Around the same time, the influential German physicist Hermann von Helmholtz wrote lectures on popular scientific topics and similarly used eighteenth-century automata as illustrations. In one of them, presented in 1854, Helmholtz recapitulates the Scientific Revolution in broad strokes on the "threshold between the Middle Ages and the modern era" and discusses contemporaneous mechanical arts. In this context, he mentions Vaucanson's duck and flute player and the Jaquet-Droz family's writer and "piano-player."[21] Helmholtz claims, like Brewster, that eighteenth-century artisans built their automata to simulate life, on the one hand, and to create machines that would replace a human's work, on the other. He draws the connection to his own time, saying, "Nowadays, we no longer aim to build machines which perform the work of *one* human but that one machine performs the work of a thousand humans."[22] Another assumption that I discussed in chapter 1, the idea that eighteenth-century automata anticipated the factory system, is thus also explicated and suggested here; it became widely disseminated from this and similar sources.

Automata were also the concern of people who studied industrial production and the organization of factories. Some of them, such as Andrew Ure, Karl Marx, and Charles Babbage, tremendously influenced nineteenth- and twentieth-century understandings of indus-

20. Huxley, *Science and Culture*, 199–245. His mention of Vaucanson is in the context of a famous thought experiment about an injured man who has a divided spinal cord and parts of his nervous system are cut off from his consciousness. The man is still irritable and can respond to simple stimuli such as tickling. He is thus paralyzed in regard to voluntary movement and cannot experience consciousness, but he moves when an irritant is applied to his foot (217–19). Huxley presents a real example later in the text, that of a wounded French soldier. He asks whether the man was "in the condition of one of Vaucanson's automata—a senseless mechanism worked by molecular changes in his nervous system" (223–27).

21. Helmholtz, *Wechselwirkung*, 3–5.

22. Ibid., 6 (Helmholtz's emphasis). On Helmholtz's work and its various connections to nineteenth-century materialism, see also Anson Rabinbach, *Human Motor*, 57.

trial economies and their historical course. For these authors, the eighteenth-century automaton served as a motif to comprehend interfaces between humans and machines in the factories of the nineteenth century. Both the workers' bodies and the factory itself were conceived as "automata."

For example, Andrew Ure, a prolific commentator of his time, provided in his 1835 book *Philosophy of Manufactures* an exposition of the "scientific, moral, and commercial economy of the factory system of Great Britain."[23] Ure uses eighteenth-century android automata, in particular Vaucanson's, but also the idea of self-moving machines more generally, to grasp the idea of the factory and its social and philosophical consequences. Early in the book, he sketches the same brief history of android automata that Brewster, Beckmann, Collinson, and others used (see notes 7 and 15), and Ure uses this brief automaton history to illustrate two historical developments: the improvement of mechanical machinery in manufactures over time and, more generally, the history of building moving objects and how doing so amounts to endowing matter with movements that "resemble," as he says, those of organized beings.[24] Ure also wrote other works in which automata figured, such as a study of Britain's cotton manufacture, textbooks, and encyclopedias.[25]

Charles Babbage, one of the leading figures in early Victorian science and culture and the inventor of a mechanical calculating machine, wrote a widely read and translated work in 1832 that was very similar in kind to Ure's *Philosophy of Manufactures*. It was entitled *On the Economy of Machinery and Manufactures*, and in it he also made a passing reference to an automaton.[26] His and Ure's understanding of the automaton in relation to the factory's mechanical machinery had a profound impact on subsequent general understandings of the relationship between humans and machines.

Much of this impact came about through Karl Marx, who in his *Cap-*

23. The quote is from the work's subtitle.

24. Machinery in factories, according to Ure, consisted of "cords, pulleys, toothed wheels, nails, screws, levers, inclined-planes," and they were used to give "inert matter precise movements resembling those of organized beings." Ure, *Philosophy of Manufactures*, 9. Throughout his book he refers to self-moving machines with the term "automaton" (331, 367, 368).

25. Ure, *Cotton Manufacture*, in which he refers to Arkwright's water-frame as an "automaton" (228); and Ure, *Dictionary*, in which his entry "automaton" is four pages long (77–81), reviewing briefly the history of automaton-making from the fourteenth century onward and mentioning the "brothers Droz of Chaux de Fonds" as "imitators" of Vaucanson. Ure also explains that "very complete automata have not been made of late years" (77).

26. Babbage, *On the Economy of Machinery and Manufactures*, 101.

ital relied a great deal on Ure and his use of eighteenth-century automata. Marx's automaton terminology occurs primarily in his chapter 13, on the factory and its machinery and their relations to the worker and the capitalist. Marx's use of the term *automaton* is similar to Ure's; Marx uses it to refer to the factory's machinery as a whole: he calls a group of machines that is run by one and the same self-moving engine (normally a steam engine) an "automaton," whether the group is "merely" a set of similar machines (as in weaving) or a combination of different types of machines (as in spinning).[27] He also points out that even when the "whole system" is driven by a steam engine, individual machines may need a worker to conduct certain motions, thus emphasizing the coexistence of humans and large machines in factories. Marx brings together his observations about the factory into a definition of the "most developed form (*Gestalt*) of production by machinery," which he bases on the "automaton": he says this most developed form of production is "a system of machines that receive their motion exclusively through transmission gear from a central automaton."[28] He further claims, in quite extravagant rhetoric, that in the place of an individual machine, as in former times, there is now a "mechanical monster" whose body fills entire factories and whose demonic power is at first hidden by the almost solemn motion of its gigantic limbs; but the monster finally breaks out in a feverish dance of its numerous actual working organs.[29] Marx's interest revolves around the factory machinery's organization, its inner order, and the forces that it exerts on the human. He observes how the factory machinery, like an "industrial perpetual motion machine," imposes its motion and rhythm on the human, and he claims, furthermore, in a move that integrates materialist with idealist categories, that "the automaton" has "consciousness and a will" and a "drive" to restrain human resistance.[30] In this context, Marx quotes Ure explicitly, calling him a bit condescendingly the "Pindar" of the automatic factory, as if Ure's philosophical work was that of a lyrical poet who writes celebratory odes.[31] Marx explains how Ure conceives of the automatic factory as, on the one hand, the cooperation of various types of workers with machines and, on the other, a gigantic automaton composed of numerous mechanical and self-conscious parts that work to-

27. Marx, *Das Kapital*, 392.
28. Ibid., 395–96.
29. Ibid., 396–97. He relies implicitly on automaton connotations here, too, as he mentions "inventions by Vaucanson, Arkwright, Watt, etc." in the following passage.
30. Ibid., 421, 422.
31. On Marx's condescension, see also Farrar, "Andrew Ure," 299.

gether in agreement. Marx points out that these two descriptions are not identical, and he pushes Ure's observations further by highlighting how, in the one case, the group of the workers is the subject and the mechanical automaton the object, while in the other, the automaton itself is the subject and workers are merely assembled as conscious organs around the nonconscious organs of the automaton. Marx argues that while the factory automaton can conceivably always be an object in any possible application of machinery in a large system, it is necessarily the subject in the capitalist order and therefore in "the modern factory system."[32]

Marx thus uses the subject status of the automaton to distinguish capitalist from noncapitalist factories. He and Ure rely on ideas of the automaton to bring into focus perceived changes in the condition of the human working in a factory, and Marx finishes this exchange on a poetic note by pointing out how Ure tends to call the factory's central machinery not only an "automaton," but also an "autocrat."[33] The German terms *Automat* (automaton) and *Autokrat* rhyme, which allows Marx to merge elegantly and poetically technical and political arguments.

Literary Movements in the Nineteenth Century

Such attempts to systematize the social and economic consequences of the Industrial Revolution took place alongside the emergence of English and German romantic theory and literature, the subsequent literary movements of Victorianism and modernism, and the appearance of the earliest works in the science fiction genre. All these intellectual projects were in productive dialogue with one another during the long nineteenth century. The android automaton proved a rich repository of motifs and symbolisms for these enterprises, and it found expression in literature, philosophy, aesthetics, and political theory. The romantic movement especially is credited with broadening and making more complex the rationalist and empiricist worldviews of the Enlightenment and integrating it with the "dark" and irrational aspects of human experience. Romantic authors were inspired by automata, chang-

32. Marx, *Das Kapital*, 440–41 (quote on 441).
33. Ibid., 441. On conceptual and historical interrelations between Marx, Babbage, and human-machine encounters, see also Schaffer, "Babbage's Intelligence"; Schaffer, "Babbage's Dancer."

ing and broadening the connotations attached to them, taking them from their earlier role as allegories for rational-mechanical orders and making them a foreboding oracle, as E. T. A. Hoffmann did in his text that included a piano-playing woman automaton. Hoffmann's text also made use of the entire list of eighteenth-century androids. (See my discussion of Hoffmann in chapter 5.)

The prolific late-nineteenth-century literary and artistic movements that succeeded the European romanticism are often grouped under the term *modernism*. They were diverse in their ideas and practices, but all included sustained efforts to grasp "the conditions of modernity."[34] Among the modern conditions that they struggled with were, of course, urban, industrial, mass-oriented culture and society and the profound and rapid change that came with the nineteenth century's industrial revolutions. The automaton—as well the marionette, which had taken on more visibility and popularity in the course of the nineteenth century as a motif that suggested an artificial human devoid of freedom—remained attractive for writers and artists to express such issues. Arthur Schnitzler was a key writer of Viennese modernism and became, incidentally, Peter Gay's historical "guide" to the European middle class from 1815 to 1914. Schnitzler revitalized the marionette motif around 1900 with three one-act puppet plays—a collection called *Marionettes*—that used the theme of marionettes and puppet-masters to pose questions about freedom and consciousness.[35] Marionettes and androids were also relevant in discussions about modernist theories of theater, the status of the actor on the stage, and, by extension, the status of the human individual in the world.[36]

In regard to the cultural history of the female automaton, the modernist period was an important moment. Jean-Marie Villiers de l'Isle-Adam's novel *L'Eve future* from 1886 is known for its role in popularizing the term *android* (*androïde*) and for putting at the novel's center a woman android. The novel also broke new ground in experimenting

34. Whitworth, *Modernism*, 3. Cathy Gere defines "modernism" as "a distinctive and often self-conscious sense of generational crisis, beginning around 1870 and persisting until just before the Second World War." Gere, *Knossos and the Prophets of Modernism*, 6.

35. *Marionetten: Drei Einakter*, in Arthur Schnitzler, *Die dramatischen Werke*, 1:838–94 (1903, 1904, and 1905). Peter Gay claims that the nineteenth century was "Schnitzler's century." Gay, *Schnitzler's Century*. See also Coen, *Vienna in the Age of Uncertainty*, 18. Dagmar Lorenz describes how Schnitzler's marionettes, unlike Kleist's from the early nineteenth century, can be described as laboring "under the burden of a modern consciousness." Lorenz, *Companion*, 16.

36. See Maurice Maeterlinck's landmark essay "Androidentheater" (original title *Un Théâtre d'Androïdes*) of 1890; and Edward Gordon Craig's equally groundbreaking text "The Actor and the Uber-marionette," of 1908, which was another argument of replacing the living actor with a puppet.

with science fiction to display the experience of the machine age.[37] In *L'Eve future*, a fictionalized Thomas Edison is approached for help by a friend who is dispirited with his fiancée's intellectual and emotional unresponsiveness. Edison builds an artificial woman for his friend and models it on the appearance of the unresponsive fiancée. The novel combines questions about machines and science in relation to the human experience with questions about sexuality, desire, and marriage.[38] Although there had been precedents for the appearance of these topics in the early nineteenth century, such as in works by E. T. A. Hoffmann and Heinrich von Kleist, interest in women dolls and automata increased only after the period of romanticism. It was in the same postromantic period that the general perception of automata changed from a spectacle to a threat. The best-known androids from the eighteenth century include a male flute player, a male writer and draftsman, a male chess-playing Turk, and the male monster created by Victor Frankenstein; the only women automata from the eighteenth century that we know of are the two women musicians who are at the center of my study. Automata blur several boundaries—between life and not-life and between nature and artifice, for example—and after the early nineteenth century, it was especially in the female versions of artificial humans that this deception operated. It was an ambivalence of attraction and threat often at the expense of the male gaze and identity. The tradition of the threatening artificial woman that began in ancient Greece with Pandora was continued with female figures, automata, and goddesses and is clearly exemplified in *L'Eve future*.[39]

The idea of a scientist's building an android in the image of a woman he loves (or used to love) found further poetic and artistic elaboration in the first half of the twentieth century. The best-known example is Fritz Lang's 1927 expressionist film *Metropolis*, which was based on a novel from the previous year. *Metropolis* is about conflicts between workers, machines, and owners in capitalist systems.[40] Both in the novel and in the film, a scientist constructs a machine android and designs it in the image of a long-lost female love interest. Literary

---

37. Other important works of this earliest phase of science fiction include two stories by H. G. Wells from the 1890s, "Lord of the Dynamos" and "When the Sleeper Wakes." William Douglas O'Connor, known as Walt Whitman, wrote a science fiction story called "The Brazen Android" in 1891, which explores the medieval legend of Friar Roger Bacon's talking head.

38. Lathers, *Aesthetics of Artifice*, 10.

39. Söntgen, "Täuschungsmanöver," 125; Lindauer, *Reconstructing Eve*, 17, 34.

40. The scriptwriter for Fritz Lang was Thea von Harbou, his wife. Bachmann, "Production and Contemporary Reception of *Metropolis*," 9.

and cultural critic Andreas Huyssen has written influential analyses of *Metropolis* that inquire in nuanced and wide-reaching ways into the relationships between male fantasies about women and sexuality on the one hand, and female androids on the other. With this analysis, Huyssen has shaped many literary critics' and historians' understandings of modernism, film, technology, and women automata. I discuss his work in detail later in this chapter.[41]

Technology in Education Curricula and Literary Criticism

At the end of the nineteenth century and the beginning of the twentieth, more ambitious and professionalized attempts at writing the history of technology emerged in Europe. Automata from the eighteenth century became an integral part of such work as they continued to be considered effective illustrations of the changing role of humans in the industrial world. Germany and France became industrial powers and leaders of the Second Industrial Revolution in Europe, and, as earlier in Britain, historical, philosophical, and literary works produced in these countries exhibited a pronounced interest in technology.[42] An influential example is the works of the German writer, collector, and historian of technology Franz Maria Feldhaus, who was a self-taught engineer and contributed critically to the establishment of the history of technology as a discipline. Feldhaus wrote encyclopedias and survey histories of technology that quickly became standard works. His frequent references to android automata helped keep the motif visible and functional through the twentieth century.[43]

Large, book-length histories of the automaton emerged at this time, too. They followed the pattern laid out by the encyclopedia entries of the late eighteenth and early nineteenth centuries (as I discussed above

41. Huyssen, "Vamp and the Machine."
42. Sanford, "Technology and Culture"; Ludwig, "Entwicklung."
43. Feldhaus founded in Berlin a private institute for primary source research on the history of technology and science and built up a massive and unique archive. *Neue Deutsche Biographie* (1961), 5:68. The professional label "engineer" was not legally protected at the time. Markus Krajewski describes Feldhaus as an autodidact. Krajewski, *Restlosigkeit*, 20, 144. A detailed entry "Automat" is found in Feldhaus's dictionary of inventions and discoveries in the natural sciences and technology. He also wrote an entry "Automat" in a history of the technology of ancient times and among indigenous peoples in a handbook for archaeologists and historians. Feldhaus, *Die Technik der Vorzeit*, 46–56. And there is an entry "artificial humans and animals" in a cultural history of technology that he wrote, Feldhaus, *Kulturgeschichte der Technik*, 1:94–96; as well as some details on automata in a 1930 history of ancient and medieval technology. Feldhaus, *Die Technik der Antike*, 69, 139–43, 268–70.

and used, opportunistically, in chapter 1), adapting it for more comprehensive and detailed texts. They typically started with the ancient Greeks, discussed the exchange between Arabic and European learned cultures during the Middle Ages, and compiled the usual list of automaton makers from the early modern period.[44] Alongside these writings on automata, histories of the mechanical clock became popular, too, tracing its history back to the fourteenth century. Such research and publishing activity occurred in conjunction with the founding of the major technical museums in the "imperial centers" Munich, London, and Paris in the late nineteenth and early twentieth centuries. All together, this period of high industrialism was characterized by a sustained public and scholarly interest in ancient and medieval technological artifacts.[45]

Another scholarly discipline that expanded and professionalized early in the twentieth century, the analysis of literary texts, also relied on the automaton. This field became prominent in the educational curricula of schools and universities and thus broadened its influence on the public. In a manner similar to that of historians of technology, literary scholars began to pay increasing attention to the industrial age in which they lived and to the role of "the machine" in literary works. The questions they raised concerned, like the literary works themselves, the period's rapid industrialization and commercialization. The automaton, the marionette, and other androids were high on the discipline's agenda, not least because of the attention that writers and artists themselves had paid to these motifs throughout the nineteenth century. Not coincidentally, there was renewed interest in La Mettrie's materialism and in the legend of the golem, which had its origins in the Middle Ages. The golem joined the android and the marionette in the stockpile of artificial beings that were perceived as threatening, in this period of high industrialism, to take control of men.[46] The automaton motif also

44. The most important example is Chapuis and Gelis, *Le monde des automates* (1928), on which I rely extensively. It became the most authoritative source on the history of automata and is to this day cited by everyone who works on automata.

45. Bassermann-Jordan, *Die Geschichte der Räderuhr*; Klemm, *Geschichte der naturwissenschaftlichen*; *Le conservatoire national*; and *Science Museum*.

46. On La Mettrie, see Bergmann, *Die Satiren des Herrn Maschine*. On the golem, see my chapter 5, note 22. Other examples are an essay by Strakosch, *Mensch und Maschine* (Strakosch was an actor); a play, *Maschinenmensch*, of 1929, by Otto Arnold; and *Der Maschinenmensch und seine Erlösung*, a novel by the journalist, philosopher, and writer Bruno Wille. On recent interpretations, see Schlemmer, "Man and Art Figure" (original title "Mensch und Kunstfigur"); Olf, "Man/ Marionette Debate in Modern Theatre," 491.

became involved in a dense exchange between English and German literature, on the one hand, and newly written dictionaries and reference works, on the other, and the motif traveled from there to other language traditions.[47]

Two of the earliest encounters in the period 1910–20 between industrialism and literary criticism occurred in two German books that studied historically the role of technology in literature and poetry. Such works provided the methodological foundation for subsequent historical surveys of androids in literature, and they also give us insight into the general relations between humanistic study and machine culture at the time. Later historical studies of the android as a literary motif became popular after each of the two world wars, and they combined the techniques of historical-literary analysis of technology developed in the second decade of the century with the material on the history of automata that was available through the surveys that began to appear around 1900. It is this genre of surveys of the android motif that has shaped so profoundly our ideas, even now, about eighteenth-century automata and their role in literature and culture.

Both authors of the historical surveys of technology in literature just mentioned take as their starting point that the reading public of the time, in their opinion, thought of technology as something that "resisted" artistic use or was "contrary" to poetry.[48] Hans-Werner Kistenmacher investigates in a 1914 review of German literature the impact that the rapid technological and social change of the nineteenth century had on literary production, asking how literary work of the time engaged key technologies such as the steam engine. He laments that many nineteenth-century poets kept their distance from "technology" and asserts that no subject should be excluded from poetic treatment or be called "unpoetic."[49] Felix Zimmermann's survey of 1913 similarly investigates "the historical relationship of German literature to technology" and ends up with a similar critical appraisal: he challenges the widespread public opinion that technology is antipodal to poetry and proposes instead that a large part of nineteenth-century poetic work already reflected the period's technologies. He did concede that, for his

---

47. Rosenberg, *Zehn Kapitel zur Geschichte der Germanistik*, 129–31; Hermand, *Geschichte der Germanistik*, 75–82; and Mauris, *French Men of Letters*.

48. Kistenmacher, *Maschine und Dichtung*, 7; Zimmermann, *Die Widerspiegelung der Technik*, 8.

49. Kistenmacher, *Maschine und Dichtung*, 7, 15. Harro Segeberg documents in the second decade of the twentieth century such efforts from the earlier part of the century to catalog literary work reflecting "technology" in literature. Segeberg, *Literatur im technischen Zeitalter*, 3–5.

own time, there was a need for new forms of aesthetic analysis of technological systems.[50]

Studies of technology in the same literary-critical tradition continued from this foundation through the twentieth century, with the automaton motif receiving increasing attention. Literary critics rediscovered the automaton repeatedly, often in tandem with renewed interest in romantic literature. Specifically, studies from after the Second World War have played a crucial role in disseminating our current ideas about the Enlightenment automaton. I discussed some of those in chapter 1: the ideas that the automaton was known to many people in the eighteenth century, that it was an embodiment of the mechanistic worldview, and that it was an oracle of the catastrophes of the industrial machine age that occurred in the twentieth century. Indeed, perhaps the most influential commentaries of all on eighteenth-century automata have come from literary critics in the past fifty years.

Some of these critics read Enlightenment and romantic texts on automata as authoritative accounts that articulated earlier than everyone else's (that is, already *before* the onset of the Industrial Revolution) the threat that the machine age poses to humans. These critics hold that Enlightenment automata, or contemporaneous texts about them, anticipated already around 1800 the catastrophes of the twentieth century.[51] Their interpretations tend to narrow dramatically the meanings of the automaton motif, treating automata as a supposedly cohesive symbol for the mechanization of the world and treating literary authors of the eighteenth and early nineteenth centuries as monolithic in their concerted and prophetic reaction to the automata of their time. Variations on this perspective include the claim that romantic authors used the automaton motif as a critique for the "empty and dead natural sciences" and the claim that the automaton motif is an expression of the "rebellion of the poet" against the "cold, mechanistic worldview."[52] Other literary critics apply more broadly conceived historical narra-

---

50. Zimmermann, *Die Widerspiegelung der Technik*, 8. For examples of later works, see Racker, "Gestalt und Symbolik," a PhD dissertation of 1932; Rosenfeld, *Die Golemsage* (1934); and Hoffmann, "Eine Untersuchung über Kapital, Industrie und Maschine von Goethe bis Immermann," a PhD dissertation of 1942.

51. An example is Gerhard Weinholz, who reads Hoffmann's work as warning us of the dangers that humans are exposed to when they follow exclusively the "mechanical-external" discourse of the "technical-scientific disciplines." He also holds that Hoffmann, in his overall work, assesses "scientific-civilizational" progress negatively. Weinholz, *"Die Automate,"* 17–18. See also Eberhard Roters, who talks about a "terror" that Hoffmann felt in the face of android automata and a "fear of the power of robots" that he harbored and aimed to communicate. Roters, *E. T. A. Hoffmann*, 97.

52. Vietta recapitulates those in a critical review. Vietta, "Das Automatenmotiv," 25.

tives, such as the idea that the aristocratic and bourgeois strata of society in the eighteenth-century had uniform political and economic interests, which were supported by the automaton motif. Or they presuppose a uniform and widely shared "dream of linear technological progress," which was supposedly inspired by the eighteenth-century automaton.[53]

Literary scholars have commented specifically on Johann Richter and E. T. A. Hoffmann, suggesting that these poets' work exposes the ways in which humans (or societies) have become machines and lost their autonomy and authenticity. These scholars also assume that the two poets resented automata and the ideas connected with them, identifying the automata as a danger threatening humans and societies.[54] I presented a more complex picture of the two poets in chapter 5, analyzing their references to contemporary automata, music-making women automata, and contemporary concerns in musical theory and pedagogy. Automata are indeed a threat, or source of discomfort or confusion, in the scenarios that Richter's and Hoffmann's texts present. But the poets explicate their critical commentary in much more specific ways, and in ways appropriate to their time: piano-playing women automata do not engage, in Richter's and Hoffmann's texts, the mechanization of humans in the industrial age. Rather, they engage issues such as sentimental society, gendered practices of music-playing, and interrelations between human agency, clockwork mechanics, and intersubjective forces.

To supplement with one more example the contrasts between my and other scholars' analyses of Richter and Hoffmann's work, I look at Andreas Huyssen's widely acknowledged analyses of the intersection of modernism, film, and women androids from the 1980s. My work intersects with his through women androids, of course, but also through the received ideas about eighteenth-century automata that I discuss in this chapter and in chapter 1, ideas on which Huyssen relies as well. I have shown that such ideas have been extensively used over long time periods but also, and maybe more importantly, that they have been consistently deemed helpful and useful as illustrations of the human condition in the industrial age. It is here that I take Huyssen's study as an opportunity to probe further the tension that my study generates: between eighteenth-century automata's obvious and widely used

---

53. Both ideas are expressed in Wawrzyn, *Der Automaten-Mensch*, 101–3.
54. See note 42 in chapter 5. Sauer, *Marionetten*, 73–74; Gendolla, *Anatomien der Puppe*, 98–99; Frank Wittig, *Maschinenmenschen*, 58, 80.

appeal to symbolize broad concerns of industrial modernity, on the one hand, and the deep roots that the automata have in a society that clearly preceded industrial modernity and is clearly distinct from it, on the other.

Huyssen was among the first to suggest connections between gender, mass culture, and industrialization and offer interpretations of technology that were based on Marxist critical theory and psychoanalysis. His work opens crucial conceptual avenues for studying industrial capitalism's ramifications in realms of sexuality, production, and consumption. He centers his discussion on Fritz Lang's *Metropolis* and claims that the film's female robot is part of a "long standing" tradition of the machine-man's being "presented as woman." Consequently, he engages in a "historical digression," as he says, to trace the roots of high industrialism's women androids back to the eighteenth century. Doing so, he operates with precisely those assumptions about eighteenth-century automata whose universality my current study challenges. He holds, for example, that philosophies such as La Mettrie's and Descartes's materialism prompted the construction of android automata in the eighteenth century, that a large number of artisans tried to build them, and that these automata were widely known and were an expression of the zeitgeist.[55] About one of the two protagonists of my study, he says: "Androids and robots such as . . . Jacquet-Droz's organ player captured the imagination of the times and seemed to embody the realization of an age-old dream."[56]

As part of his historical digression, Huyssen analyzes La Mettrie's materialism in *L'homme machine*, saying that such materialism "denies" emotion and subjectivity and that it "ultimately led to the notion of a blindly functioning world machine, a gigantic automaton." He claims, furthermore, that eighteenth-century materialism indicates that the "age of modern technology and its legitimatory apparatuses had begun."[57] Huyssen casts La Mettrie's eighteenth-century materialism in terms—"emotion" and "subjectivity"—that had their origins in the nineteenth century. Indeed, he uses no eighteenth-century sources other than La Mettrie, thus projecting a late nineteenth-century cul-

---

55. Huyssen, "Vamp and the Machine," 224, 225. Huyssen says, "It is no coincidence that in the [Enlightenment] age literally hundreds of mechanics attempted to construct human automata who could walk and dance, draw and sing, play the flute or the piano, and whose performance became a major attraction in the courts and cities of 18th-century Europe."
56. Ibid., 225.
57. Ibid.

tural and technical modernity onto a text that preceded not only the idea of the factory but even James Watt's steam engine. Huyssen transforms La Mettrie's materialist, medical, and theological scholarship into an anticipation of industrial technology. Furthermore, rhetoric about "blindly functioning world machines" and "gigantic automata" also has its roots in the mid-nineteenth century, as suggested by my readings of Marx and Carlyle, who wrote almost a century after La Mettrie's materialism and the construction of the eighteenth century's two women automata, in a distinctly more industrial period.

There is another question that Huyssen tries to answer with the help of his historical digression and the assumption that *Metropolis*'s female robot (named Maria) "stands in a tradition" that reaches back to the women automata of the eighteenth century, and that is the question of why robot Maria, who was "intended to replace the human workers" in the production process, would actually appear as a woman. This question, I would argue, is difficult to answer with the help of a "historical digression" to the eighteenth century. The "intention" to replace workers in the production process does not have its main roots in eighteenth-century women automata, since such production processes as shown in *Metropolis* did not exist then; even production processes that existed at the time of Thomas Carlyle's essay "Signs of the Times," the 1830s, did not exist when La Mettrie's *L'homme machine* was written or when the women automata were built. And to assume that industrial production processes were anticipated in La Mettrie's work or in the automata is problematic historically. Huyssen's strategy of looking to the eighteenth century to explain the convergence of female gendering of androids and the mechanization of the production processes in the industrial age is conceptually extremely fruitful, and yet historically it is difficult to place coherently into protoindustrial and industrial eras. The replacement of humans with machines on a large scale is a historical phenomenon—but also a key imagination and a fantasy—of the industrial age: it was particularly in the factory era that rapid increases in productivity and efficiency were accomplished by replacing human physical labor with machines. A question that remains standing as I close my study is why mechanical androids produced sixty to seventy years earlier would be taken to anticipate this phenomenon.

Huyssen presents his argument with a great deal of subtlety, and he never claims that eighteenth-century androids are the cause of Western industrialization. Furthermore, his analysis of *Metropolis* brings together two sets of themes that are at the center of my analysis: industrialization and gender. But he consistently lets his attention become captivated by the "industrial condition" and anachronistically renders Enlightenment automata as objects that somehow signify the origins of the Industrial Revolution, whether in ideological or mechanical-practical ways. In contrast to Huyssen's and others' analyses of industrial modernity through female androids—which have contributed a great deal to our understanding of gender, desire, and technology—I emphasize, through my findings, the very boundary between the industrial and the preindustrial economies and raise questions about how we choose objects and symbols to help us understand the significance of the process of the industrialization of modernity.[58]

The Automaton in Twentieth-Century Science Fiction

Two additional cultural conduits through which the automaton or robot became widely known in the late twentieth century were the science fiction novel and the popularization of scientific and engineering research in robotics and artificial intelligence. To finish my overview of the journey that eighteenth-century automata took in the two hundred years after their production, I now look at a few more channels through which current ideas about them emerged. I trace the term *robot*, which in the twentieth century soon eclipsed *automaton* as the standard name for a mechanical human or animal, the genre of science fiction and a few prominent texts and films, and three rather distinct academic disciplines, critical theory, cybernetics, and history.

The Czech writer Karel Čapek coined the word *robot* just after the First World War in his play *R.U.R.* (*Rossum's Universal Robots*), and Isaac Asimov underscored its profound influence when he said that Čapek's play was "immortal for that one word."[59] In the following decades, the

58. The case of Vaucanson's automated loom is an example that comes closer to the idea of "anticipating" the Industrial Revolution (ibid., 224). Huyssen contributes a third argument to this debate. He makes a connection between gender and mass culture, investigating the vexed relationship between women and "real" or "high" culture and art in the age of industrialization and mass culture. The traditional exclusion of women, he argues, took on new connotations in this period, and mass culture was gendered as feminine, even by critics such as Horkheimer and Adorno and Siegfried Kracauer. Huyssen, "Mass Culture as Woman," esp. 47.

59. Asimov, *Asimov on Science Fiction*, 71. There is a wide consensus about this origin story: virtually every encyclopedia or textbook etymology of the word *robot* mentions the play *R.U.R.* It was actually not Karel Čapek but his brother Josef, also a respected Czech writer, who coined

distinction between robots and automata became increasingly blurred, and *automaton* started referring to entities that were markedly different from mechanical androids.[60] The novel and distinct idea of joining human bodies with machines, for example, first appeared in science fiction writing in 1923 and then went on to become an essential element of twentieth-century science fiction.[61] Leading science fiction writers picked up the automaton motif, in the new form of the robot, and it profoundly shaped the productive period of the genre between the 1930s and the 1950s.[62] The examples I single out here are Isaac Asimov and Stanislaw Lem, perhaps the masters of the genre. They started publishing stories that regularly featured artificial humans in the late 1930s. Philip K. Dick's works, first published in the early 1950s, construct worlds inhabited by both humans and androids, and thus he engaged preoccupations of the Cold War era. His novel *Do Androids Dream of Electric Sheep?* was the basis of the later Ridley Scott film *Blade Runner,* which became well known in the 1990s. A favorite example, finally, for scholars in the history and philosophy of technology, to illustrate the key process in the industrialization of the modern world—the replacement of working humans with automated machines—is Kurt Vonnegut's 1950 short story "epicac," which features a fictional computer that also appears in his 1952 novel *Player Piano.* In a subplot in the novel, a mechanic is replaced by an automated machine (the epicac computer). Historians and critics such as David Mindell and Langdon Winner, for example, use the player piano to illustrate the replacement of human labor by machines in the industrial age.[63]

---

the word. The science fiction play premiered in Prague in 1921, and it was successful in Britain and the United States soon afterward. Čapek's Robots (always capitalized) are not mechanical entities, however, but biological ones. In the play, Čapek deals with the technocratic visions, revolutionary utopias, and war horrors of his age. See the introduction by Ivan Klíma in Čapek, *R.U.R.*, x–xvi.

60. An example here is Grässe, *Zur Geschichte des Puppenspiels* (1856), which is hardly at all about automata. See also Schücking, "Die Marionette"; Rapp, *Die Marionette* (1924); and Boehn, *Puppets and Automata* (1929).

61. The first mention of such an entity, supposedly, was in E. V. Odle's novel *The Clockwork Man* (1923). See Brian Stableford's entry "cyborg" in *Science Fact and Science Fiction*, 114–16; and Clute, "Identity of E.V. Odle," 343.

62. On the "golden age" of science fiction, see Asimov, *Before the Golden Age.* John W. Campbell, editor of the prominent magazine *Astounding Science Fiction*, wrote stories such as "The Last Evolution" in 1932, in which aliens invade the earth and humanity defends itself by developing machine minds—minds outside of the body. Another example is Neil R. Jones, who is known for coining the term *astronaut* and for his use of robotic characters.

63. Gunn, *Isaac Asimov*; Davis, "Metaphysics of Philip K. Dick"; Lem's "Smart Robots," in Swirski, *Art and Science of Stanislaw Lem*, 13–21; Mindell, *War, Technology, and Experience*, 135;

Anxieties about human-machine relations reached new heights in the mid-twentieth century in response to distinctively new technologies of warfare in the Cold War and cooperation between governments, the military, and corporations on an unprecedented scale (the "military-industrial complex").[64] New forms of social commentary expressed these concerns, and eighteenth-century androids continued to be part of the symbolism to articulate the increasingly troubling relationships between technology and humanity. The preindustrial mechanical objects of my study thus persisted as valid metaphorical tools, even though the size and complexity of late twentieth-century technological systems increased steadily.

The social commentary of the time included the emergent countercultures, whose participants asked, among other things, about the foundations of what they called the "technological society."[65] They couched their critique in terms of stipulating "the organic instead of the mechanical," thus reviving rhetoric from the nineteenth and the eighteenth centuries.[66] The renewed discussion about technology in a new geopolitical order gained wide attention in the public spheres of the industrialized nations. Among the best-known authors were Lewis Mumford, Jacques Ellul, E. F. Schumacher, Sigfried Giedion, and Marshall McLuhan, all of whom, each in his own way, pronounced their society "mechanized" and investigated a variety of technical and social phenomena in the mid-twentieth century.[67] Members of the first two generations of the Frankfurt School were equally prominent in their theoretical investigations of an increasingly mechanized modernity,

---

Mindell, *Between Human and Machine*, 316; Winner, *Autonomous Technology*, 33. See also Reed, *Short Fiction of Kurt Vonnegut*; Marvin, *Kurt Vonnegut*, 25–42.

64. This term was famously coined by Dwight D. Eisenhower in his farewell address in January of 1961. Johnson, "From Boeing to Berkeley."

65. *The Technological Society* is the title of Jacques Ellul's influential book of 1964. See also Hughes, *American Genesis*, 444. Tom Hughes brings this counterculture into focus by describing how critics raised questions about the wisdom of organizing nations "into massive military, production, and communication systems" and attacked the systematizing and administrative aspects of modern technology (3). Hughes also analyzes historically the emergence of a new distrust of technology (444).

66. Hughes, *American Genesis*, 12.

67. See Mumford, *Technics and Civilization* (1934); and his two-volume work *The Myth of the Machine*, of 1967 and 1970. His *The City in History* (1961), was awarded the National Book Award. Hughes calls Mumford "a window on characteristic problems of [the twentieth] century." Hughes, *American Genesis*, 447. See also Hughes, *Lewis Mumford*. David Mindell provides a beautiful interpretation of the relations between Mumford's "neotechnic machine worlds" and Norbert Wiener's cybernetics, and a critique of their common perceptions. Mindell, *Between Human and Machine*, 4.

evaluating the accomplishments of industrial society at the level of ideology, economics, and culture.[68]

In this vein, historians as well turned their attention, once again, to the early modern period and its supposedly characteristic "mechanistic worldview." They asked whether this worldview, in connection with materialism, was an ultimate historical cause of the horrors of the First and the Second World War and the new political and military order of the Cold War. Such studies were, because of the coincidence of materialist works and automaton-making, another occasion for engaging with seventeenth- and eighteenth-century androids in the mid-twentieth century.[69] The growing skepticism toward industrial civilization in the second half of the Cold War generated, not least, yet another set of surveys of the history of the automaton and the robot, and I have relied on some of them in my own study. They made the spectacular eighteenth-century automata widely visible once again.[70]

Advanced scientific and engineering research during and after the Second World War and its popularization in the Cold War contributed to increasing visibility of the digital computer and of cybernetics as a field of interdisciplinary study of both physical and social systems. This made an impact on the ways the terminology of "automata" was used. Norbert Wiener's first book, *Cybernetics*, for example, was published in 1948, and he uses "automaton" in ways similar to the uses of other authors that I have discussed, namely as a motif to illustrate key moments in the modern history of ideas such as Newtonianism, Cartesianism, Darwinism, and industrialism, as well as for self-moving

68. They included Walter Benjamin in the 1930s, Horkheimer and Adorno in the 1940s, and Herbert Marcuse and Jürgen Habermas in the 1960s. On the relationships specifically between technology and the countercultures, see also Wisnioski, "'Liberal Education Has Failed.'"

69. The Enlightenment as a period also received renewed attention after the Second World War for the potential answers it could give to questions about the origins of modernity's destructive powers. James Schmidt has analyzed how scholars and critics in the twentieth and early twenty-first centuries have mistakenly appropriated the eighteenth century and the "Enlightenment Project" as explanations for the catastrophes of the twentieth century. See Schmidt, "What Enlightenment Project?"

70. For example, in 1947, 1949, and 1952, Alfred Chapuis published works on automata: a survey, a work specifically on the relationship between automata and literary imagination, and a work on the watch from the late eighteenth to the first half of the twentieth century. Later examples include Cohen's *Human Robots*; Swoboda's *Der künstliche Mensch*; Simmen's *Der mechanische Mensch* (1967); Aram Vartanian's "Man-Machine" (1973); Matt's *Die Augen der Automaten* (1971); Baruzzi's *Mensch und Maschine* (1973); Wawrzyn's *Der Automaten-Mensch* (1978); Völker's anthology *Künstliche Menschen* (1976); Sauer's *Marionetten* (1983); Beyer's *Faszinierende Welt der Automaten* (1983); Bammé's collection of essays *Maschinen-Menschen* (1983); Hildenbrock's *Das andere Ich*; Drux's *Marionette Mensch* (1986); and Richter's *Wunderbares Menschenwerk* (1989).

machines and self-organizing systems. In subsequent editions, he keeps up, as he says, with novel "automata," warning against "social dangers" that accompany them.[71] His 1954 work *The Human Use of Human Beings* uses "automaton" in relation to the replacement of human labor in the factory and to life-imitating objects.[72] In *God and Golem, Inc.*, published in 1964, finally, Wiener merges the ideas of monster and automaton, proclaiming on the back cover that the golem is "a monster, an automaton."[73]

Even though in the twentieth century the term *automaton* became increasingly severed from its eighteenth-century context, some historians and philosophers of the automaton continued to see a more or less unbroken legacy from the construction of android automata in the Enlightenment (or even in classical antiquity) to the age of the factory in the nineteenth century, and from there to the age of the computer in the Cold War. The survey works and critical analyses mentioned above contributed their share to this idea and its variations. The coinage of the term *robot* and its successful career in popular usage and literature in the twentieth century were factors in the emerging paradox that although the automaton was becoming increasingly visible, at the same time its meaning and referents were becoming increasingly blurred. The political, military, and economic conditions in late-twentieth-century industrial nations were of a kind that encouraged commentary and critique in terms of artificial or mechanical humans. Cultural critics were concerned in particular with the conditions of democratic government in the industrial and postindustrial societies, as well as with the seemingly rapid progress in artificial intelligence and artificial-life research, and they employed the android automaton as an attractive and useful motif to express these concerns. It was, as this chapter shows, transmitted through a period of two hundred years and retained its value as a metaphor or symbol for human-machine relations.[74] In this long period, in the discussions between cultural and literary criticism, political theory, philosophy, and computing history, the android automaton was increasingly construed as a coherent category, and the "industrial

71. The full title of the first edition of Wiener's first book is *Cybernetics; or, Control and Communication in the Animal and the Machine.* The quote is in the edition of 1961, "Preface to the Second Edition," vii.

72. He uses the term often, as the index reveals.

73. See also Langdon Winner's comments on Norbert Wiener in 1947 and the connection to Greek tragedy. Winner, *Autonomous Technology,* 17.

74. See, for example, the essays in Day, Beiner, and Masciulli, *Democratic Theory and Technological Society,* 43, 276, 179. The "automaton" is here transmitted mostly through the work of Karl Marx and E. T. A. Hoffmann.

anxieties" attached to it were increasingly projected back onto the time that the automata were made. In contrast, I found in this same time period unusually prolific artisan worlds in which android automata mimicking body practices of bourgeois emancipation were designed, and I found poets who used the metaphorical and symbolic value of these automata to raise questions, in very different ways, about subject-formation in an era that preceded the industrial age.

Conclusion

The two woman automata under consideration in this study bring together two features of the early modern world: pre- and protoindustrial artisanship and nascent sentimental bourgeois culture. Both features are indispensable for explaining the automata's design, construction, and interpretation. There is an obvious tension between the early modern, preindustrial, court society in which the automata were made and first interpreted and our modern, industrial age that so keenly appropriates them as symbols for its own discontents and crises. This tension has been the organizing theme of my study and my attempts to cast in a new light mechanical androids and their relevance to our understanding of both pre-industrial and industrial human-machine relations.

I explored the economic histories of Switzerland and the Rhineland to understand the environments in which such sophisticated artifacts were built, and then I analyzed mechanically the music-playing automata's internal clockworks to explain their design and function through a cultural history of musical and sentimental practices of bourgeois selfhood in Enlightenment Europe. I traced how piano-playing women emerged as a literary motif between the 1780s and 1820s and became integrated into wide-reaching poetic agendas that influence interpretations of modern technology and culture to this day. The two literary texts that I looked at engaged the ambiguous status of the bourgeois practices replicated in the automata's mechanisms, and they also laid the foundation for the eighteenth-century android's later use as a token for the perils of the industrial age. In chapter 6 I outlined the multiple modes and manners in which the idea of the "Enlightenment automaton" traveled through the nineteenth and twentieth centuries, explaining how it could so successfully transform itself, in multiple steps, into one of the key allegories of later, industrial modernities. In the process, I traced ideas attached to the boundaries between preindustrial and industrial worlds and how they have entered into our own understanding, in the postindustrial age, of the relationship between humans and machines, also bringing into sharper focus the technologies that we single out to help us understand this relationship.

Mechanical humans, and questions about unstable human-machine boundaries, certainly preceded the industrial age, and they operated in numerous and influential philosophical, cultural, and artifactual ways in many earlier periods. At the same time, it is industrial machinery—not artisan production—that is at the center of the most influential critiques of the mechanical age. Our current accounts of modernity were developed in, and usually refer to, the steam-engine era of factories, not the age of artisan clock-, textile-, and furniture-making. As a result, such accounts often take questions about modernity, implicitly, to be questions about *industrial* modernity. When pondering problems about "the modern," we tend to distinguish insufficiently between the preindustrial and the industrial "modern," and this might be, not least, a reason why the eighteenth-century automaton—the product of decidedly pre-steam-engine manufactures—appears so easily applicable to crises of later, and very different, modernities. Our insufficient distinctions between preindustrial and industrial modernity entail not only misunderstandings of many of modernity's most foundational aspects but also misconceptions of its most important promises and crises, before, during, and after the factory age.

My case of the two women automata integrates the history of protoindustrialism and early bourgeois culture with the history and criticism of modern industrial production and selfhood. It revisits preindustrial roots of the industrial revolutions and early modern roots of the bourgeois subject and clarifies the role that the Enlightenment plays in our assumptions about the interdependencies between technology and modernity. I found ample evidence for the idea—in itself not very radical—that concerns about unstable human-machine boundaries take on different forms (and different technologies) in different time periods and that, in the context of my two women automata, such concerns carry connotations that are different from the ones they carry nowadays. However, my study also brought out a more specific and less obvious finding. While concerns about unstable human-machine boundaries often take on a generic form (say, the generic fear that humans have become "like" machines), historically more specific forms of this concern capture in broader and more complete ways the promises and crises of the modern age in both its industrial and its preindustrial elements. One of those historically more specific forms of unstable human-machine boundaries sits at the core of my analysis of the two women automata. The motions that the automata perform mimic an eighteenth-century sentimental, self-reflexive, subject-forming experience of music-making. The automata's mechanical replication of this experience makes the human-machine boundary unstable, in the first instance, in a generic way: it becomes problematic for a spectator to distinguish clearly between human and machine when witnessing the androids' musical play. The mechanical performance raises, in the second instance, however, the question of whether the process of reflexive, sentimental, musical subject-formation—claimed and cultivated by Enlightenment theorists and practitioners—is reproduced in visible and credible ways in the automata. And that raises the two questions of, first, whether the android itself experiences self-reflexively its own subject-forming music-playing, and, second, whether a spectator could "see" the android's encounter with itself during its sentimental-musical performance. The women automata's performance thus asks whether the process of subject-formation is "real" or merely "mechanical" and fake.

The bodily practices that accompanied the constitution of sentimental and rational selfhood in the eighteenth century, and their reproduction in mechanical artifacts, created shared experiences of sociability and at the same time cast doubt on the reality and reliability of such sociability. The two women automata (as well as Richter's and Hoffmann's texts about them) cause uncertainty not only about the human-machine boundary as such, but also about the idea that sentimental body practices could serve as the foundation of social order. The doubt arising from the mechanical replication of sentimental practices in machines illustrates both the robustness and the potential failures of this particular moment of the making of modern selfhood in late-eighteenth-century Europe. Destabilizing the human-machine boundary by building piano-playing women automata makes two types of uncertainties of this historical moment coincide: the uncertainty over distinguishing humans from machines and the uncertainty over the possibility of creating modern society by mass-producing sentimental selves.

The scenarios replicated mechanically in the two women automata bring into focus the problem of self-expression and its multiple and reliable reproduction. The dissemination and circulation of this scenario (both in the human and in the mechanical version), the organized way in which it was accomplished, and the promise of social order attached to it all point to a contradiction typical for the modern age, the contradiction that underlies all attempts to accomplish refined and reliable individuality by mass-producing it. Historically speaking, my case is a curious concurrence of preindustrial, custom-made production of technological artifacts with a cultural mass-production of bourgeois selves. Indeed, it is the *technologies* in my case that are unique, individual, and artisan-produced, while it is the *human individuals* that are being mass-produced in uniform, identical ways and in large numbers. The artisans, the sentimental music-making, and the human-machine boundaries in my case reorganize narratives of androids, subject-formation, rationality, machine-culture, and the disciplining of individuals in the context of making society modern. Eighteenth-century android automata have provided for us, sometimes maybe too conveniently, molds in which to cast questions about the ever-changing and complex relationship between humans and machines. And historical studies about them often seek to identify the ultimate origins of our contemporary machine culture and human-machine boundaries. The search for antecedents, however, often fails to yield a sufficient understanding beyond integrating current robots into an imagined, uninterrupted genealogy.[1] John Cohen expresses this historiographical problem aptly when he says that the "automaton refuses to be taken hold of and put in a cage with a label attached. It pops up here, there, and everywhere."[2]

I have provided a fresh set of questions and answers about what it means to reproduce mechanically techniques of human subject-formation. The metaphor of the machine-man typically characterizes the ambiguities of the technical and social accomplishments and failures of the modern age. The fact that the man-machine brings this ambiguity into focus so effectively has been used by a significant number of critics of the modern age. At the center of my analyses of machine-

---

1. Daston makes a similar argument about the history of probability. Daston, *Classical Probability in the Enlightenment*, 8. The publishing of books about automata between 1950 and 2005 (as I analyze in chapter 6) neatly went hand-in-hand with the emergence of the electronic computer and of artificial intelligence and with the evolution of the literary genre of science fiction, during the Cold War.

2. Cohen, *Human Robots*, 7.

women, artisan worlds, clockwork, bourgeois practices, and poetic texts have been the social production and mechanical reproduction of musical and sentimental selfhood. And in my case, the unstable human-machine boundary has raised the question not only of whether we are becoming more and more like machines, but also whether we can assume that stable and reliable selfhood is possible even in the absence of machines.

Bibliography

Manuscript Materials

Archiv der Brüder-Unität, Herrnhut/Oberlausitz (UA)
    UA R.22.36 68–82 A. (Roentgen's biography)
    UA R. 2. A. 1–4. A. Synoden und Conferenzen der
    Zinzendorf'schen Zeit (protocols of the synods and
    conferences of the Brotherhood's governing bodies during
    Zinzendorf's lifetime)
    UA R.2.A. no. 2, 2a, R.2.A. no. 2, 2b (Abraham's missions)
    UA R. 8. 35 a. Herrnhaagische Kirchenbücher (Abraham's
    marriage to Susanne Bausch in Marienborn)
    UA R. 7. G. b. no. 1. c. Diarien, Berichte, Memorabilien,
    1782–1793 (Prussian king Friedrich Wilhelm II, visit in 1792)
    UA R 7. G. a. 2, 53. Akta, das Etablissement in Neuwied betref-
    fend, 1750–1756, Haus- und Tischzinsbücher (Arts and
    trades and their accounting in the Brotherhood's Neuwied
    settlement)
    UA R. 7. G. b. 3 a. Kirchenbuch-Extrakte, Neuwied, 1750–1827
    UA R. 7. G. a. 1. Briefwechsel Zinzendorfs mit dem Grafen von
    Wied (Negotiations about privileges for a Herrnhut settle-
    ment in Neuwied) (counterparty of this is in the correspon-
    dence in FWA 65–11–4)
    UA R. 7. G. a. 3. Akta, das Etablissement in Neuwied betref-
    fend, 1750–1756 (Negotiations about privileges for a Herrn-
    hut settlement in Neuwied)
    UA R. 7. G. b. 3. a. Kirchenbuch-Extrakte, Neuwied, 1750–1827
    (arrival of the Brotherhood group in Neuwied, 1750)
    UA R. 7. G. b. 4., June 1787. Protokolle der Ältesten-Conferenz
    (David Roentgen's request for readmission into the
    Brotherhood)

UA R. 7. B. b. 7 a. Protokolle des Aufseher-Collegiums, 1779–1803 (David Roentgen's request for readmission into the Brotherhood)

UA R. 7. G. b. 9. a. Briefwechsel der UAC mit Neuwied und Andernach, 1765–1814 (David Roentgen's request for readmission into the Brotherhood)

UA R. 7. G. a. 9. Akta, gräfl. Zusicherungs-Decret für Anbau eines neuen Quarré (1781–82) (Building permissions for the Brotherhood)

Archiv des Unitätsvorsteher-Collegs, Herrnhut (UVC)

UVC R5 I K. Der Brüdergemeinorte Oeconomicum oder Rechnungswesen, Jahresrechnungen Neuwied (accounting, billing, loans)

UVC R6 I K. Der Brüdergemeinen Civile oder bürgerliche Verfassung (civil code, constitution)

UVC. IX. 23. (loan negotiations with Abraham Roentgen)

Archive of the Deutsches Museum, Munich

Kaufmann papers

Archives d'Etat, Genève

Archives privées no. 77: Papiers Leschot et Jaquet-Droz 1716–1957

Jur. Civ. F, no. 825 (1791) (Inventaires)

Coll. Fatio 16.17

Archives of the Académie des Sciences, Paris

Letter of J. M. F. Lassone, 4 March 1785, to the Académie

Procès Verbaux, 5 mars 1785, folio 46

Plumitif, 16 mars 1785

Procès Verbaux, 20 avril 1785, folio 67

Archives of the Conservatoire national des arts et métiers (CNAM), Paris

Folder "07501–0000–Joueuse de tympanon"

Archives of the Technisches Museum, Vienna

Folder "Friedrich von Knaus, Allesschreibende Wundermaschine"

Bibliothèque de la Ville La Chaux-de-Fonds, Switzerland

"Journal de A. L. Sandoz": Nb36, Nb37, Nb38, Nb39

"Journal pour Abram Louis Sandoz; Justicier de la Chaux-de-Fonds; 1757–59"

"Journal du Voyage d'Espagne de Jaquet-Droz, 1758–1759": D 1853

"P. Jaquet-Droz; Lettres d'Espagne, 1758–1759": Nb 60

Bibliothèque de Genève

Manuscript Collection "History," vol. 5:

No. 954: Notes diverses (mécanique, etc.) de Jean-Frédéric Leschot.

No. 955: Comptabilité de Jean-Frédéric Leschot pour les années 1786–1824.

No. 956–960: Comptabilité de Jean-Frédéric Leschot et Henry-Louis Jaquet-Droz pour les années 1791–1812.

Manuscript Collection "Theology, Literature, Sciences," vol. 6: nos. 961–64: Copie de lettres de Jean-Frédéric Leschot 1792–1822

Manuscript Collection "French Manuscripts," vol. 19: Ms. fr. 4498–4500: Papiers Leschot

Léonard Bourdillon, Bibliothèque publique de Genève, manuscrits, vol. 27: 206–7

Fürstlich-Wiedisches Archiv, Neuwied, Germany (FWA) (call numbers: Schrank-Gefach-Fascikel-Seite)

FWA 65-11-13 (privilege for the so-called Inspirierten)

FWA 61-6-2 (privilege for the Freemasons, 1752)

FWA 26-10-6 (general princely privilege for immigrants with building permit, 1733)

FWA 65-11-13 (privilege for the so-called Inspirierten, 1739)

FWA 80-2-4 (general concession: tax exemption for ten years to everyone who built a house in the city of Neuwied, 1747)

FWA 65-11-6 (privilege for the Augsburg branch of the Moravian Church (Brüder-Unität Augsburgischer Konfession) from 1756

FWA 65-11-4 (correspondence between the County of Neuwied's administration and the Brotherhood's authorities in the early 1750s)

FWA 65-11-6 (General-Concession for the Herrnhut Brotherhood: exemption from military service and from an oath of allegiance to the Count, 1751)

Renteirechnungen 1761, S. 215, no. 1279

Renteirechnungen 1761, S. 223, no. 1388

Renteirechnungen 1769, S. 300, no. 1022 bill for Roentgen furniture:

Renteirechnungen 1769, S. 300, no. 1023 bill for Roentgen furniture:

(*Renteirechnungen*: lists of payments for deliveries, receipts)

Mathematisch-Physikalischer Salon, Dresden

Metropolitan Museum of Art, New York:

Manuscript collection MS-44, Part I, folders 1 and 2 ("Correspondence, David Roentgen")

Musée d'art et d'histoire, Neuchâtel, Switzerland

Musée International d'Horlogerie, La Chaux-de-Fonds

Jaquet-Droz Papers: D 166, D 172, D 476, D 571, D 2292, D 3767, D 3768, D 3839, D 3840, D 3895

Périodique: Pb99
Leschot Papers: D 158, D 1053
Biographies – marques de fabriques "Jaquet-Droz": B 384, B 783

Staatsarchiv, Stadt Basel, Switzerland
Straf- und Polizeiakten, F4 Schaustellungen, 1667–1865.
Manuel du Conseil, 1742–43 (7 September 1743)

Published Materials

Aeschlimann, W. "Pierre Jaquet-Droz et Jean-Frédéric Leschot." *Almanach du vieux Genève* 29 (1954): 51–56.

Affiches, annonces, et avis divers, ou, Journal général de France. Paris: Bureau du Journal général de France, ou Affiches (1761–89), 1781.

Alder, Ken. *Engineering the Revolution: Arms and Enlightenment in France, 1763–1815.* Princeton, NJ: Princeton University Press, 1997.

Alexander, John T. *Catherine the Great: Life and Legend.* New York: Oxford University Press, 1989.

Alford, C. Fred. *Science and the Revenge of Nature: Marcuse and Habermas.* Gainesville: University Presses of Florida, 1985.

Allen, Robert C. *The British Industrial Revolution in Global Perspective.* Cambridge: Cambridge University Press, 2009.

Allgemeine deutsche Biographie. Berlin: Duncker & Humblot, 1967–71.

Almanach de Gotha contenant diverse connoissances curieuses et utiles. Gotha: C. G. Ettinger, 1789.

Altick, Richard Daniel. *The Shows of London.* Cambridge, MA: Harvard University Press, 1978.

———. *Victorian People and Ideas: A Companion for the Modern Reader of Victorian Literature.* London: J. M. Dent, 1973.

Apostolidès, Jean-Marie. *Le roi-machine: Spectacle et politique au temps de Louis XIV.* Paris: Editions de minuit, 1981.

Arnim, Ludwig Achim von. *Armuth, Reichthum, Schuld und Busse der Gräfin Dolores; eine wahre Geschichte zur lehrreichen Unterhaltung armer Fräulein.* Berlin: Realschulbuchhandlung, 1810.

———. *Isabella von Aegypten.* Berlin: Realschulbuchhandlung, 1812.

Ashby, Sir Eric. "Education for an Age of Technology." In *The Late Nineteenth Century, c. 1850–c. 1900,* vol. 5 of *A History of Technology,* 8 vols., edited by Charles Joseph Singer, 776–99. Oxford: Clarendon Press, 1954–84.

Asimov, Isaac, ed. *Asimov on Science Fiction.* Garden City, NY: Doubleday, 1981.

———. *Before the Golden Age: A Science Fiction Anthology of the 1930s.* Garden City, NY: Doubleday, 1974.

Augarde, Jean-Dominique. *Les ouvriers du temps: La pendule à Paris de Louis XIV à Napoléon Ier.* Geneva: Antiquorum Editions, 1996.

Auslander, Leora. *Taste and Power: Furnishing Modern France*. Berkeley: University of California Press, 1996.

"Auszug eines Schreibens aus Neuwied, den 10. November, 1785." *Literattur und Völkerkunde* 4, nos. 7–8 (1786): 686–96.

Baasner, Rainer. *Georg Christoph Lichtenberg*. Darmstadt: Wissenschaftliche Buchgesellschaft, 1992.

Babbage, Charles. *On the Economy of Machinery and Manufactures*. London: C. Knight, 1832.

Bach, Adolf. *Goethes Rheinreise mit Lavater und Basedow im Sommer 1774: Dokumente*. Zürich: Seldwyla, 1923.

Bach, Carl Philipp Emanuel. *Six sonates pour le Clavecin: A l'usage des dames*. [c. 1727]. Microfiche manuscript. Original in Universitäts- und Stadtbibliothek Köln.

———. *Versuch über die wahre Art, das Clavier zu spielen*. Berlin: Printed by author, 1753.

Bachaumont, Louis Petit de. *Mémoires secrets pour servir à l'histoire de la république des lettres en France depuis MDCCLXII jusqu'à nos jours, ou, Journal d'un observateur.* . . . 36 vols. London: Chez John Adamson, 1780–89.

Bachelin, Auguste. *L'horlogerie neuchateloise*. Neuchâtel, Switzerland: Attinger Frères, 1888.

Bachmann, Holger. "The Production and Contemporary Reception of *Metropolis*." In *Fritz Lang's Metropolis: Cinematic Visions of Technology and Fear*, edited by Michael Minden, Michael Bachmann, and Holger Bachmann, 3–46. Rochester, NY: Camden House, 2000.

Bachmann-Medick, Doris. *Die ästhetische Ordnung des Handelns: Moralphilosophie und Ästhetik in der Popularphilosophie des 18. Jahrhunderts*. Stuttgart: Metzler, 1989.

Bahr, Erhard. "Aufklärung." In *Geschichte der deutschen Literatur: Kontinuität und Veränderung vom Mittelalter bis zur Gegenwart*, edited by Erhard Bahr, 1–132. Tübingen: A. Francke Verlag, 1998.

Bailly, Christian. *Automata: The Golden Age, 1848–1914*. London: P. Wilson for Sotheby's, 1987.

Baker, F. Grenfell. *The Model Republic: A History of the Rise and Progress of the Swiss People*. London: H. S. Nichols, 1895.

Balet, Leo, and E. Gerhard. *Die Verbürgerlichung der deutschen Kunst, Literatur und Musik*. Dresden: VEB Verlag der Kunst, 1979.

Bammé, Arno, ed. *Maschinen-Menschen, Mensch-Maschinen: Grundrisse einer sozialen Beziehung*. Reinbek: Rowohlt, 1983.

Baruzzi, Arno. *Mensch und Maschine: Das Denken sub specie machinae*. Munich: Wilhelm Fink, 1973.

Bassermann-Jordan, Ernst. *Die Geschichte der Räderuhr unter besonderer Berücksichtigung der Uhren des bayerischen Nationalmuseums*. Frankfurt am Main: P. Keller, 1905.

————. *Uhren: Ein Handbuch für Sammler und Liebhaber.* Berlin: R. C. Schmidt, 1914.

Baumgärtel, Ehrfried. *Die Almanache, Kalender und Taschenbücher (1750–1860) der Landesbibliothek Coburg.* Wiesbaden: Otto Harrassowitz, 1970.

Bayerl, Günter. "Die Anfänge der Technikgeschichte bei Johann Beckmann und Johann Heinrich Moritz von Poppe." In König and Schneider, *Die technikhistorische Forschung,* 13–34.

Beci, Veronika. *Musikalische Salons: Blütezeit einer Frauenkultur.* Düsseldorf, Germany: Artemis und Winkler, 2000.

Becker-Cantarino, Barbara. "Vorwort." In *Tagebuch einer Reise durch Holland und England,* by Sophie von La Roche. Karben, Germany: P. Wald, 1997.

Beckmann, Johann. *Beyträge zur Geschichte der Erfindungen.* 5 vols. Leipzig: Paul Gotthelf Kummer, 1782–1805.

————. *Beyträge zur Oekonomie, Technologie, Polizey und Cameralwissenschaft.* 12 vols. Göttingen: Vandenhoeck, 1779–91.

Bedini, Silvio. "The Role of Automata in the History of Technology." *Technology and Culture* 5 (1964): 24–42.

Beer, Gavin R. de. *Travelers in Switzerland.* London: Oxford University Press, 1949.

Beiser, Frederic. *Enlightenment, Revolution, and Romanticism: The Genesis of German Political Thought, 1790–1800.* Cambridge, MA: Harvard University Press, 1992.

Beiträge der Schweiz zur Technik: Festschrift zum Jubiläum 700 Jahre Eidgenossenschaft. Edited by Stiftung Eisenbibliothek, Schaffhausen, Georg Fischer AG, Schaffhausen, SIG Schweizerische Industrie-Gesellschaft Holding AG, Lorenz Häfliger, et al. Oberbözberg, Switzerland: Olynthus, 1991.

Benjamin, Walter. "Das Kunstwerk im Zeitalter seiner technischen Reproduzierbarkeit." *Zeitschrift für Sozialforschung* 5, no. 1 (1936).

Bennett, James. A. "The Mechanics' Philosophy and the Mechanical Philosophy." *History of Science* 24 (1986): 1–27.

Benthien, Claudia, Anne Fleig, and Ingrid Kasten. *Emotionalität: Zur Geschichte der Gefühle.* Weimar: Böhlau Verlag, 2000.

Bentley, Richard. *A Confutation of Atheism from the Structure and Origin of Humane Bodies.* London: Parkhurst, 1692.

Berend, Eduard. "Prolegomena zur historisch-kritischen Gesamtausgabe von Jean Pauls Werken." *Abhandlungen der Preussischen Akademie der Wissenschaften* 1 (1927): 3–43.

Berg, Maxine. *The Age of Manufactures, 1700–1820: Industry, Innovation, and Work in Britain.* 2nd ed. London: Routledge, 1994.

Berg, Maxine, Pat Hudson, and Michael Sonenscher, eds. *Manufacture in Town and Country before the Factory.* Cambridge: Cambridge University Press, 1983.

Bergier, Jean-François. *Die Wirtschaftsgeschichte der Schweiz: Von den Anfängen bis zur Gegenwart.* Zürich: Benziger, 1983.

———. *Zu den Anfängen des Kapitalismus—Das Beispiel Genf.* Cologne: Selbstverlag, 1972.

Bergmann, Ernst. *Die Satiren des Herrn Maschin: Ein Beitrag zur Philosophie- und Kulturgeschichte des 18. Jahrhunderts.* Leipzig: E. Wiegandt, 1913.

Bernoulli, Johann. *Johann Bernoulli's Sammlung kurzer Reisebeschreibungen. Erster überzähliger Band. Beschreibung des Fürstenthums Welsch-Neuenburg und Vallengin.* Berlin: Printed by the editor, 1783.

———. *Sammlung kurzer Reisebeschreibungen und anderer zur Erweiterung der Länder- und Menschenkenntniß dienender Nachrichten.* Berlin: Printed by the editor, 1781–87.

Berthoud, Ferdinand. *Essai sur l'horlogerie dans lequel on traite de cet art, relativement à l'usage civil, à l'astronomie et à la navigation, en établissant des principes confirmés par l'expérience.* Paris: J. C. Jombert, 1763.

Bertola de' Giorgi, Aurelio. *Malerische Rhein-Reise von Speyer bis Düsseldorf.* Mannheim: Schwan und Götz, 1796.

Bertuch, Friedrich Justin, and Georg Melchior Kraus. "Einleitung." *Journal der Moden* 1 (1786): 3–30.

Beyer, Annette. *Faszinierende Welt der Automaten: Uhren, Puppen, Spielereien.* Munich: Callwey, 1983.

Beyreuther, Erich. *Nikolaus Ludwig von Zinzendorf: Selbstzeugnisse und Bilddokumente: Eine Biographie.* Giessen: Brunnen, 2000.

Biagioli, Mario. "From Print to Patents: Living on Instruments in Early Modern Europe." *History of Science* 44 (2006): 139–86.

Biagioli, Mario, and Peter Galison. *Scientific Authorship: Credit and Intellectual Property in Science.* New York: Routledge, 2003.

Biographie universelle (Michaud) ancienne et moderne. Paris, Madame C. Desplaces, 1854–65.

Biucchi, Basilio M. "The Industrial Revolution in Switzerland." In *The Fontana Economic History of Europe,* vol. 4, *The Emergence of Industrial Societies,* edited by Carlo M. Cipolla, 627–54. London: Collins/Fontana, 1972.

Blair, Ann. "Humanist Methods in Natural Philosophy: The Commonplace Book." *Journal of the History of Ideas* 53 (1992): 541–15.

Blanning, Timothy. C. W. *The Culture of Power and the Power of Culture: Old Regime Europe, 1660–1789.* Oxford: Oxford University Press, 2002.

———, ed. *The Eighteenth Century: Europe, 1688–1815.* Oxford: Oxford University Press, 2000.

Bloch, Oscar, and Walther von Wartburg. *Dictionnaire étymologique de la langue française.* Paris: Presses Universitaires de France, 1968. Originally published in 1932.

Blondel, Christine, and Bernadette Bensaude-Vincent, eds. *Science and Spectacle in the European Enlightenment.* Aldershot: Ashgate, 2008.

Bodmer, Walter. *Der Einfluß der Refugianteneinwanderung von 1550–1700 auf die schweizerische Wirtschaft.* Zürich: Leemann, 1946.

Boehn, Max von. *Puppen und Puppenspiele.* 2 vols. Munich: Bruckmann, 1929.

———. *Puppets and Automata.* New York: Dover, 1972.

Boit, Johann Peter. *Fassliche Beschreibung der gemeinnützlichsten Künste und Handwerke für junge Leute.* 2 vols. Nürnberg: Weigel und Schneider, 1790.

Bonaventura. *Die Nachtwachen des Bonaventura.* Munich: Kurt Desch, [1947?].

Bonds, Mark Evan. *Wordless Rhetoric: Musical Form and the Metaphor of the Oration.* Cambridge, MA: Harvard University Press, 1991.

Böning, Thomas. *Widersprüche: Zu den "Nachtwachen" von Bonaventura und zur Theoriedebatte.* Freiburg im Breisgau: Rombach, 1996.

Borries, Ernst von, and Erika von Borries. *Deutsche Literaturgeschichte.* Vol. 4, *Zwischen Klassik und Romantik: Hölderlin, Kleist, Jean Paul.* Munich: Deutscher Taschenbuch Verlag, 1997.

Bouziane, Dagmar, Heike Krems, and Ruth Weiß. *". . . und die Lust und Trieb zu arbeiten unbeschreiblich . . .": Johann Georg Krünitz und seine Oekonomisch-technologische Encyklopädie.* Wiesbaden, 1996.

Brackett, Oliver. *English Furniture Illustrated: Le mobilier anglais illustré: Englands Möbelwerk in Bildern.* London, Spring Books, [1958].

Braun, Rudolf. *Das ausgehende Ancien Régime in der Schweiz: Aufriß einer Sozial- und Wirtschaftsgeschichte des 18. Jahrhunderts.* Göttingen: Vandenhoeck und Ruprecht, 1984.

Brecht, Martin, and Klaus Deppermann, eds. *Geschichte des Pietismus.* Vol. 2, *Der Pietismus im achtzehnten Jahrhundert.* Göttingen: Vandenhoeck und Ruprecht, 1995.

Brentano, Clemens. *Entweder wunderbare Geschichte von Bogs dem Uhrmacher, wie er zwar das menschliche Leben längst verlassen, nun aber doch, nach vielen musikalischen Leiden zu Wasser und zu Lande, in die bürgerliche Schützenge-sellschaft aufgenommen zu werden Hoffnung het. . . .* [Heidelberg?], 1807.

Brewster, David. *Letters on Natural Magic: Addressed to Sir Walter Scott.* New York: J. & J. Harper, 1832.

Brieger, Lothar. *Das Genrebild: Die Entwicklung der bürgerlichen Malerei.* Munich: Delphin-Verlag, 1922.

Buchner, Alexander. *Mechanische Musikinstrumente.* Hanau, Germany: Werner Dausien, 1992.

Bunzel, Wolfgang. "Almanache und Taschenbücher." In *Von Almanach bis Zeitung: Ein Handbuch der Medien in Deutschland, 1700–1800,* edited by E. Fischer, W. Haefs, and Y.-G. Mix. Munich: C. H. Beck, 1999.

Burri, Adolf. *Johann Rudolf Sinner von Ballaigues 1730–1787: Ein Beitrag zur Kultur- und Geistesgeschichte des 18. Jahrhunderts.* Bern: A. Francke, 1913.

Busch, Gabriel Christoph Benjamin. *Almanach der Fortschritte neuesten Erfin-dungen und Entdeckungen in Wissenschaften, Künsten, Manufakturen und Handwerken. . . .* Erfurt, Germany: Georg Adam Keyser, 1798–1807.

———. *Versuch eines Handbuchs der Erfindungen.* 8 vols. Eisenach: J. G. E. Witte-kind, 1790–98.

Busch-Salmen, Gabriele. "Die Frau am Tasteninstrument: Thesen zur Interpretation eines Bildtopos." In *Frauen- und Männerbilder in der Musik: Festschrift für Eva Rieger,* edited by Freia Hoffmann, Jane Bowers, and Ruth Heckmann, 41–46. 1, Germany: BIS Verlag, 2000.

Caetano da Rosa, Catarina. "Androiden im 18. Jahrhundert." Master's thesis, Technical University of Berlin, 2006.

Calhoun, Craig. *Habermas and the Public Sphere.* Cambridge, MA: MIT Press, 1992.

Campbell-Kelly, Martin, and William Aspray. *Computer: A History of the Information Machine.* New York: Basic Books, 1996.

Campe, Joachim Heinrich. "Sensation, Sensibilität, Sentiment, sentimental, sentimetalisieren u.s.w." In *Wörterbuch zur Erklärung und Verdeutschung der unserer Sprache aufgedrungenen fremden Ausdrücke.* Braunschweig: Vieweg, 1813.

Canguilhem, Georges. "Organisme et modèles mécaniques: Reflexions sur la biologie cartésienne." *Revue philosophique de la France et de l'Etranger,* 1955, 281–99.

Čapek, Karel. *R.U.R. (Rossum's universal robots).* Translated by Claudia Novack. Introduction by Ivan Klíma. New York: Penguin Books, 2004.

Cardinal, Catherine, and François Mercier. *Musées d'horlogerie La Chaux-de-Fonds, Le Locle.* Geneva: Banque Paribas, 1993.

Carlyle, Thomas. *Critical and Miscellaneous Essays.* 6 vols. London: Chapman and Hall, 1869.

Carrera, Roland, Dominique Loiseau, and Olivier Roux. *Androiden: Die Automaten von Jaquet-Droz.* Lausanne: Scriptar, 1979.

Casey, Timothy J. *Jean Paul: A Reader.* Baltimore: Johns Hopkins University Press, 1992.

Cassirer, Ernst. *The Philosophy of the Enlightenment.* Princeton, N.J.: Princeton University Press, 2009.

Catalogue du musée, section Z: automates et mécanismes à musique. Paris: Conservatoire national des artes et métiers, 1973.

Caus, Salomon de. *Les raisons des forces mouvantes.* Frankfurt am Main, 1615.

Cayrou, Gaston. *Le français classique: Lexique de la langue du dix-septième siècle expliquant d'après les dictionnaires du temps et les remarques des grammairiens: Le sens et l'usage des mots aujourd'hui vieillis ou différemment employés.* Paris: H. Didier, 1923.

Chapuis, Alfred. *Histoire de la boîte á musique et de la musique mécanique.* Lausanne: Editions du *Journal suisse d'horlogerie et de bijouterie,* 1955.

———. *Histoire de la pendulerie neuchâteloise: (Horlogerie de gros et de moyen volume).* Paris: Attinger Frères, [1917?].

———. *Les automates, figures artificielles d'hommes et d'animaux: Histoire et technique.* Neuchâtel, Switzerland: Editions du Griffon, 1949.

———. *Les automates dans les œuvres d'imagination.* Neuchâtel, Switzerland: Editions du Griffon, 1947.

————. *Relations de l'horlogerie suisse avec la Chine: La montre "chinoise."* Neuchâtel, Switzerland: Attinger frères, [1918?].

Chapuis, Alfred, and Alphonse Bernoud. *Automates, machines automatiques et machinisme.* Geneva: S.A. Des Publications Techniques, 1928.

Chapuis, Alfred, and Edmond Droz. *Automata: A Historical and Technological Study.* Neuchâtel, Switzerland: Editions du Griffon, 1958.

————. *Les automates des Jaquet-Droz.* Booklet of the Musée d'art et d'histoire in Neuchâtel. N.p, n.d.

Chapuis, Alfred, and Edouard Gelis. *Le monde des automates: Etude historique et technique.* 2 vols. Paris: Printed by the authors, 1928.

Chippendale, Thomas. *The gentleman and cabinet-maker's director: Being a large collection of the most elegant and useful designs of houshold furniture in the Gothic, Chinese and modern taste . . . and other ornaments. . . .* London: Printed for the author, 1754.

Cipolla, Carlo M. *Clocks and Culture, 1300–1700.* New York: Norton, 2003.

Cloot, Julia. *Geheime Texte: Jean Paul und die Musik.* Berlin: Walter de Gruyter, 2001.

Clute, John. "The Identity of E.V. Odle." *Science Fiction Studies* 8, no. 3 (November 1981): 343.

Coen, Deborah R. *Vienna in the Age of Uncertainty: Science, Liberalism, and Private Life.* Chicago: University of Chicago Press, 2007.

Cohen, I. Bernhard, ed. *Puritanism and the Rise of Modern Science: The Merton Thesis.* New Brunswick, NJ: Rutgers University Press, 1990.

Cohen, John. *Human Robots in Myth and Science.* London: Allen & Unwin, 1966.

Collins, Harry. M. *Artificial Experts: Social Knowledge and Intelligent Machines.* Cambridge, MA: MIT Press, 1990.

Condillac, Etienne Bonnot de. *Essai sur l'origine des connoissances humaines: Ouvrage où l'on réduit à un seul principe tout ce qui concerne l'entendement humain.* Amsterdam: Chez Pierre Mortier, 1746.

————. *Traité des animaux, où on entreprend d'expliquer leurs principales facultés: On a joint à cet ouvrage un extrait raisonné du Traité des Sensations.* Amsterdam, 1755.

————. *Traité des sensations.* London, 1754.

Conrad, William Cooke. *Automata Old and New.* London, 1893.

Correll, Ernst H. *Das schweizerische Täufermennonitentum.* Tübingen: Mohr, 1925.

Coski, R. Christopher. "Condillac: Language, Thought, and Morality in the Man and Animal Debate." *French Forum* 28, no. 1 (Winter 2003): 57–75.

Couperin, François. *L'art de toucher le clavecin.* Paris: Printed by the author, 1717.

Coxe, William. *Travels in Switzerland in a Series of Letters to William Melmoth, Esq.* London: T. Cadell, 1789.

Craig, Edward Gordon. "The Actor and the Uber-marionette." In *Modernism: An Anthology of Sources and Documents,* edited by Vassiliki Kolocotroni,

Jane Goldman, and Olga Taxidou, 150–54. Chicago: University of Chicago Press, 1998.

Cumming, Alexander. *The elements of clock and watch-work: Adapted to practice, in two essays*. London: Printed for the author, 1766.

Darnton, Robert. *The Great Cat Massacre and Other Episodes in French Cultural History*. New York: Basic Books, 1984.

Daston, Lorraine. "Afterword: The Ethos of Enlightenment." In *The Sciences in Enlightened Europe*, edited by William Clark, Jan Golinski, and Simon Schaffer, 495–504. Chicago: University of Chicago Press, 1999.

———. *Classical Probability in the Enlightenment*. Princeton, NJ: Princeton University Press, 1988.

———, ed. *Things That Talk: Object Lessons from Art and Science*. New York: Zone Books, 2004.

Davis, Erik. "The Metaphysics of Philip K. Dick." *Wired* 11, no. 12 (December 2003).

Davis, Natalie Zemon. *The Return of Martin Guerre*. Cambridge, MA: Harvard University Press, 1983.

Day, Richard B., Ronald Beiner, and Joseph Masciulli, eds. *Democratic Theory and Technological Society*. Armonk, NY: M. E. Sharpe, 1988.

Dear, Peter. "A Mechanical Microcosm: Bodily Passions, Good Manners, and Cartesian Mechanism." In *Science Incarnate: Historical Embodiments of Natural Knowledge*, edited by Christopher Lawrence and Steven Shapin, 51–82. Chicago: University of Chicago Press 1998.

De Man, Paul. *The Rhetoric of Romanticism*. New York: Columbia University Press, 1984.

Derham, William. *The artificial clock-maker: A treatise of watch, and clock-work, wherein the art of calculating numbers for most sorts of movements is explained to the capacity of the unlearned: Also, the history of clock-work, both ancient and modern, with other useful matters*. London: Printed for James Knapton, 1696.

Descartes, René. *Meditations on First Philosophy: With Selections from the Objections and Replies*. Translated and edited by John Cottingham. Cambridge: Cambridge University Press, 1996.

Deutsche Literaturgeschichte: Von den Anfängen bis zur Gegenwart. Edited by Wolfgang Beutin. Stuttgart: Metzler, 2001.

De Vries, Jan. *The Industrious Revolution: Consumer Behavior and the Household Economy, 1650 to the Present*. Cambridge: Cambridge University Press, 2008.

Dictionary of Literary Biography. Vol. 94, *German Writers in the Age of Goethe: Sturm und Drang to Classicism*. Edited by James Hardin and Christoph E. Schweitzer. Detroit: Gale Research, 1990.

Dietmann, Carl Gottlob, and Johann G. Haymann. *Neue europäische Staats- und Reisegeographie: Worinnen kürzlich alles was zur geographischen, physikalischen, politischen und topographischen Kenntniss eines jeden Staats gehöret, nach und nach vorgestellet, und mit nöthigen Landkarten, auch andern zur Historie dientlichen Kupfern, versehen werden soll.* 16 vols. Leipzig: Richter, 1750–70.

Dijck, Lucas van, and Ton Koopman. *Het klavecimbel in de Nederlandse kunst tot 1800* Zutphen, Netherlands: Walburg Pers, 1987.

Dijksterhuis, Eduard Jan. *The Mechanization of the World Picture.* Translated by C. Dikshoorn. Oxford, Clarendon Press, 1961.

Dixon, C. Scott, and Luise Schorn-Schütte. *The Protestant Clergy of Early Modern Europe.* New York: Palgrave Macmillan, 2003.

Dixon, Simon. *Catherine the Great.* Harlow, UK: 2001.

Dohrn-van Rossum, Gerhard. *History of the Hour: Clocks and Modern Temporal Orders.* Translated by Thomas Dunlap. Chicago: University of Chicago Press, 1996.

Dolan, Emily. "E. T. A. Hoffmann and the Ethereal Technologies of 'Nature Music.'" *Eighteenth-Century Music* 5, no. 1 (2008): 7–26.

Donnert, Erich. *Katharina II. die Grosse (1729–1796): Kaiserin des Russischen Reiches.* Regensburg: F. Pustet, 1998.

Dotzler, Bernhard, Peter Gendolla, and Jörgen Schäfer, eds. *Maschinen-Menschen: Eine Bibliographie.* Bibliographien zur Literatur- und Mediengeschichte, vol. 1. Frankfurt am Main: Peter Lang, 1992.

Drux, Rudolf, ed. *Der Frankenstein-Komplex: Kulturgeschichtliche Aspekte des Traums vom künstlichen Menschen.* Frankfurt am Main: Suhrkamp, 1999.

———. *Marionette Mensch: Ein Metaphernkomplex und sein Kontext von Hoffmann bis Büchner.* Munich: Wilhelm Fink, 1986.

Du Bois-Reymond, Emil. "La Mettrie." In *Vorträge über Philosophie und Gesellschaft,* by Emil Du Bois-Reymond, 79–103. Hamburg: Meiner, 1974.

Duden, Barbara. "Das schöne Eigentum: Zur Herausbildung des bürgerlichen Frauenbildes an der Wende vom 18. zum 19. Jahrhundert." *Kursbuch* 47 (1977): 125–40.

Düll, Sigrid, and Walter Pass, eds. *Frau und Musik im Zeitalter der Aufklärung.* Sankt Augustin, Germany: Academia Verlag, 1998.

Dülmen, Richard van. *Die Gesellschaft der Aufklärer: Zur bürgerlichen Emanzipation und aufklärerischen Kultur in Deutschland.* Frankfurt am Main: Fischer Taschenbuch Verlag, 1986.

———. *Kultur und Alltag in der Frühen Neuzeit,* vol. 3, *Religion, Magie, Aufklärung, 16.-18. Jahrhundert.* Munich: Beck, 1994.

———. *Poesie des Lebens: Eine Kulturgeschichte der deutschen Romantik.* Vol. 1, *Lebenswelten.* Cologne: Böhlau, 2002.

Duplessis, Georges. *Mémoires et journal de J.-G. Wille, graveur du roi, pub. d'après les manuscrits autographes de les manuscrits autographes de la Bibliothèque impériale.* 2 vols. Paris: Ve J. Renouard, 1857.

Ebel, Johann. G. *Anleitung, auf die nützlichste und genussvollste Art die Schweitz zu bereisen. Dritter Theil. Zweyte, ganz umgearbeitete und sehr vermehrte Auflage.* Zürich: Orell, Füssli, 1805.

Ehrmann, Marianne. "'Uber weibliche Beschäftigungen': Musik—Spiel—Tanz." In *Maria Anna Mozart: Die Künstlerin und ihre Zeit*, edited by Sigrid Düll and Otto Neumaier, 297–308. Möhnesee, Germany: Bibliopolis, 2001.

Elias, Norbert. *Die höfische Gesellschaft: Untersuchungen zur Soziologie des Königtums und der höfischen Aristokratie, mit einer Einleitung: Soziologie und Geschichtswissenschaft.* Neuwied: Luchterhand, 1969.

———. *Über den Prozess der Zivilisation: Soziogenetische und psychogenetische Untersuchungen.* Frankfurt am Main: Suhrkamp, 1978.

Ellul, Jacques. *The Technological Society.* Translated from the French by John Wilkinson. Introduction by Robert K. Merton. New York: Knopf, 1964.

Encyclopédie, ou, Dictionnaire raisonné des sciences, des arts et des métiers: Par une société de gens de lettres; mis en ordre & publié par M. Diderot . . . & quant à la partie mathematique, par M. d'Alembert. 28 vols. Geneva, 1754–72.

Engelmann, Max. "Aus der Geschichte einer Uhrmacherfamilie: Die 'Akustiker' Kaufmann und ihr Werk." *Deutsche Uhrmacher-Zeitung* 31, no. 40 (1907): 797–800; no. 42 (1907): 841–44.

———. *Leben und Wirken des württembergischen Pfarrers und Feintechnikers Philipp Matthäus Hahn.* Berlin: Richard Carl Schmidt, 1923.

Engels, Friedrich. *Die Lage der arbeitenden Klasse in England: Nach eigner Anschauung und authentischen Quellen.* Leipzig, 1845.

Epstein, Stephan R. "Craft Guilds in the Pre-Modern Economy: A Discussion," *Economic History Review* 61, no. 1 (2008): 155–74.

Europäische Schlüsselwörter. Vol. 2, *Kurzmonographien I. Wörter im geistigen und sozialen Raum.* Munich: Hueber, 1964.

Evans, Robert J. W., and Alexander Marr, eds. *Curiosity and Wonder from the Renaissance to the Enlightenment.* Aldershot: Ashgate, 2006.

Fabian, Dietrich. *Abraham und David Roentgen: Das noch aufgefundene Gesamtwerk ihrer Möbel- und Uhrenkunst in Verbindung mit der Uhrmacherfamilie Kinzing in Neuwied.* Bad Neustadt/Saale: Internationale Akademie der Kulturwissenschaften, 1996.

———. *Kinzing und Roentgen: Uhren aus Neuwied.* Bad Neustadt/Saale: Internationale Akademie der Kulturwissenschaften, 1983.

Faessler, François. "250e anniversaire de Pierre Jaquet-Droz, génial pendulier et créateur d'automates." In F. Faessler, S. Guye, and Edmond Droz, *Pierre Jaquet-Droz et son temps*, 9–25.

Faessler, François, Samuel Guye, and Edmond Droz, eds. *Pierre Jaquet-Droz et son temps.* La Chaux-de-Fonds: Courvoisier, 1971.

Farrar, W. V. "Andrew Ure, F.R.S., and the Philosophy of Manufactures." *Notes and Records of the Royal Society of London* 27, no. 2. (February 1973): 299–324.

Favre, Henry. *Neuenburgs Union mit Preussen und seine Zugehörigkeit zur Eidgenossenschaft: Ein Beitrag zur Verfassungsgeschichte von Neuenburg bis zu seinem Aufgehen in der Eidgenossenschaft.* Leipzig: Theodor Weicher, 1932.

Favre, Louis. "Journal de David Sandoz." *Musée neuchatelois* 11 (1874): 228.

Favre, Maurice. "Daniel JeanRichard, horloger (1665–1741)." In *Biographies Neuchâteloises*, vol. 1, *De saint Guillaume à la fin des Lumières*, edited by Michel Schlup, 149–53. Neuchâtel, Switzerland: Editions Gilles Attinger, 1996.

Feldhaus, Franz Maria. *Die Technik der Antike und des Mittelalters*. Potsdam, Germany: Athenaion 1930.

———. *Die Technik der Antike und des Mittelalters*. Hildesheim: Georg Olms Verlag, 1971.

———. *Die Technik der Vorzeit, der geschichtlichen Zeit und der Naturvölker: Ein Handbuch für Archäologen und Historiker, Museen und Sammler, Kunsthändler und Antiquare, mit 873 Abbildungen*. Leipzig: W. Engelmann, 1914.

———. *Geschichte des technischen Zeichnens*. Wilhelmshaven: Franz Kuhlmann, 1967.

———. *Kulturgeschichte der Technik*. 2 vols. Berlin: O. Salle, 1928.

———. *Lexikon der Erfindungen und Entdeckungen auf den Gebieten der Naturwissenschaften und Technik in chronologischer Übersicht mit Personen- und Sachregister*. Heidelberg: Carl Winter's Universitätsbuchhandlung, 1904.

Fetzer, John F. *Romantic Orpheus: Profiles of Clemens Brentano*. Berkeley: University of California Press, 1974.

Feulner, Adolf. *Die Zick: Deutsche Maler des 18. Jahrhunderts*. Munich: Hübschmann, 1920.

Fischer, Ernst, Wilhelm Haefs, and York-Gothart Mix. "Einleitung: Aufklärung, Öffentlichkeit und Medienkultur in Deutschland im 18. Jahrhundert." In *Von Almanach bis Zeitung: Ein Handbuch der Medien in Deutschland, 1700–1800*, edited by Ernst Fischer, Wilhelm Haefs, and York-Gothart Mix, 9–23. Munich: C. H. Beck, 1999.

Fischer, Hermann. "Beiträge zur Geschichte der Holzbearbeitungsmaschinen." *Beiträge zur Geschichte der Technik und Industrie* 1 (1909): 176–82; 3 (1911): 61–80.

Flaad, Peter. *Untersuchungen zur Kulturgeographie der Neuenburger Hochjuratäler von La Brévine und Les Ponts*. Basel: Komm, Helbling & Lichtenhahn, 1974.

Fleckenstein, Joachim Otto. *Johann und Jakob Bernoulli*. Basel, Birkhäuser, 1949.

Foucault, Michel. "Technologies of the Self." In Martin, Gutman, and Hutton, *Technologies of the Self*.

Frasca-Spada, Marina, and Nick Jardine, eds. *Books and the Sciences in History*. Cambridge: Cambridge University Press, 2000.

Frenzel, Herbert A. *Daten deutscher Dichtung: Chronologischer Abriss der deutschen Literaturgeschichte*. Munich: Deutscher Taschenbuch Verlag, 2001.

Freud, Sigmund. "Das Unheimliche." In *Gesammelte Werke*, edited by Anna Freud, 12:227–68. Frankfurt am Main: Suhrkamp, 1947.

Frevert, Ute. "Bürgerliche Familie und Geschlechterrollen: Modell und Wirklichkeit." In *Bürgerliche Gesellschaft in Deutschland: Historische Einblicke, Fragen, Perspektiven*, 90–98. Frankfurt am Main: Fischer, 1990.

———. "Bürgerliche Meisterdenker und das Geschlechterverhältnis: Konzepte, Erfahrungen, Visionen an der Wende vom 18. zum 19. Jahrhundert." In *Bürgerinnen und Bürger: Geschlechterverhältnisse im 19. Jahrhundert*, edited by Ute Frevert, 17–48. Göttingen: Vandenhoeck und Ruprecht, 1988.

———. *Mann und Weib, und Weib und Mann: Geschlechter-Differenzen in der Moderne*. Munich: C. H. Beck, 1995.

———. *Women in German History: From Bourgeois Emancipation to Sexual Liberation*. New York: St. Martin's Press, 1989.

Frieling, Kirsten. *Ausdruck macht Eindruck: Bürgerliche Körperpraktiken in sozialer Kommunikation um 1800*. Frankfurt: Lang, 2003.

Fröhner, Annette. *Technologie und Enzyklopädismus im Übergang vom 18. zum 19. Jahrhundert*. Mannheim, Palatium Verlag, 1994.

Funk, Julika. "Tiere und Zähne—Geheimschrift Handschrift." In *Hand: Medium—Körper—Technik*, edited by Ulrike Bergermann, Andrea Sick, and Andrea Klier, 47–59. Bremen: Thealit Frauen.Kultur.Labor, 2001.

Furger, Fridolin. *Zum Verlagssystem als Organisationsform des Frühkapitalismus im Textilgewerbe*. Stuttgart: W. Kohlhammer, 1927.

Gall, Lothar. *Von der ständischen zur bürgerlichen Gesellschaft*. Munich: Oldenbourg, 1993.

Gärtner, Christoph, "Remuer l'Ame or Plaire à l'Oreille? Music, Emotions, and the Mind—Body Problems in French Writings of the Later Eighteenth Century." In Gouk and Hills, *Representing Emotions*, 173–88.

Gaukroger, Stephen. *The Soft Underbelly of Reason: The Passions in the Seventeenth Century*. London: Routledge, 1998.

Gay, Peter. *Schnitzler's Century: The Making of Middle-Class Culture, 1815–1914*. New York: Norton, 2002.

Gehler, J. S. T. *Johann Samuel Traugott Gehler's Physikalisches Wörterbuch*. Edited by Brandes, Gmelin, Horner, Muncke, and Pfaff. 11 vols. Leipzig: E. B. Schurckert, 1825–45.

———. *Physikalisches Wörterbuch oder Versuch einer Erklärung der vornehmsten Begriffe und Kunstwörter in der Naturlehre*. 6 vols. Leipzig: Schwickert, 1798.

Geißler, J. G. *Alexander Cummings's Elemente der praktischen Groß- und Klein-Uhrmacherkunst*. Leipzig: Baumgärtner, 1802.

Geminiani, Francesco. *The art of playing on the violin: containing all the rules necessary to a perfection on that instrument, with great variety of compositions, which will also be very useful to those who study the violoncello, harpsichord &c.* London: J. Johnson, 1751.

Gendolla, Peter. *Anatomien der Puppe: Zur Geschichte des MaschinenMenschen bei Jean Paul, E. T. A. Hoffmann, Villiers d'Ísle-Adam und Hans Bellmer*. Heidelberg: Carl Winter Universitätsverlag, 1992.

Gere, Cathy. *Knossos and the Prophets of Modernism*. Chicago: University of Chicago Press, 2009.

Gersdorf, Lilo. "Die Vogelorgel oder Dame, die Abwechslung in ihren Zeit-
vertreib bringt." In *Die Künstlerin und ihre Zeit*, by Maria Anna Mozart,
edited by Sigrid Düll and Otto Neumaier, 197–203. Möhnesee, Germany:
Bibliopolis, 2001.

Giedion, Sigfried. *Mechanization Takes Command: A Contribution to Anonymous
History.* New York: Oxford University Press, 1948.

Ginzburg, Carlo. *The Cheese and the Worms: The Cosmos of a Sixteenth-Century
Miller.* Translated by John Tedeschi and Anne Tedeschi. London: Penguin,
1992.

———. "Microhistory: Two or Three Things That I Know about It." *Critical
Inquiry* 20, no. 1 (Autumn 1993): 10–35.

Gladt, Karl. *Almanache und Taschenbücher aus Wien.* Vienna: Jugend und Volk,
1971.

Goebel, Ralf. *Der handschriftliche Nachlaß Jean Pauls und die Jean-Paul-Bestände
der Staatsbibliothek zu Berlin—Preußischer Kulturbesitz. Teil 1 (Fasz. I-XV).*
Wiesbaden: Harrassowitz, 2002–.

Goethe, Johann Wolfgang von. *Sämtliche Werke, Briefe, Tagebücher und
Gespräche.* Frankfurt am Main: Deutscher Klassiker Verlag, 1985–.

———. "Tag- und Jahres-Hefte." In Goethe, *Sämtliche Werke*, vol. 17, p. 141,
§ 450.

Goldsmith, Elizabeth C., and Dena Goodman. *Going Public: Women and
Publishing in Early Modern France.* Ithaca, NY: Cornell University Press,
1995.

Gondorf, Bernhard. "Der Kunsttischler Abraham Roentgen: Eine biografische
Skizze." In *Möbel von Abraham und David Roentgen*, 10–15. Schriften des
Kreismuseums Neuwied. Neuwied, Germany: Landkreis, 1990.

Goodman, Dena, ed. *Marie-Antoinette: Writings on the Body of a Queen*, edited by
Dena Goodman, 1–15. New York: Routledge, 2003.

———. "The Secrétaire and the Integration of the Eighteenth-Century Self." In
Goodman and Norberg, *Furnishing the Eighteenth Century*, 183–204.

Goodman, Dena, and Kathryn Norberg, eds. *Furnishing the Eighteenth Century:
What Furniture Can Tell Us about the European and American Past.* New York:
Routledge, 2007.

Gouk, Penelope, and Helen Hills. *Representing Emotions: New Connections in the
Histories of Art, Music, and Medicine.* Aldershot: Ashgate, 2005.

Grässe, Johann Georg Theodor. *Zur Geschichte des Puppenspiels und der Automaten.* Berlin, 1856.

Greber, Josef M. *Abraham und David Roentgen: Möbel für Europa: Werdegang,
Kunst und Technik einer deutschen Kabinett-Manufaktur.* Starnberg: Josef
Keller, 1980.

Grimm, Friedrich Melchior. *Correspondance littéraire.* 3 vols. Ferney-Voltaire,
France: Centre international d'étude du XVIIIe siècle, 2006–7.

———. *Lettres de Grimm à l'Imperatrice Catherine II: Sous les auspices de la Societe
imperiale d'histoire russe.* Saint Petersburg: Tip. I. Akademii nauk, 1885.

Grimminger, Rolf, ed. *Deutsche Aufklärung bis zur Französischen Revolution: 1680–1789.* Hansers Sozialgeschichte der deutschen Literatur vom 16. Jahrhundert bis zur Gegenwart, vol. 3. Munich: Hanser, 1980.

Groenewegen, Peter D. *Eighteenth-Century Economics: Turgot, Beccaria, and Smith and Their Contemporaries.* London: Routledge, 2002.

Grosz, Elizabeth A. *Volatile Bodies: Toward a Corporeal Feminism.* Bloomington: Indiana University Press, 1994.

Grotjahn, Rebecca, and Freia Hoffmann. "Einleitung: Geschlechterpolaritäten in der Musikgeschichte des 18. bis 20. Jahrhunderts." In *Geschlechterpolaritäten in der Musikgeschichte des 18. bis 20. Jahrhunderts,* edited by Rebecca Grotjahn and Freia Hoffmann, 1–6. Herbolzheim, Germany: Centaurus Verlag, 2002.

Gülich, Gustav von. *Geschichtliche Darstellung des Handels, der Gewerbe und des Ackerbaus: Der bedeutendsten handeltreibenden Staaten unsrer Zeit.* Jena, Germany: F. Frommann, 1830–45.

Gunn, James E. *Isaac Asimov: The Foundations of Science Fiction.* Oxford: Oxford University Press, 1982.

Günther, Walther. "*Herrnhaagisches Kirchenbuch* und ihre Ökonomie, ihr geistliches Selbstverständnis, ihre Lebensformen und ihre Spiritualität." In *Herrnhuter Architektur am Rhein und an der Wolga,* edited by Landkreis Neuwied/Rhein, with Reinhard Lahr and Bernd Willscheid, 19–26. Neuwied: Landkreis Neuwied, 2001.

———. "Die Wurzeln der Brüderunität in Böhmen und die erneuerte Brüderunität in Herrnhut." In *Herrnhuter Architektur am Rhein und an der Wolga,* edited by Landkreis Neuwied/Rhein, with Reinhard Lahr and Bernd Willscheid, 27–57. Neuwied: Landkreis Neuwied, 2001.

Gütle, Johann Conrad. *Beschreibung eines mathematisch-physikalischen Maschinen und Instrumenten-Kabinets: Mit zugehörigen Versuchen zum Gebrauch für Schulen.* Leipzig: Adam Gottlieb Schneiderschen Kunst-und Buchhandlung, 1790.

———. *Nüzliche Versuche und Erfahrungen für Fabrikanten, Künstler und Oekonomen: Zur Beförderung der Oekonomie, der Künste, Gewerbe und Handlung, der Fabriken und Manufacturen.* Nürnberg: Monath und Kußler, 1796.

———. *Zaubermechanik oder Beschreibung mechanischer Zauberbelustigungen mit darzu gehörigen Maschinen für Liebhaber belustigende Künste.* Nürnberg: J. C. Monath und J. F. Kußler, 1794.

Guyot. *Nouvelles récréations physiques et mathematiques, contenant, toutes celles qui ont été découvertes & imaginées dans ces derniers temps, sur l'aiman, les nombres, l'optique, la chymie, &c., & quantité d'autres qui n'ont jamais été rendues publiques.* 4 vols. Paris: Gueffier, 1769–70.

Habermas, Jürgen. *Strukturwandel der Öffentlichkeit: Untersuchungen zu einer Kategorie der bürgerlichen Gesellschaft.* New edition with a new foreword. Frankfurt am Main: Suhrkamp, 1990. Originally published in 1962.

Hafter, Daryl M. *Women at Work in Preindustrial France*. University Park: Pennsylvania State University, 2007.

Hahn, Roger. *The Anatomy of a Scientific Institution: The Paris Academy of Sciences, 1666–1803*. Berkeley: University of California Press, 1971.

Handbuch der Schweizer Geschichte. Vol. 2. Zürich: Verlag Berichthaus, 1977.

Hankins, Thomas. *Science and the Enlightenment*. Cambridge: Cambridge University Press, 1985.

Hankins, Thomas L., and Robert J. Silverman. *Instruments and the Imagination*. Princeton, NJ: Princeton University Press, 1995.

Haraway, Donna. "A Cyborg Manifesto: Science, Technology, and Socialist-Feminism in the Late Twentieth Century." In *Simians, Cyborgs, and Women: The Reinvention of Nature*, edited by Donna Haraway, 149–81. New York: Routledge, 1991.

Haspels, Jan Jaap. *Automatic Musical Instruments: Their Mechanics and Their Music, 1580–1820*. Koedijk, Netherlands: Nirota, Muziekdruk C.V., 1987.

Hauser, Albert. *Schweizerische Wirtschafts- und Sozialgeschichte*. Zürich: Eugen Rentsch Verlag, 1961.

Heal, Ambrose. *The London Furniture Makers, from the Restoration to the Victorian Era, 1660–1840: A Record of 2500 Cabinet-Makers, Upholsterers, Carvers, and Gilders, with Their Addresses and Working Dates, Illustrated by 165 Reproductions of Makers's Trade-cards*. London: Batsford, [1953?].

Heckmann, Herbert. *Die andere Schöpfung: Geschichte der frühen Automaten in Wirklichkeit und Dichtung*. Frankfurt am Main: Umschau Verlag Breidenstein, 1982.

Heckmann, Ruth. "Mann und Weib in der 'musikalischen Republick': Modelle der Geschlechterpolarisierung in der Musikanschauuung 1750–1800." In *Geschlechterpolaritäten in der Musikgeschichte des 18. bis 20. Jahrhunderts*, edited by Rebecca Grotjahn and Freia Hoffmann, 19–30. Herbolzheim, Germany: Centaurus Verlag, 2002.

Heckscher, Eli F. *Mercantilism*. With a new introduction by Lars Magnusson. London: Routledge, 1994. Originally published in 1935.

Heesen, Anke te. "News, Papers, Scissors." In *Things That Talk*, edited by L. Daston and Anke te Heesen. Berlin: MPI für Wissenschaftsgeschichte, 2003.

Hegel, Georg Wilhelm Friedrich. *Ästhetik*. Berlin: Aufbau-Verlag, 1955.

Helmholtz, Hermann von. *Popular Lectures on Scientific Subjects*. Translated by E. Atkinson. New York: D. Appleton, 1873.

———. *Ueber die Wechselwirkung der Naturkräfte und die darauf bezüglichen neuesten Ermittelungen der Physik*. Königsberg: Gräfe & Unzer, 1872.

Hentschel, Uwe. *Mythos Schweiz: Zum deutschen literarischen Philhelvetismus zwischen 1700–1850*. Tübingen: Max Niemeyer Verlag, 2002.

Hepplewhite, A., and Company. *The cabinet-maker and upholsterer's guide; or, Repository of designs for every article of household furniture microform: in the newest and most approved taste . . . also, the plan of a room, shewing the proper distribution of the furniture . . . from drawings.* London: [I. and J. Taylor, 1789?].

Hermand, Jost. *Geschichte der Germanistik.* Reinbek bei Hamburg: Rowohlt, 1994.

Herrnhuter Architektur am Rhein und an der Wolga. Edited by Landkreis Neuwerd/ Rhein, with Reinhard Lahr and Bernd Willscheid. Neuwied: Landkreis Neuwied, 2001.

Hettche, Walter. "Goethes Sommerreise 1805." *Jahrbuch des Freien Deutschen Hochstifts,* 2005, 56–81.

Heuson, Hans-Velten. *Büdingen gestern und heute: Arbeiten zur Geschichte der Stadt und ihres Umfeldes: 1956–2000.* Collected and edited by Volkmar Stein. Büdingen: Magistrat der Stadt Büdingen, Amt für Jugend, Kultur und Soziales, 2004.

Heyd, Michael. "Changing Emotions? The Decline of Original Sin on the Eve of the Enlightenment." In Gouk and Hills, *Representing Emotions,* 123–36.

Hildenbrock, Aglaja. *Das andere Ich: Künstlicher Mensch und Doppelgänger in der deutsch- und englischsprachigen Literatur.* Tübingen: Stauffenburg, 1986.

Hill, David, ed. *Literature of the Sturm und Drang.* Rochester, NY: Camden House, 2003.

Histoire de l'academie royale des sciences. 1785.

Historisch-biographisches Lexikon der Schweiz. 7 vols. Neuenburg: Administration des Historisch-Biographischen Lexikons der Schweiz, 1921–34.

Hitchcock, Susan Tyler. *Frankenstein: A Cultural History.* New York: W. W. Norton, 2007.

Hochadel, Oliver. *Öffentliche Wissenschaft: Elektrizität in der deutschen Aufklärung.* Göttingen: Wallstein Verlag, 2003.

Hoffmann, E. T. A. *Briefwechsel.* Collected and annotated by H. von Müller und F. Schnapp, edited by F. Schnapp. Munich: Winkler, 1967–69.

———. *Die Automate.* In *Die Serapions-Brüder,* edited by Wulf Segebrecht, 396–417. Frankfurt am Main: Deutscher Klassiker Verlag, 2008.

———. *Tagebücher.* Munich, Winkler-Verlag, [1971?].

Hoffmann, Freia. *Instrument und Körper: Die musizierende Frau in der bürgerlichen Kultur.* Frankfurt am Main: Insel Verlag, 1991.

Hoffmann, Friedrich G., and Herbert Rösch. *Grundlagen, Stile, Gestalten der deutschen Literatur: Eine geschichtliche Darstellung.* Berlin: Cornelsen, 1996.

Hoffmann, Henriette. "Eine Untersuchung über Kapital, Industrie und Maschine von Goethe bis Immermann." PhD diss., University of Vienna, 1942.

Hohendahl, Peter Uwe. *Der europäische Roman der Empfindsamkeit.* Wiesbaden: Athenaion, 1977.

Holme, Bryan. *Princely Feasts and Festivals: Five Centuries of Pageantry and Spectacle.* London: Thames and Hudson, 1988.

Horloges et automates: Musée du Conservatoire National des Arts et Métiers, septembre–novembre 1954. Paris: Les Presses Artistiques, 1954.

Horn, Jeff, Leonard N. Rosenband, and Merritt Roe Smith, eds. *Reconceptualizing the Industrial Revolution*. Cambridge, MA. MIT Press, 2010.

Hosler, Bellamy. *Changing Aesthetic Views of Instrumental Music in 18th-Century Germany*. Ann Arbor, MI: UMI Research Press, 1978.

Hubbard, Frank. *Three Centuries of Harpsichord Making*. Cambridge, MA: Harvard University Press, 1965.

Huber, Friedrich. *Daniel Bernoulli (1700–1782) als Physiologe und Statistiker*. Basel: Bennop Schwabe, 1959.

Hudson, Nicholas, and Aaron Santesso, eds. *Swift's Travels: Eighteenth-Century British Satire and Its Legacy*. Cambridge: Cambridge University Press, 2008.

Hughes, Thomas Parke. *American Genesis: A Century of Invention and Technological Enthusiasm, 1870–1970*. Chicago: University of Chicago Press, 2004.

———. *Lewis Mumford: Public Intellectual*. New York: Oxford University Press, 1990.

Hull, Isabel. *Sexuality, State, and Civil Society in Germany, 1700–1815*. Ithaca, NY: Cornell University Press, 1996.

Hunt, Lynn Avery. "The Many Bodies of Marie-Antoinette: Political Pornography and the Problem of the Feminine in the French Revolution." In *Marie-Antoinette: Writings on the Body of a Queen*, edited by Dena Goodman, 117–38. New York: Routledge, 2003.

Huth, Hans. *Abraham und David Roentgen und ihre Neuwieder Möbelwerkstatt*. Berlin: Deutscher Verein für Kunstwissenschaft, 1928.

Hutton, Charles. *Mathematical and philosophical dictionary: containing an explanation of the terms, and an account of the several subjects, comprized under the heads mathematics, astronomy, and philosophy both natural and experimental; with an historical account of the rise progress and present state of these sciences; also memoirs of the lives and writings of the most eminent authors. . . .* 2 vols. London: Johnson and Robinson, 1795–96.

Huxley, Thomas Henry. *Science and Culture, and other Essays*. 1st ed. London: Macmillan, 1881.

Huyssen, Andreas. "Mass Culture as Woman: Modernism's Other." In *After the Great Divide: Modernism, Mass Culture, Postmodernism*, edited by Andreas Huyssen, 44–62. Bloomington: Indiana University Press, 1986.

———. "The Vamp and the Machine: Technology and Sexuality in Fritz Lang's Metropolis." *New German Critique* 24–25 (1981–82): 221–37.

Iffland, August Wilhelm. *Die Marionetten: Lustspiel in einem Aufzuge*. Edited and with an afterword by Gunhild Berg. Hannover: Wehrhahn, 2009.

Impey, Oliver, and Arthur MacGregor. *The Origins of Museums: The Cabinet of Curiosities in Sixteenth and Seventeenth-Century Europe*. New York: Oxford University Press, 1985.

Ince, William, and Jonathan Mayhew. *The Universal System of Houshold Furniture*. London: Printed by the authors, 1760.

Innes, William. *Social Concern in Calvin's Geneva*. Allison Park, PA: Pickwick, 1983.

Jackson, Barbara Garvey. "Musical Women of the Seventeenth and Eighteenth Centuries." In *Women and Music: A History,* edited by Karin Pendle, 97–144. Bloomington: Indiana University Press, 2001.

Jackson, Myles W. *Harmonious Triads: Physicists, Musicians, and Instrument Makers in Nineteenth-Century Germany.* Cambridge, MA: MIT Press, 2006.

Jacobsson, Johann Karl Gottfried. *Technologisches Wörterbuch, oder Alphabetische Erklärung aller nützlichen mechanischen Künste, Manufakturen, Fabriken und Handwerker.* . . . 8 vols. With a preface by Johann Beckmann. Berlin: F. Nicolai, 1781–1802.

Jacquat, Marcel. "Abraham Gagnebin: Médecin (1707–1800)." In *Biographies Neuchâteloises,* vol. 1, *De saint Guillaume à la fin des Lumières,* edited by Michel Schlup, 97–102. Neuchâtel, Switzerland: Editions Gilles Attinger, 1996.

Jauch, Ursula Pia. *Jenseits der Maschine: Philosophie, Ironie und Ästhetik bei Julien Offray de La Mettrie.* Munich: Hanser, 1998.

———. "Maschinentraum und Traummaschine bei Julien Offray de La Mettrie." In *Julien Offray de La Mettrie: Ansichten und Einsichten,* edited by Hartmut Hecht, 49–61. Berlin: Berliner Wissenschaftsverlag, 2004.

Jeanneret, Frédéric Alexandre M. *Biographie Neuchâteloises.* 2 vols. Le Locle, Switzerland: Eugène Courvoisier, 1863.

———. *Etrennes Neuchâteloises.* Le Locle, Switzerland: Eugène Courvoisier, 1862.

Jean Paul. *Jean Paul's sämmtliche Werke.* Berlin: G. Reimer, 1840–42.

———. *Jean Pauls sämtliche Werke: Historisch-kritische Ausgabe.* Weimar: H. Böhlaus Nachfolger, 1927–

———. *Werke.* Edited by Norbert Miller and Wilhelm Schmidt-Biggemann. Munich: Carl Hanser, 1976.

Johnson, Ann. "From Boeing to Berkeley: Civil Engineers, the Cold War, and the Origins of Finite Element Analysis." In *Growing Explanations: Historical Perspectives on Recent Science,* edited by M. Norton Wise, 133–58. Durham, NC: Duke University Press, 2004.

Journal de politique et de littérature: Contenant les principaux Evènemens de toutes les Cours; les Nouvelles de la République des Lettres, &c. No. 3, 25 January. Brussels, 1775.

Jullien, Adolphe. *La ville et la cour au XVIII siècle: Mozart—Marie Antoinette—Les Philosophes.* Paris: Rouveyere, 1881.

Jung-Stilling, Johann Heinrich. *Versuch eines Lehrbuchs der Fabrikwissenschaft: Zum Gebrauch academischer Vorlesungen.* Nürnberg: Grattenauerischen Buchhandlung, 1785.

Justi, Johann Heinrich Gottlob von. *Vollständige Abhandlung von denen Manufacturen und Fabriken.* 2 vols. Kopenhagen: Rothensche Buchhandlung, 1758–61.

Jüttemann, Herbert. *Mechanische Musikinstrumente: Einführung in Technik und Geschichte.* Frankfurt am Main: Erwin Bochinsky, 1987.

Kaiser, Gerhard. *Aufklärung, Empfindsamkeit, Sturm und Drang.* Munich: Francke, 1976.

Kämmerer, Harald. *Nur um Himmels willen keine Satyren . . . Deutsche Satire und Satiretheorie des 18. Jahrhunderts im Kontext von Anglophilie, Swift-Rezeption und ästhetischer Theorie.* Heidelberg: Universitätsverlag C. Winter, 1999.

Kang, Minsoo. *Sublime Dreams of Living Machines: The Automaton in the European Imagination.* Cambridge, MA: Harvard University Press, 2011.

Karmarsch, Karl. *Geschichte der Technologie seit der Mitte des 18. Jahrhunderts.* Munich, Oldenbourg, 1872.

Katritzky, Linde. *A Guide to Bonaventura's Nightwatches.* New York: P. Lang, 1999.

Kellenbenz, Herrmann. "Rural Industries in the West from the End of the Middle Ages to the Eighteenth Century." In *Essays in European Economic History, 1500–1800,* edited by Peter Earle, 45–88. Oxford: Clarendon, 1974.

Kempelen, Wolfgang von. *Wolfgangs von Kempelen k.k. wirklichen Hofraths Mechanismus der menschlichen Sprache: Nebst der Beschreibung seiner sprechenden Maschine.* Vienna: J. V. Degen, 1791.

Kircher, Athanasius. *Musurgia universalis.* Rome, 1650.

Kistenmacher, Hans Werner. *Maschine und Dichtung: Ein Beitr. z. Geschichte d. Deutschen Literatur im 19. Jh.* Greifswald: Hartmann, 1914.

Klappert, Annina. *Die Perspektiven von Link und Lücke: Sichtweisen auf Jean Pauls Texte und Hypertexte.* Bielefeld, Germany: Aisthesis, 2006.

Klemm, Friedrich. *Geschichte der naturwissenschaftlichen und technischen Museen.* Munich: VDI-Verlag, 1973.

Klussmann, Paul Gerhard, and York-Gothart Mix, eds. *Literarische Leitmedien: Almanach und Taschenbuch im kulturwissenschaftlichen Kontext.* Wiesbaden: Harrassowitz, 1998.

Knott, Sarah, and Barbara Taylor, eds. *Women, Gender, and Enlightenment.* New York: Palgrave Macmillan, 2005.

Koch, Heinrich Christoph. *Versuch einer Anleitung zur Composition.* 3 vols. Leipzig: A. F. Böhme, 1782.

Kocka, Jürgen, "Bürgertum und Bürgerlichkeit als Probleme der deutschen Geschichte vom späten 18. bis zum frühen 20. Jahrhundert." In *Bürger und Bürgerlichkeit im 19. Jahrhundert,* edited by Jürgen Kocka, 21–64. Göttingen: Vandenhoeck und Ruprecht, 1987.

König, Wolfgang, and Helmuth Schneider. *Die technikhistorische Forschung in Deutschland von 1800 bis zur Gegenwart.* Kassel, Germany: Kassel University Press, 2007.

Körner, Martin. "Town and Country in Switzerland, 1450–1750." In *Town and Country in Europe, 1300–1800,* edited by Stephan R. Epstein, 229–49. Cambridge: Cambridge University Press, 2001.

Kotte, Andreas. "Iffland, Kleist und das Marionettentheater." *Figura* 23 (1998): 4–8.

———. *Theaterwissenschaften: Eine Einführung.* Cologne: Böhlau, 2005.

Kowar, Helmut. *Die Wiener Flötenuhr: "Sie spielt besser als das Orchester im Kärntnertor."* Vienna: Technisches Museum Wien, 2001.

Krajewski, Markus. *Restlosigkeit: Weltprojekte um 1900.* Frankfurt am Main: Fischer Taschenbuch Verlag, 2006.

Krätz, Otto. "'Der makabre Android . . . sein Spiegelbild in der Literatur': Kultur und Technik." *Zeitschrift des Deutschen Museums München* 2 (1978): 22–27.

Krause, Christian Gottfried. *Von der musikalischen Poesie.* Berlin, J. F. Voss, 1752.

Kremer, Detlef. *Prosa der Romantik.* Stuttgart: J. B. Metzler, 1996.

Kriedte, Peter. *Taufgesinnte und großes Kapital: Die niederrheinisch-bergischen Mennoniten und der Aufstieg des Krefelder Seidengewerbes (Mitte des 17. Jahrhunderts–1815).* Göttingen: Vandenhoeck & Ruprecht, 2007.

Kriedte, Peter, Hans Medick, and Jürgen Schlumbohm. *Industrialization before Industrialization: Rural Industry in the Genesis of Capitalism.* Translated by Beate Schempp. Cambridge: Cambridge University Press, 1981.

Krieg, Dieter. "Das alte Herrnhuter Viertel zu Neuwied." *Neuwied Kalender,* 1963.

———. "Das Brüderhaus in Neuwied—einstige Stätte soliden Handwerkertums." *Neuwied Kalender,* 1959.

———. "Im Chorhaus der ledigen Brüder zu Neuwied Ende des 18. Jahrhunderts." *Neuwied Kalender,* 1961.

Krüger, Hans-Jürgen. "Religiöse Toleranz aus religiöser Gleichgültigkeit? Die Herrnhuter Brüdergemeine in Neuwied." *Neuwied Kalender,* 2002.

Krüger, Renate. *Das Zeitalter der Empfindsamkeit: Kunst und Kultur des späten 18. Jahrhunderts.* Vienna: Anton Schroll, 1972.

Krünitz, Johann Georg. *Oeconomisch-technologische Encyklopädie, oder allgemeines System der Staats-, Stadt-, Haus- und Landwirthschaft und der Kunstgeschichte in alphabetischer Ordnung.* Berlin: Paul, 1773–1858.

Kuijken, Barthold. "Einführung." In *Versuch einer Anweisung, die flute traversière zu spielen,* by Johann Joachim Quantz, xi–xv. Facsimile of the 1752 edition. Wiesbaden: Breitkopf & Härtel, 1988.

Kunisch, Johannes. *Absolutismus: Europäische Geschichte vom Westfälischen Frieden bis zur Krise des Ancien Régime.* Göttingen: Vandenhoeck und Ruprecht, 1999.

Küster, Karl Daniel. "Empfindsam." In *Sittliches Erziehungs-Lexicon, oder Erfahrungen und geprüfte Anweisungen: Wie Kinder von hohen und mittlern Stande, zu guten Gesinnungen und zu wohlanständigen Sitten können angeführet werden: Ein Handbuch für edelempfindsame Eltern, Lehrer und Kinder-Freunde, denen die sittliche Bildung ihrer Jugend am Herzen liegt,* 47–51. Magdeburg: Scheidhauersche Buchhandlung, 1774.

Lacan, Jacques. *The Seminar of Jacques Lacan.* Edited by Jacques-Alain Miller. Translated by John Forrester. Cambridge: Cambridge University Press, 1988.

La Mettrie, Julien Offray de. *L'homme machine.* Leyde, Netherlands: Elie Luzac, 1748.

———. *L'homme machine.* Paris: Frédéric Henry, 1865.

———. *Man a Machine.* Chicago: Open Court, 1912.

Lanckorońska, Maria Gräfin, and Arthur Rümann. *Geschichte der deutschen Taschenbücher und Almanache aus der klassisch-romantischen Zeit.* Munich: Ernst Heimeran, 1954.

Landes, David. *Revolution in Time: Clocks and the Making of the Modern World.* Cambridge, MA: Harvard University Press, 1983.

———. *The Unbound Prometheus: Technological Change and Industrial Development in Western Europe from 1750 to the Present.* Cambridge: Cambridge University Press, 2003 (1969).

Landes, Joan. *Women and the Public Sphere in the Age of the French Revolution.* Ithaca, NY: Cornell University Press, 1988.

Lang, Joseph Gregor. *Reise auf dem Rhein von Mainz bis Düsseldorf.* 2 vols. Koblenz, 1789–90.

Lange, Friedrich Albert. *Geschichte des Materialismus und Kritik seiner Bedeutung in der Gegenwart.* 2 vols. Iserlohn, Germany: J. Baedeker, 1866–73.

Lanier, Jacques François. *L'abbé Michel Servan, ou, De Servan: prêtre, historien, ingénieur: Romans 1746–Lyon 1837.* Valence: J. F. Lanier, 2000.

La restauration musicale de la Joueuse de tympanon. Booklet with audio CD. Conservatoire national des arts et métiers. Paris, 1995.

La Roche, Sophie von. *Tagebuch einer Reise durch Holland und England.* Offenbach am Main: Ulrich Weiss und Carl Ludwig Brede, 1788.

Lathers, Marie. *The Aesthetics of Artifice: Villiers's "L'Eve future."* Chapel Hill: Department of Romance Languages, University of North Carolina, 1996.

Le Blanc, Charles. *Catalogue de l'oeuvre de Jean Georges Wille, graveur: Avec une notice biographique.* Leipsic: R. Weigel, 1847.

Le conservatoire national des arts et métiers au coeur de Paris: 1794–1994: [Sous la] dir. de Michel le Moël et Raymond Saint-Paul [et al.]. La Délégation à l'action artistique de la ville de Paris. Paris: Conservatoire national des arts et métiers: Délégation à l'action artistique de la ville de Paris, 1994.

Le Corbeiller, Clare. "James Cox and His Curious Toys." *Metropolitan Museum of Art Bulletin,* n.s., 18, no. 10 (June 1960): 318–24.

Leibniz, Gottfried Wilhelm. *Remarques sur le discours de M. H. Sully.* Vienna, 1714.

Le Paute, Jean André. *Traité d'horlogerie contenant tout ce qui est nécessaire pour bien connoitre et pour regler les pendules et les montres, la description des pieces d'horlogerie les plus utiles, des repétitions. . . .* Paris: J. Chardon, 1755.

Leppert, Richard. *Arcadia at Versailles: Noble Amateur Musicians and Their Musettes and Hurdy-Gurdies at the French Court: A Visual Study.* Amsterdam: Swets & Zeitlinger B. V., 1978.

Le Roy, Julien. *Avis contenant les vrais moyens de régler les montres tant simples qu'à répétition.* Paris: Chardon-père, 1719.

Les androides Jaquet-Droz: La grande histoire d'une petite famille mécanique. TALIA films, Philippe Sayous. Paris: T. I. L. Productions, 2001. Videocassette (VHS), Musée internationale d'horlogerie.

Lespinasse, Julie de. *Correspondance entre Mademoiselle de Lespinasse et le comte de Guibert: Publiée pour la première fois d'après le texte original par le comte de Villeneuve-Guibert.* Paris: Calman-Lévy, 1906.

Leupold, Jacob. *Theatrum machinarum generale: Schau-Platz des Grundes mechanischer Wissenschaften.* Leipzig: Christoph Zunkel, 1724.

Levy, René. *The Social Structure of Switzerland: Outline of a Society.* 5th ed. Zürich: Pro Helvetia, 1991.

Lichtenberg, Georg Christoph. *Briefe an Dieterich, 1770–1798: Zum hundertjährigen Todestage Lichtenberg's herausgegeben von Eduard Grisebuch.* Leipzig: Dieterich'sche Verlagsbuchhandlung, T. Weicher, 1898.

———. *Taschenbuch zum Nutzen und Vergnügen.* Göttingen: Johann Christian Dieterich, 1777–99.

Lindauer, Tanja. *Reconstructing Eve: Automatenmenschen in Literatur und Film.* Marburg, Germany: Tectum Verlag, 2008.

Lindner, Burkhardt. "Jean Paul als J. P. F. Hasus." In Schweikert, *Jean Paul,* 411–50.

Lister, Martin. *New Media: A Critical Introduction.* 2nd ed. New York: Routledge, 2009.

Loesser, Arthur. *Men, Women, and Pianos: A Social History.* New York: Simon and Schuster, 1954.

Lorenz, Dagmar C. G. *A Companion to the Works of Arthur Schnitzler.* Rochester, NY: Camden House, 2003.

Ludwig, Karl-Heinz. "Entwicklung, Stand und Aufgaben der Technikgeschichte." *Archiv für Sozialgeschichte* 18 (1978): 502–23.

Lynch, Michael, and Harry Collins. "Introduction: Humans, Animals, and Machines." *Science, Technology, and Human Values* 23, no. 4 (Autumn 1998): 371–83.

Maeterlinck, Maurice. "Androidentheater" [1890]. In *Texte zur Theorie des Theaters,* edited by Klaus Lazarowicz and Christopher Balme, 364–73. Stuttgart: P. Reclam, 1991.

Maier, Anneliese. *Die Mechanisierung des Weltbildes im 17. Jahrhundert.* Leipzig: Meiner, 1983.

Marpurg, Friedrich Wilhelm. *Anleitung zum Clavierspielen.* Berlin: Haude und Spener, 1755.

Martens, Wolfgang. *Die Botschaft der Tugend: Die Aufklärung im Spiegel der Moralischen Wochenschriften.* Stuttgart: Metzler, 1968.

Martin, Luther H., Huck Gutman, and Patrick H. Hutton. *Technologies of the Self: A Seminar with Michel Foucault.* Amherst: University of Massachusetts Press, 1988.

Martius, Johann Nikolaus. *Unterricht in der natürlichen Magie, oder, Zu allerhand belustigenden und nützlichen Kunststücken: Völlig umgearbeitet von Johann Christian Wiegleb.* 20 vols. Berlin: Friedrich Nicolai, 1779–1805.

Marvin, Thomas F. *Kurt Vonnegut: A Critical Companion.* Westport, CT: Greenwood Press, 2002.

Marx, Karl. *Das Kapital: Kritik der politischen Oekonomie*. 2 vols. 2nd ed. Hamburg: O. Meissner, 1872.

Matérialistes français du XVIIIe siècle: La Mettrie, Helvétius, d'Holbach. Paris: Presses universitaires de France, 2006.

Matt, Peter von. *Die Augen der Automaten: E. T. A. Hoffmanns Imaginationslehre als Prinzip seiner Erzählkunst*. Tübingen: Niemeyer, 1971.

Matthes, Dieter. "Goethes Reise nach Helmstedt und seine Begegnung mit Gottfried Christoph Beireis: Eine Untersuchung zum Bildstil der 'Tag- und Jahreshefte.'" *Braunschweigisches Jahrbuch* 49 (1968): 121–201.

Mattheson, Johann. *Der vollkommene Capellmeister, das ist, Gründliche Anzeige aller derjenigen Sachen, die einer wissen, können, und vollkommen inne haben muss, der eine Kapelle mit Ehren und Nutzen vorstehen will*. Hamburg: Herold, 1739.

Maunder, Charles Richard Francis. *Keyboard Instruments in Eighteenth-Century Vienna*. New York: Oxford University Press, 1998.

Mauris, Maurice. *French Men of Letters*. New York: D. Appleton, 1902. Originally published in 1880.

Mautner, Franz H. *Lichtenberg: Geschichte seines Geistes*. Berlin: Walter de Gruyter, 1968.

Mayr, Otto. *Authority, Liberty, and Automatic Machinery in Early Modern Europe*. Baltimore. Johns Hopkins University Press, 1986.

Meadow, Mark A. "Merchants and Marvels: Hans Jacob Fugger and the Origins of the Wunderkammer." In *Merchants and Marvels: Commerce and the Representation of Nature*, edited by Paula Findlen and Pamela Smith, 182–200. London: Routledge, 2001.

Meighörner, Jeannine. *Sophie von La Roche: "Was ich als Frau dafür halte": Deutschlands erste Bestsellerautorin*. Erfurt, Germany: Sutton, 2006.

Meiners, Christoph. *Briefe über die Schweiz*. Berlin: Spener, 1790.

Meinhardt, Albert. "Der Werdegang Neuwieds." In *1653–1953: 300 Jahre Neuwied. Ein Stadt- und Heimatbuch*, edited by Albert Meinhardt, 67–332. Neuwied: Neuwieder Verlagsgesellschaft, 1953.

Mendels, Franklin. "Proto-Industrialization: The First Phase of the Process of Industrialization." *Journal of Economic History* 32 (1972): 241–61.

Merton, Robert King. *Science, Technology, and Society in Seventeenth Century England*. New York: H. Fertig, 1970.

Mestral, Aymon de. *Daniel JeanRichard, Founder of the Jura Watch Industry, 1672–1741*. Zürich: Institute of Economic Research, 1957.

Meusel, Johann Georg. *Miscellaneen artistischen Inhalts*. Vol. 22. Erfurt, Germany: Heft, 1785.

Meyer, Dietrich. "Zinzendorf und Herrnhut." In Brecht and Deppermann, *Geschichte des Pietismus*, 2:5–105.

Meyer, Torsten. "Die Anfänge technikhistorischen Arbeitens in Deutschland: Johann Beckmanns 'Beyträge zur Geschichte der Erfindungen.'" *Technikgeschichte* 64, no. 3 (1997): 161–79.

Milham, Willis. *Time and Timekeepers: The History, Construction, Care, and Accuracy of Clocks and Watches.* New York: Macmillan, 1923.

Mindell, David A. *Between Human and Machine: Feedback, Control, and Computing before Cybernetics.* Baltimore: Johns Hopkins University Press, 2002.

———. *Digital Apollo: Human and Machine in Spaceflight.* Cambridge, MA: MIT Press, 2008.

———. *War, Technology, and Experience aboard the USS Monitor.* Baltimore: Johns Hopkins University Press, 2000.

Mittendorfer, Monika. "Unterdrückte Kreativität: Cornelia Goethe und Nannerl Mozart." In *Maria Anna Mozart: Die Künstlerin und ihre Zeit,* edited by Sigrid Düll and Otto Neumaier, 135–54. Möhnesee, Germany: Bibliopolis, 2001.

Mix, York-Gothart. *Die deutschen Musenalmanache des 18. Jahrhunderts.* Munich: C. H. Beck, 1987.

Mizler, Lorenz Christoph von Kolof. *Neu eröffnete musikalische Bibliothek, oder Gründliche Nachricht. . . .* 4 vols. Leipzig, 1739–54.

Mokyr, Joel, ed. *The British Industrial Revolution: An Economic Perspective.* 2nd ed. Boulder, CO.: Westview Press, 1999.

Möller, Horst. *Fürstenstaat oder Bürgernation: Deutschland, 1763–1815.* Berlin: Wolf Jobst Siedler Verlag, 1989.

Möller, Uwe. *Rhetorische Überlieferung und Dichtungstheorie im frühen 18. Jahrhundert: Studien zu Gottsched, Breitinger und G. Fr. Meier.* Munich: W. Fink, 1983.

Mörikofer, Johann Kaspar. *Geschichte der evangelischen Flüchtlinge in der Schweiz.* Leipzig: S. Hirzel, 1876.

Mozart, Leopold. *Versuch einer gründlichen Violinschule.* Augsburg: J. J. Lotter, 1756.

Müller, Götz. *Jean Paul im Kontext: Gesammelte Aufsätze.* Würzburg: Königshausen & Neumann, 1996.

———. *Jean Pauls Exzerpte.* Würzburg: Könighausen und Neumann, 1988.

Müller-Tamm, Pia, and Katharina Sykora. *Puppen, Körper, Automaten: Phantasmen der Moderne.* Düsseldorf, Germany: Oktagon, 1999.

Mulvey, Laura. *Visual and Other Pleasures.* Basingstoke, Hampshire: Macmillan, 1989.

Mumford, Lewis. *The Myth of the Machine.* 2 vols. New York: Harcourt, Brace & World, 1967–70.

———. *Technics and Civilization.* New York: Harcourt, Brace, 1934.

Munz, Alfred. *Philipp Matthäus Hahn: Pfarrer, Erfinder und Erbauer von Himmelsmaschinen, Waagen, Uhren und Rechenmaschinen.* Sigmaringen, Germany: Jan Thorbecke Verlag, 1977.

Nabholz, Hans, Leonhard von Muralt, Richard Feller, and Edgar Bonjour. *Geschichte der Schweiz.* 2 vols. Zürich: Schulthess, 1938.

Nedoluha, Alois. *Kulturgeschichte des technischen Zeichnens.* Vienna: Springer-Verlag, 1960.

Nemnich, Philipp Andreas. *Tagebuch einer der Kultur und Industrie gewidmeten Reise*. Tübingen, J. G. Cotta, 1809.

Neubauer, John. *The Emancipation of Music from Language: Departure from Mimesis in Eighteenth-Century Aesthetics*. New Haven, CT: Yale University Press, 1986.

Neue deutsche Biographie. Berlin: Duncker & Humblot, 1953–.

Neuls-Bates, Carol. *Women in Music: An Anthology of Source Readings from the Middle Ages to the Present*. Boston: Northeastern University Press, 1996.

Neumayr, Franz. *Frag: Ob der Mensch weiter nichts seye, als eine Machine*. Munich: Frantz Xaveri Crätz und Thomas Summer, 1761.

Niemann, Walter. "Vorwort des Herausgebers." In *Versuch über die wahre Art das Klavier zu spielen*, by Karl Philipp Emanuel Bach, iii–v. Leipzig: Kahnt Nachfolger, 1906.

Niethammer, Lutz. "Bürgerliche Gesellschaft als Projekt." In *Bürgerliche Gesellschaft in Deutschland: Historische Einblicke, Fragen, Perspektiven*, 17–40. Frankfurt am Main: Fischer Taschenbuch Verlag, 1990.

Nouvelle biographie générale depuis les temps les plus reculés jusqu'à nos jours. Paris: Firmin Didot Frères, 1857.

Ogilvie, Sheilagh C., and Markus Cerman, eds. *European Proto-Industrialization*. Cambridge: Cambridge University Press, 1996.

Olf, Julian. "The Man/Marionette Debate in Modern Theatre." *Educational Theatre Journal* 26, no. 4 (December 1974): 488–94.

Opitz, Claudia, Ulrike Weckel, and Elke Kleinau, eds. *Tugend, Vernunft und Gefühl: Geschlechterdiskurse der Aufklärung und weibliche Lebenswelten*. Münster: Waxmann, 2000.

Ord-Hume, Arthur W. J. G. *Clockwork Music: An Illustrated History of Mechanical Musical Instruments from the Musical Box to the Pianola, from Automaton Lady Virginal Players to Orchestrion*. London: George Allen and Unwin, 1973.

Ostervald, Samuel Frédéric. *Descriptions des montagnes et des vallées qui font partie de la principauté de Neuchâtel et Valangin*. 2nd ed. Neuchâtel, Switzerland: Samuel Fauche, 1766. Originally published in 1765.

Outram, Dorinda. "The Enlightenment Our Contemporary." In *The Sciences in Enlightened Europe*, edited by William Clark, Jan Golinski, and Simon Schaffer, 32–42. Chicago: University of Chicago Press, 1999.

Pearl, Mildred. "The Suite in Relation to Baroque Style." PhD diss., New York University, 1957.

Perregaux, Charles, and F.-Louis Perrot. *Les Jaquet-Droz et Leschot*. Neuchâtel, Switzerland: Attinger Frères, 1916.

Perrin, Charles. "Un solliciteur Loclois au XVIIIme siècle." *Musée neuchatelois*, 1906, 61–70.

Petersen-Mikkelsen, Birger. *Die Melodielehre des vollkommenen Capellmeisters von Johann Mattheson: Eine Studie zum Paradigmenwechsel in der Musiktheorie des 18. Jahrhunderts*. Eutin: [Birger Petersen-Mikkelsen], 2002.

Pfeisinger, Gerhard. *Arbeitsdisziplinierung und frühe Industrialisierung 1750–1820.* Vienna: Böhlau, 2006.

Pfleghart, A. *Die schweizerische Uhrenindustrie, ihre geschichtliche Entwicklung und Organisation.* Leipzig: Duncker & Humblot, 1908.

Philipp, Guntram. "Wirtschaftsethik und Wirtschaftspraxis in der Geschichte der Herrnhuter Brüdergemeine." In *Unitas Fratrum: Herrnhuter Studien,* edited by Mari P. van Buijtenen, Cornelis Dekker, and Huib Leeuwenberg, 401–63. Utrecht: Rijksarchief, 1975.

Poggi, Gianfranco. *Calvinism and the Capitalist Spirit: Max Weber's Protestant Ethic.* Amherst: University of Massachusetts Press, 1983.

Poppe, Johann Heinrich Moritz. *Ausführliche Geschichte der theoretisch-praktischen Uhrmacherkunst, seit der ältesten Art den Tag einzutheilen, bis an das Ende des achtzehnten Jahrhunderts.* Leipzig: Roch, 1801.

———. *Ausführliche Volks-Gewerbslehre, oder, Allgemeine und besondere Technologie zur Belehrung und zum Nutzen für alle Stände nach dem neuesten Zustande der technischen Gewerbe und deren Hilfswissenschäften bearbeitet.* Stuttgart: C. Hoffmann, 1833.

———. *Encyclopädie des gesammten Maschinenwesens, oder vollständiger Unterricht in der praktischen Mechanik und Maschinenlehre, mit Erklärungen der dazu gehörigen Kunstwörter in alphabetischer Ordnung: ein Handbuch für Mechani-ker, Kameralisten, Baumeister und Jeden, dem Kenntnisse des Maschinenwesens nöthig und nützlig sind.* Leipzig: Voss, 1803–27.

———. *Geschichte aller Erfindungen und Entdeckungen im Bereiche der Gewerbe, Künste und Wissenschaften von der frühesten Zeit bis auf unsere Tage.* Stuttgart, Hoffmann'sche Verlags-Buchhandlung, 1837.

———. *Geschichte der Technologie seit der Wiederherstellung der Wissenschaften bis an das Ende des achtzehnten Jahrhunderts.* Göttingen: J. F. Röwer, 1807–11.

———. *Lehrbuch der Maschinenkunde: Nach einem neuen, umfassenderen Plane und ohne Voraussetzung höherer analytischer Kenntnisse, hauptsächlich für angehende Kameralisten, Oekonomen, Baumeister und jeden Liebhaber der Mechanik.* Tübingen: Osiander, 1821.

———. *Versuch einer Geschichte der Entstehung und Fortschritte der theoretisch-praktischen Uhrmacherkunst.* Göttingen: Vandenhoeck and Ruprecht, 1797.

———. *Wunder der Mechanik: oder, Beschreibung und Erklärung der berühmten Tendlerschen Figuren, der Vaucansonschen, Kempeleschen, Drozschen, Maillar-detschen und anderer merkwürdiger Automaten und ähnlicher bewunderungs-würdiger mechanischer Kunstwerke.* Tübingen: C. F. Osiander, 1824.

Poritzky, Jakob Elias. *Julien Offray de Lamettrie: Sein Leben und seine Werke.* Berlin: F. Dümmler, 1900.

Proß, Wolfgang. *Jean Pauls geschichtliche Stellung.* Tübingen: Max Niemeyer, 1975.

Protz, Albert. *Mechanische Musikinstrumente.* Kassel: Bärenreiter, 1939.

Quantz, Johann Joachim. *Versuch einer Anweisung, die flute traversière zu spielen.* 3rd ed. Breslau: Johann Friedrich Korn, 1789.

Rabinbach, Anson. *The Human Motor: Energy, Fatigue, and the Origins of Modernity*. New York: Basic Books, 1990.

Racker, Miriam. "Gestalt und Symbolik des künstlichen Menschen in der Dichtung des 19. und 20. Jahrhunderts." PhD diss., University of Vienna, 1932.

Rankl, Maximilian. *Jean Paul und die Naturwissenschaft*. Frankfurt am Main: Peter Lang, 1987.

Raphael, David Daiches. *The Impartial Spectator: Adam Smith's Moral Philosophy*. Oxford: Clarendon Press, 2007.

Rapp, Eleonore. *Die Marionette in der deutschen Dichtung vom Sturm und Drang bis zur Romantik*. Leipzig: Engelhardt, 1924.

Reed, Peter J. *The Short Fiction of Kurt Vonnegut*. Westport, CT: Greenwood Press, 1997.

Rheinische Kunstwerke von der Renaissance bis zum Expressionismus. Düsseldorf, Germany: Rheinland-Verlag, 1970.

Richter, Klaus. "Aus der Baugeschichte der Herrnhuter Brüdergemeine." In *Herrnhuter Architektur am Rhein und an der Wolga*, edited by Landkreis Neuwerd/Rhein, with Reinhard Lahr and Bernd Willscheid, 27–57. Koblenz: Görres-Druckerei, 2001.

Richter, Siegfried. *Wunderbares Menschenwerk: Aus der Geschichte der mechanischen Automaten*. Leipzig: Edition Leipzig, 1989.

Rieger, Eva. *Frau, Musik und Männerherrschaft: Zum Ausschluß der Frau aus der deutschen Musikpädagogik, Musikwissenschaft und Musikausübung*. Frankfurt: Ullstein, 1981.

———. *Nannerl Mozart: Leben einer Künstlerin im 18. Jahrhundert*. Frankfurt am Main: Insel Verlag, 1990.

Riley, Matthew. *Musical Listening in the German Enlightenment: Attention, Wonder, and Astonishment*. Burlington, VT: Ashgate, 2004.

Riskin, Jessica. "The Adventures of Mr. Machine, with Morals." In *A Cultural History of the Human Body*, vol. 4, *A History of the Human Body in the Age of Enlightenment, 1650–1800*, edited by Carole Reeves, 73–92. Oxford: Berg, 2010.

———. "The Defecating Duck; or, The Ambiguous Origins of Artificial Life." *Critical Inquiry* 20 (2003): 599–633.

———. "Eighteenth-Century Wetware," *Representations* 83 (2003): 97–125.

———. "Mr. Machine and the Imperial Me." In "The Super-Enlightenment: Daring to Know Too Much," edited by Dan Edelstein, special issue, *Studies on Voltaire and the Eighteenth Century* 1 (2010): 75–94.

———. *Science in the Age of Sensibility: The Sentimental Empiricists of the French Enlightenment*. Chicago: University of Chicago Press, 2002.

Röder, Birgit. *A Study of the Major Novellas of E.T.A. Hoffmann*. Rochester, NY: Camden House, 2003.

Roentgen, Ludwig. *Das erste Buch meines Lebens*. Rotterdam: J. Van Baalen, 1845.

Rose, Gillian. *Visual Methodologies: An Introduction to the Interpretation of Visual Materials*. London: Sage, 2007.

Rosenband, Leonard N. *Papermaking in Eighteenth-Century France: Management, Labor, and Revolution at the Montgolfier Mill, 1761–1805.* Baltimore: Johns Hopkins University Press, 2000.

Rosenberg, Rainer. *Zehn Kapitel zur Geschichte der Germanistik: Literaturgeschichtsschreibung.* Berlin: Akademie-Verlag, 1981.

Rosenfeld, Beate. *Die Golemsage und ihre Verwertung in der deutschen Literatur.* Breslau, H. Priebatsch, 1934.

Rosenfield, Leonora Cohen. *From Beast-Machine to Man-Machine: Animal Soul in French Letters from Descartes to La Mettrie.* New York: Oxford University Press, 1941.

Rössig, Carl Gottlob. *Lehrbuch der Technologie für den angehenden Staatswirth und den sich bildenden oder reisenden Technologen.* Jena: Akademische Buchhandlung, 1790.

———. *Versuch einer pragmatischen Geschichte der Oekonomie-Polizey- und Cameralwissenschaften seit dem sechzehnten Jahrhunderte bis zu unsern Zeiten.* 2 vols. Leipzig, 1781–82.

Roters, Eberhard. *E. T. A. Hoffmann.* Berlin: Stapp, 1984.

Roubo, André Jacques. *L'art du menuisier ébéniste.* 3 vols. Paris: Académie Royale des Sciences, 1774.

Sabel, Charles, and Jonathan Zeitlin. "Historical Alternatives to Mass Production: Politics, Markets, and Technology in Nineteenth-Century Industrialization." *Past and Present* 108 (1985): 133–76.

———, eds. *World of Possibilities: Flexibility and Mass Production in Western Industrialization.* Cambridge: Cambridge University Press/Editions de la Maison des Sciences de l'Homme, 1997.

Sadie, Julie Anne. "*Musiciennes* of the Ancien Régime." In *Women Making Music: The Western Art Tradition*, edited by Jane Bowers and Judith Tick, 191–223. Urbana: University of Illinois Press, 1986.

Safley, Thomas Max, and Leonard N. Rosenband. *The Workplace before the Factory: Artisans and Proletarians, 1500–1800.* Ithaca. NY: Cornell University Press, 1993.

Sanford, Charles L. "Technology and Culture at the End of the Nineteenth Century: The Will to Power." In *Technology in Western Civilization*, edited by Melvin Kranzberg and Carroll W. Pursell Jr., 1:726–39. New York: Oxford University Press, 1967.

Sargentson, Carolyn. "Looking at Furniture Inside Out: Strategies of Secrecy and Security in Eighteenth-Century French Furniture." In *Furnishing the Eighteenth Century: What Furniture Can Tell Us about the European and American Past*, edited by Dena Goodman and Kathryn Norberg, 205–36. New York: Routledge, 2007.

Sauder, Gerhard. "Die empfindsamen Tendenzen in der Musikkultur nach 1750." In *Carl Philipp Emanuel Bach und die europäische Musikkultur des mittleren 18. Jahrhunderts*, edited by Hans Joachim Marx, 41–64. Göttingen, Vandenhoeck und Ruprecht, 1990.

————. *Empfindsamkeit.* Vol. 1, *Voraussetzungen und Elemente.* Stuttgart: Metzler, 1974.

Sauer, Lieselotte. *Marionetten, Maschinen, Automaten: Der künstliche Mensch in der deutschen und englischen Romantik.* Bonn: Bouvier, 1983.

Saunders, E. Stewart. "The Archives of the Académie des Sciences." *French Historical Studies* 10 (Fall 1978): 696–702.

Sautermeister, Gert. *Georg Christoph Lichtenberg.* Munich: C. H. Beck, 1993.

Sawday, Jonathan. *Engines of the Imagination: Renaissance Culture and the Rise of the Machine.* London: Routledge, 2007.

Schaffer, Simon. "Babbage's Dancer and Impresarios of Mechanism." In *Cultural Babbage, Technology, Time, and Invention,* edited by Francis Spufford and Jenny Uglow, 53–80. London: Faber & Faber, 1997.

————. "Babbage's Intelligence: Calculating Engines and the Factory System." *Critical Inquiry* 21, no. 1 (Autumn 1994): 203–27.

————. "Deus et machina: Human Nature and Eighteenth Century Automata." *La Lettre de la Maison Française d'Oxford* 9 (1998): 9–18.

————. "Enlightened Automata." In *The Sciences in Enlightened Europe,* edited by William Clark, Jan Golinski, and Simon Schaffer, 126–65. Chicago: University of Chicago Press, 1999.

Schardin, Joachim, "Musikuhren und Musikautomaten der Dresdner Instrumentenbaur- und 'Akustiker'-Familie Kaufmann." *Uhren und Schmuck* 18 (1981): 17–19.

Schindler, Georg. "Die legendären Automaten von Jaquet-Droz, Vater und Sohn." *Schriften des Fachkreises alter Uhren in der Deutschen Gesellschaft für Chronometrie* 13 (1974): 11–22.

Schlaffer, Heinz. "Jean Paul." In Schweikert, *Jean Paul,* 389–410.

Schlemmer, Oskar. "Man and Art Figure." In *The Theater of the Bauhaus,* edited by Walter Gropius and translated by Arthur S. Wensinger. Middletown: Wesleyan University, 1961.

Schmidt, James. "What Enlightenment Project?" *Political Theory* 28, no. 6 (December 2000): 734–57.

Schmidt-Biggemann, Wilhelm. *Maschine und Teufel: Jean Pauls Jugendsatiren nach ihrer Modellgeschichte.* Freiburg: Karl Alber, 1975.

Schmidt-Hannisa, Hans-Walter: "Lesarten. Autorschaft und Leserschaft bei Jean Paul." *Jahrbuch der Jean-Paul-Gesellschaft* 37 (2002): 35–52.

Schmitz-Emans, Monika. "Georg Christoph Lichtenberg und der Maschinenmann: Zur Interferenz von literarischer Phantasie und naturwissenschaftlicher Modellbildung." *Jahrbuch der Jean-Paul-Gesellschaft* 25 (1990), 74–111.

————. *Seetiefen und Seelentiefen: Literarische Spiegelungen innerer und äusserer Fremde.* Würzburg: Königshausen & Neumann, 2003.

Schneider, Falko. *Öffentlichkeit und Diskurs: Studien zu Entstehung, Struktur und Form der Öffentlichkeit im 18. Jahrhundert.* Bielefeld, Germany: Aisthesis Verlag, 1992.

Schnitzler, Arthur. *Die dramatischen Werke.* 2 vols. Frankfurt: S. Fischer, 1962.

Schott, Caspar. *Mechanica Hydraulico-pneumatica*. Würzburg, 1657.

———. *Technica curiosa*. Nürnberg, 1664.

Schubert, Gotthilf Heinrich von. *Ansichten von der Nachtseite der Naturwissen-schaft*. Dresden: Arnoldische Buchhandlung, 1808.

Schücking, J. L. "Die Marionette bei E. T. A. Hoffmann und Heinrich von Kleist." *Puppentheater* 1 (1923): 63–70.

Schultz, Wilhelm. "Der Mechanismus der Drozschen Androiden." *Deutsche Uhrmacher-Zeitung* 31, no. 8 (1907): 128–29; no. 10:158–61; no. 12:188–89; no. 14:222–24.

Schulz, Andrew. "Spaces of Enlightenment: Art, Science, and Empire in Eighteenth-Century Spain." In *Spain in the Age of Exploration, 1492–1819*, edited by Chiyo Ishikawa, 189–204. Lincoln: University of Nebraska Press, 2004.

Schulz, Heinrich. *Rudolph Zacharias Becker als Volkserzieher*. Berlin: W. Pilz, 1926.

Schütz, Rosemarie. "Der 'Königliche Kabinettmacher' aus Neuwied." In *Möbel von Abraham und David Roentgen: Schriften des Kreismuseums Neuwied*, 17–25. Neuwied: Landkreis, 1990.

Schweikert, Uwe, ed. *Jean Paul*. Darmstadt: Wissenschaftliche Buchgesellschaft, 1974.

The Science Museum: The First Hundred Years. London: H. M. Stationery Office, 1957.

Scott, David. "Leibniz and the Two Clocks." *Journal of the History of Ideas* 58, no. 3 (1997): 445–63.

Segeberg, Harro. *Literatur im technischen Zeitalter: Von der Frühzeit der deutschen Aufklärung bis zum Beginn des Ersten Weltkriegs*. Darmstadt: Wissenschaftli-che Buchgesellschaft, 1997.

Segel, Harold B. *Pinocchio's Progeny: Puppets, Marionettes, Automatons and Robots in Modernist and Avant-Garde Drama*. Baltimore: Johns Hopkins University Press, 1995.

Seibert, Peter. *Der literarische Salon: Literatur und Geselligkeit zwischen Aufklärung und Vormärz*. Stuttgart: Metzler, 1993.

Shank, John Bennett. *The Newton Wars and the Beginning of the French Enlighten-ment*. Chicago: University of Chicago Press, 2008.

Shapin, Steven. "Of Gods and Kings: Natural Philosophy and Politics in the Leibniz-Clarke Disputes." *ISIS* 72, no. 2 (1981): 187–215.

———. "Understanding the Merton Thesis," *ISIS* 79 (1988): 594–605.

Shelley, Mary Wollstonecraft. *Frankenstein*. Edited by Joseph Pearce. San Fran-cisco: Ignatius Press, 2008.

Sheraton, Thomas. *The Cabinet-Maker and Upholsterer's Drawing-Book, in Three Parts*. London: T. Bensley, 1793–94.

Shieber, Stuart, ed. *The Turing Test: Verbal Behavior as the Hallmark of Intelligence*. Cambridge, MA: MIT Press, 2004.

Sick, Birgit: "Jean Pauls unveröffentlichte Satiren- und Ironienhefte (1782–1803)." *Jahrbuch der Jean-Paul-Gesellschaft* 35–36 (2001): 205–17.

Simmen, René. *Der mechanische Mensch: Texte und Dokumente über Automaten, Androiden und Roboter: Eine Sammlung.* Zürich: Simmon, 1967.

Simon, Ernst. *Mechanische Musikinstrumente früherer Zeit und ihre Musik.* Wiesbaden: Breitkopf und Härtel, 1960.

Sinner de Ballaigues, Jean-Richard. *Voyage historique et littéraire dans la Suisse occidentale.* Neuchâtel, Switzerland: Société Typographique, 1781.

Smith, Adam. *The Theory of Moral Sentiments.* London: Printed for A. Millar, A. Kincaid, and J. Bell, in Edinburgh, 1759.

Smith, Pamela H. *The Body of the Artisan: Art and Experience in the Scientific Revolution.* Chicago: University of Chicago Press, 2004.

Solla Price, Derek de. "Automata and the Origins of Mechanism and Mechanistic Philosophy." *Technology and Culture* 5 (1964): 9–23.

Söntgen, Beate. "Täuschungsmanöver: Kunstpuppe—Weiblichkeit—Malerei." In Müller-Tamm and Sykora, *Puppen, Körper, Automaten,* 125–39.

Sørensen, Bengt Algot. *Geschichte der deutschen Literatur.* Vol. 1, *Vom Mittelalter bis zur Romantik.* Munich: Beck, 1997.

Stableford, Brian M. *Science Fact and Science Fiction: An Encyclopedia.* New York: Routledge, 2006.

Standage, Tom. *The Turk: The Life and Times of the Famous Eighteenth-Century Chess-Playing Machine.* New York: Berley, 2002.

Stollberg-Rilinger, Barbara. *Der Staat als Maschine: Zur politischen Metaphorik des absoluten Fürstenstaats.* Berlin: Duncker & Humblot, 1986.

Stosch, Samuel Johann Ernst. "Empfindsam. Empfindlich." In *Kritische Anmerkungen über die gleichbedeutenden Wörter der deutschen Sprache,* 4:206–8. Vienna: J. T. Edler v. Trattnern, 1786.

Strakosch, Alexander. *Mensch und Maschine: Ausblicke auf eine neue Stellung des Menschen zur Technik.* Stuttgart: Waldorfschul-Spielzeug & Verlag, 1928.

Strauss, Linda Marlene. "Automata: A Study in the Interface of Science, Technology, and Popular Culture, 1730–1885." PhD diss., University of California–San Diego, 1987.

Sturdy, David. *Fractured Europe, 1600–1721.* Oxford: Blackwell, 2002.

Stürmer, Michael. "Die Roentgen-Manufaktur in Neuwied (I)." *Kunst und Antiquitäten* 5 (1979): 24–36.

———. *Handwerk und höfische Kultur: Europäische Möbelkunst im 18. Jahrhundert.* Munich: C. H. Beck, 1982.

———. *Luxus, Leistung und die Liebe zu Gott: David Roentgen: Kgl. Kabinettmacher 1743–1807.* Munich: Bayrische Vereinsbank, 1993.

———. *Scherben des Glücks: Klassizismus und Revolution.* Berlin: W. J. Siedler, 1987.

Sully, Henri. *Regle artificielle du tems, pour apprendre la division naturelle et artificielle du tems, et connvître toutes sortes d'horloges et de montres, et la maniere de s'en servir adroitement.* Vienna: André Heyinger, 1714.

Sulzer, Johann Georg. *Allgemeine Theorie der schönen Künste: In einzeln, nach alphabetischer Ordnung der Kunstwörter auf einander folgenden, Artikeln abgehandelt.* 2 vols. Leipzig: Weidmann Reich, 1771–74.

Sutter, Alex. *Göttliche Maschinen: Die Automaten für Lebendiges bei Descartes, Leibniz, La Mettrie und Kant.* Frankfurt am Main: Athenäum, 1988.

Swann, Julian. "Politics and the State in Eighteenth-Century Europe." In Blanning, *Eighteenth Century*, 11–51.

Swirski, Peter, ed. and trans. *The Art and Science of Stanislaw Lem.* Montreal: McGill-Queen's University Press, 2006.

Swoboda, Helmut. *Der künstliche Mensch.* Munich: Ernst Heimeran, 1967.

Terrall, Mary. *The Man Who Flattened the Earth: Maupertuis and the Sciences in the Enlightenment.* Chicago: University of Chicago Press, 2002.

Thiout, Antoine. *Traité de l'horlogerie, méchanique et pratique: Approuvé par l'Academie royale des sciences.* 2 vols. Paris: C. Moette, 1741.

Tissot, André. *Voyage de Pierre Jaquet-Droz à la cour du roi d'Espagne, 1758–1759, d'après le journal d'Abraham-Louis Sandoz son beau-père.* Neuchâtel, Switzerland: Editions de la Baconnière, 1982.

Toynbee, Arnold. *Lectures on the Industrial Revolution in England: Popular Addresses, Notes, and Other Fragments.* London: Rivingtons, 1884.

Turgot, Anne-Robert-Jacques. *Reflections on the Formation and Distribution of Wealth.* Translated from the French. London: J. Good, 1793.

Trepp, Anne-Charlott. "Diskurswandel und soziale Praxis. Zur These von der Polarisierung der Geschlechter seit dem 18. Jahrhundert." In *Geschlechterpolaritäten in der Musikgeschichte des 18. bis 20. Jahrhunderts*, edited by Rebecca Grotjahn and Freia Hoffmann, 7–18. Herbolzheim, Germany: Centaurus Verlag, 2002.

Trevor-Roper, Hugh R. *The Crisis of the Seventeenth Century: Religion, the Reformation, and Social Change.* New York, Harper & Row, 1967.

Tribe, Keith. *Strategies of Economic Order: German Economic Discourse, 1750–1950.* New York: Cambridge University Press, 1995.

Troeltsch, Ernst. *Aufsätze zur Geistesgeschichte und Religionssoziologie.* Vol. 4 of *Gesammelte Schriften.* Tübingen: J. C. B. Mohr, 1924.

Tromlitz, J. G. *Ausführlicher und gründlicher Unterricht die Flöte zu spielen.* Leipzig: Adam Friedrich Böhme, 1791.

Tullius, Wilhelm. *Die wechselvolle Geschichte des Hauses Wied.* Neuwied: Peter Kehrein, 2002.

Türk, Daniel Gottlob. *Klavierschule, oder, Anweisung zum Klavierspielen für Lehrer und Lernende: Mit kritischen Anmerkungen.* Leipzig: Schwickert, 1789.

Ueding, Gert. *Jean Paul.* Munich: C. H. Beck, 1993.

———. *Klassik und Romantik: Deutsche Literatur im Zeitalter der Französischen Revolution 1789–1815.* Munich: Hanser, 1987.

Ure, Andrew. *The Cotton Manufacture of Great Britain Systematically Investigated . . . with an Introductory View of Its Comparative state in Foreign Countries. . . .* 2 vols. London, C. Knight, 1836.

———. *A Dictionary of Arts, Manufactures, and Mines, Containing a Clear Exposition of Their Principles and Practice.* London: Longman, Orme, Brown, Green, & Longmans, 1839.

――――. *The Philosophy of Manufactures; or, An Exposition of the Scientific, Moral, and Commercial Economy of the Factory System of Great Britain*. London: C. Knight, 1835.

――――. *Recent Improvements in Arts, Manufactures, and Mines, Being a Supplement to His Dictionary*. New York: D. Appleton, 1845.

Vanden Berghe, Marc. "Henri-Louis Jaquet-Droz: Horloger mécanicien (1752–1791)." In *Biographies Neuchâteloises*, vol. 1, *De saint Guillaume à la fin des Lumières*, edited by Michel Schlup, 149–53. Neuchâtel, Switzerland: Editions Gilles Attinger, 1996.

――――. "Pierre Jaquet-Droz: Horloger mécanicien (1721–1790)." In *Biographies Neuchâteloises*, vol. 1, *De saint Guillaume à la fin des Lumières*, edited by Michel Schlup, 154–58. Neuchâtel, Switzerland: Editions Gilles Attinger, 1996.

Vartanian, Aram. "Biographical Sketch of La Mettrie." In *La Mettrie's L'homme machine: A Study in the Origins of an Idea*, edited by Aram Vartanian, 1–138. Princeton, NJ: Princeton University Press, 1960.

――――. *Diderot and Descartes: A Study of Scientific Naturalism in the Enlightenment*. Princeton, NJ: Princeton University Press, 1953.

――――. "Man-Machine from the Greeks to the Computer." In *Dictionary of the History of Ideas*, edited by Philip P. Wiener, 3:131–46. New York: Scribner, 1973.

Vaucanson, Jacques de. *Le Mécanisme du fluteur automate: presenté a messieurs de l'Académie royale des sciences/Par m. Vaucanson, auteur de cette me machine; Avec la description d'un canard artificiel, mangeant, beuvant, digerant & se vuidant, épluchant ses aîles & ses plumes, imitant en diverses manieres un canard vivant. Inventé par la mesme. Et aussi celle d'une autre figure, également merveilleuse, jouant du tambourin & de la flute, suivant la relation, qu'il en a donnée dépuis son mémoire écrit*. Paris: J. Guerin, 1738.

Vietta, Silvio. "Das Automatenmotiv und die Technik der Motivschichtung im Erzählwerk E. T. A. Hoffmanns." *Mitteilungen der E. T. A. Hoffmann-Gesellschaft* 26 (1980): 25–33.

Vila, Anne C. *Enlightenment and Pathology: Sensibility in the Literature and Medicine of Eighteenth-Century France*. Baltimore: Johns Hopkins University Press, 1998.

Vincent-Buffault, Anne. *The History of Tears: Sensibility and Sentimentality in France*. London: Macmillan Press, 1991.

Vogel, Christian. *Ferdinand Berthoud's (eines berühmten Uhrmachers zu Paris) Anweisung zur Kenntniß, Gebrauch und guten Haltung der Wand- und Taschenuhren*. Meissen: Karl Friedrich Wilhelm Erbstein, 1791.

Vogel, Steven. *Against Nature: The Concept of Nature in Critical Theory*. Albany: State University of New York Press, 1996.

Volk, Stefan. "Peuplierung und religiöse Toleranz: Neuwied von der Mitte des 17. bis zur Mitte des 18. Jahrhunderts." *Rheinische Vierteljahrsblätter* 55 (1991), 205–31.

Völker, Klaus, ed. *Künstliche Menschen: Dichtungen und Dokumente über Golems, Homunculi, Androiden und liebende Statuen.* Munich: Hanser, 1976.

Voltaire chez lui: Ferney 1758–1778. Yens sur Morges, Switzerland: Cabédita, 1999.

Vuilleumier, Marc. *Immigrants and Refugees in Switzerland: An Outline History.* Zürich: Pro Helvetia, 1987.

Wäber, A. *Landes- und Reisebeschreibungen: Ein Beitrag zur Bibliographie der schweizerischen Reiseliteratur.* Bern: Wyss, 1899.

Wackernagel, Hans Georg. *Die Matrikel der Universität Basel.* 5 vols. Basel: Verlag der Universitätsbibliothek, 1951–80.

Wakefield, Andre. *The Disordered Police State: German Cameralism as Science and Practice.* Chicago: University of Chicago Press, 2009.

Wakkerbart, Freiherr von. *Rheinreise.* Halberstadt: Buchhandlung der Gross-schen Erben, 1794.

Walter, Emil. *Soziale Grundlagen der Entwicklung der Naturwissenschaften in der alten Schweiz.* Bern: Francke Verlag, 1958.

Wartburg, Walther von. *Französisches etymologisches Wörterbuch: Eine Darstellung des galloromanischen Sprachschatzes.* 25 vols. Bonn: Schroeder, 1922–2003.

Wartburg, Wolfgang von. *Geschichte der Schweiz.* Munich: Oldenbourg, 1951.

Watt, Jeffrey R. *The Making of Modern Marriage: Matrimonial Control and the Rise of Sentiment in Neuchâtel, 1550–1800.* Ithaca, NY: Cornell University Press, 1992.

Wawrzyn, Lienhard. *Der Automaten-Mensch: E. T. A. Hoffmanns Erzählung vom "Sandmann."* Berlin: Klaus Wagenbach, 1978.

Weber, Carl Maria von. "Der Trompeter, eine Maschine von der Erfindung des Mechanicus, Hrn. Friedrich Kaufmann, in Dresden." *Allgemeine musikalische Zeitung* 14, no. 41 (7 October 1812): 663–66.

Weber, Max. *The Protestant Ethic and the Spirit of Capitalism.* Translated by Talcott Parsons. London: Routledge, 1992.

Weber, Otto. "Physik und Technologie am Darmstädter Hof: Von der Drechslerwerkstatt zum Physikalischen Kabinett." In *Darmstadt in der Zeit des Barock und Rokoko.* Darmstadt: Magistrat der Stadt, 1980.

Wegmann, Nikolaus. *Diskurse der Empfindsamkeit: Zur Geschichte eines Gefühls in der Literatur des 18. Jahrhunderts.* Stuttgart: Metzler 1988.

Wehler, Hans-Ulrich. *Deutsche Gesellschaftsgeschichte.* Vol. 1, *Vom Feudalismus des Alten Reiches bis zur defensiven Modernisierung der Reformära, 1700–1815.* Munich: C. H. Beck, 1987.

Weigl, Engelhard. *Aufklärung und Skeptizismus: Untersuchungen zu Jean Pauls Frühwerk.* Hildesheim: Gerstenberg, 1980.

Weinholz, Gerhard. *E.T.A. Hoffmanns Erzählung "Die Automate": Eine Kritik an einseitiger naturwissenschaftlich-technischer Weltsicht vor zweihundert Jahren.* Essen: Verlag Die Blaue Eule, 1991.

Wernle, Paul. *Der schweizerische Protestantismus im XVIII. Jahrhundert.* 3 vols. Tübingen: Mohr, 1923.

White, Lynn T. *Medieval Technology and Social Change*. Oxford: Clarendon Press, 1962.

Whitebook, Joel. "The Problem of Nature in Habermas." *Telos* 40 (1979): 41–69.

Whitworth, Michael H., ed. *Modernism*. Malden, MA: Blackwell, 2007.

Wiegleb, Johann Christian [Johann Nikolaus Martius]. *Unterricht in der natürlichen Magie, oder zu allerhand belustigenden und nüzlichen Kunststücken*. Berlin: F. Nicolai, 1779–1805.

Wiegmann, Hermann. *Die ästhetische Leidenschaft: Texte zur Affektenlehre im 17. und 18. Jahrhundert*. Hildesheim: Georg Olms Verlag, 1987.

Wiener, Norbert. *Cybernetics; or, Control and Communication in the Animal and the Machine*. Cambridge, MA: Technology Press, 1948.

———. *Cybernetics: or, Control and Communication in the Animal and the Machine*. 2nd ed. New York: MIT Press, 1961.

———. *God and Golem, Inc.: A Comment on Certain Points where Cybernetics Impinges on Religion*. Cambridge, MA: MIT Press, 1964.

———. *The Human Use of Human Beings: Cybernetics and Society*. 2nd ed. rev. Garden City, NY: Doubleday, 1954.

Wilke, Jürgen. *Literarische Zeitschriften des 18. Jahrhunderts (1688–1789). Teil I, Grundlegung*. Stuttgart: Metzler, 1978.

———. *Literarische Zeitschriften des 18. Jahrhunderts (1688–1789). Teil II, Repertorium*. Stuttgart: Metzler, 1978.

Wille, Bruno. *Der Maschinenmensch und seine Erlösung: Roman*. Pfullingen in Württemberg: J. Baum, 1930.

Williams, Rosalind. *Dream Worlds: Mass Consumption in Late Nineteenth-Century France*. Berkeley: University of California Press, 1982.

Willscheid, Bernd. "Der Kundenkreis von Abraham und David Roentgen." In *Möbel von Abraham und David Roentgen*, 27–39. Schriften des Kreismuseums Neuwied. Neuwied: Landkreis, 1990.

Wilson, Arthur M. "Sensibility in France in the Eighteenth Century." *French Quarterly* 13 (1931): 35–45.

Winner, Langdon. *Autonomous Technology: Technics-out-of-Control as a Theme in Political Thought*. Cambridge, MA: MIT Press, 1977.

Winzer, Fritz. *Emigranten: Geschichte der Emigration in Europa*. Frankfurt am Main: Ullstein, 1986.

Wisnioski, Matthew H. "'Liberal Education Has Failed': Reading Like an Engineer in 1960s America." *Technology and Culture* 50, no. 4 (October 2009): 753–82.

Wittig, Frank. *Maschinenmenschen: Zum Wandel eines literarischen Motivs im Kontext von Philosophie, Natur und Technik*. Würzburg: Königshausen & Neumann, 1997.

Wolf, Rebecca. *Friedrich Kaufmanns Trompeterautomat: Ein musikalisches Experiment um 1810*. Stuttgart: Franz Steiner Verlag, 2011.

Wolf, Rudolf. *Biographien zur Kulturgeschichte der Schweiz*. 4 vols. Zürich: Orell, Füssli, 1858–62.

Wolff, Christian. *Der vernünfftigen Gedancken von Gott, der Welt und der Seele des Menschen, auch allen Dingen überhaupt; anderer Theil bestehend in ausführlichen Anmerckungen und zu besserem Verstande und bequemerem Gebrauche derselben.* . . . 4th ed. Frankfurt am Main: J. B. Andrea und H. Hort, 1733.

Wood, Gaby. *Living Dolls: A Magical History of the Quest for Mechanical Life.* London: Faber and Faber, 2002.

Wunder und Wissenschaft: Salomon de Caus und die Automatenkunst in den Gärten um 1600. Edited by Stiftung Schloss and Park Benrath. Düsseldorf, Germany: Grupello Verlag, 2008.

Zemanek, Heinz. "Die Automaten des 18. Jahrhunderts eingebettet in eine Kompaktform der Automaten-Geschichte." In Kowar, *Die Wiener Flötenuhr,* 237–71.

Zimmermann, Felix. *Die Widerspiegelung der Technik in der deutschen Dichtung von Goethe bis zur Gegenwart.* Dresden: Druck von W. Ulrich, 1913.

Index

Page numbers in italics indicate figures.

absolutism, 26n41, 89n4, 164–65; and mercantilism, 92; and nation-state, 92, 164. *See also* Louis XIV; mercantilism

Académie des Sciences (Paris), 5, 30, 116, 118–20, 122, 131, 132

Academy of Sciences (St. Petersburg), 35

Adorno, Theodor, 220n58, 223n68

Alder, Ken, 2n1, 26n39

Alexander, Johann Friedrich (Count, later Prince, of Neuwied), 96n30, 97–99, 103n59, 103n60, 105n69, 109, 118n110

Almanach de Gotha, 69, 76, 77n133, 82–83

Altick, Richard Daniel, 63n86, 205n11

anatomy and physiology, study of, 23, 35, 41n9, 154n61, 155n70, 204, 207. *See also under* automata, eighteenth-century

ancien régime, 25, 53, 76n129, 91. *See also* court and estate society

android. *See under* automata, eighteenth-century

Apostolidès, Jean-Marie, 26n41

Aquinas, Thomas, 28

Arabic culture, 28, 214

Arnim, Ludwig Achim von, 175n16, 176

artificial humans. *See* artificial intelligence; automata; cybernetics; dolls; golems; marionettes; puppets; robots

artificial intelligence, 6n6, 14, 194, 220–21, 224, 229n1

artisan culture, 5–8, 15–18, 25–26, 128–29, 139n16, 226–30; Augsburg and Nürnberg, as centers for, 26, 28–29, 89; clock-making, 139–40 (*see also* La Chaux-de-Fonds: clock-making, as center for); Enlightenment economics and, 87, 96–98 (*see also* Moravian Brothers: economics of); furniture-making, 89–90 (*see also* Roentgen manufacture); guilds and, 12, 26n40, 45–47, 91, 97–98, 100, 105, 109, 117; London, as center for, 50, 60–62, 79, 89–92, 104, 139; Paris, as center for, 50, 55, 72, 79, 89–90, 107–9, 116–18; post-1800 writing and, 173, 175, 207, 218. *See also* Jaquet-Droz company; natural philosophy: and automata; protoindustrial era; Roentgen manufacture

Asimov, Isaac, 220–21

automata, before 1700, 27–29, 213–14

automata, eighteenth-century: anatomy and physiology, relation to study of, 17, 20, 21n28,

Wiener, Norbert, 201, 222n67, 223–24
Williams, Rosalind, 15n14, 26n39
Wisnioski, Matthew H., 223n68
Wittig, Frank, 16n18, 172n3, 182n42,
 183n44, 217n54
Wolf, Rebecca, 32n58, 33n63, 33n64,
 34n67, 132n11
Wolff, Christian, 151n49, 173

Wood, Gaby, 16n18, 19n24
Wunder und Wissenschaft (catalog), 27n43,
 140n17

Zinzendorf, Nikolaus von, 93, 95, 97n35,
 98n40, 100. *See also* Moravian Brothers
Zwingli, Ulrich, 44